THE NEW GROVE
NORTH EUROPEAN BAROQUE
MASTERS

THE NEW GROVE
DICTIONARY OF MUSIC AND MUSICIANS

Editor: Stanley Sadie

The Composer Biography Series

BACH FAMILY

BEETHOVEN

EARLY ROMANTIC MASTERS 1

EARLY ROMANTIC MASTERS 2

HANDEL

HAYDN

HIGH RENAISSANCE MASTERS

ITALIAN BAROQUE MASTERS

LATE ROMANTIC MASTERS

MASTERS OF ITALIAN OPERA

MODERN MASTERS

MOZART

NORTH EUROPEAN BAROQUE MASTERS

SCHUBERT

SECOND VIENNESE SCHOOL

TURN OF THE CENTURY MASTERS

WAGNER

THE NEW GROVE®

North European Baroque Masters

SCHÜTZ FROBERGER
BUXTEHUDE
PURCELL TELEMANN

Joshua Rifkin
Colin Timms
George J. Buelow
Kerala J. Snyder
Jack Westrup
Martin Ruhnke

M
MACMILLAN

First published in
The New Grove Dictionary of Music and Musicians®,
edited by Stanley Sadie, 1980

The New Grove and *The New Grove Dictionary of Music and Musicians* are registered trademarks of Macmillan Publishers Limited, London

First published in UK in paperback with additions 1985 by
PAPERMAC
a division of Macmillan Publishers Limited
London and Basingstoke

First published in UK in hardback with additions 1985 by
MACMILLAN LONDON LIMITED
4 Little Essex Street London WC2R 3LF
and Basingstoke

British Library Cataloguing in Publication Data

North European baroque masters.—
(The composer biography series)
1. Music—Europe—History and criticism
2. Composers
I. Rifkin, Joshua II. The New Grove dictionary of music and musicians III. Series
780′.92′2 ML240

ISBN 0-333-39017-2 (hardback)
ISBN 0-333-39018-0 (paperback)

First American edition in book form with additions 1985 by
W. W. NORTON & COMPANY
New York and London

ISBN 0-393-01695-1 (hardback)
ISBN 0-393-30099-4 (paperback)

Printed in Great Britain by
Redwood Burn Limited, Trowbridge, Wiltshire,
and bound by Pegasus Bookbinding, Melksham, Wiltshire.

Contents

List of illustrations

Illustration acknowledgments

We are grateful to the following for permission to reproduce illustrative material: Ratsschulbibliothek, Zwickau (fig.1); Staatsarchiv, Dresden (fig.2); Musikinstrumenten-Museum, Karl-Marx-Universität, Leipzig, photo (fig.3); Gesamthochschul-Bibliothek (Landesbibliothek und Bibliothek der Stadt), Kassel (fig.4); Herzog August Bibliothek, Wolfenbüttel (fig.5); Österreichische Nationalbibliothek, Vienna (fig.6); Museum für Hamburgische Geschichte, Hamburg (fig.7, cover); Universitetsbiblioteket, Uppsala (fig.8); National Portrait Gallery, London (figs.9, 11); Dean and Chapter of Exeter Cathedral (fig.10); British Library, London (figs.12, 14, 16 right); Hessische Landes- und Hochschulbibliothek, Darmstadt (fig.16 left); Deutsche Staatsbibliothek, Berlin (fig.17)

General abbreviations

A	alto, contralto [voice]
adds	additions
Anh.	Anhang [appendix]
aut.	autmn
B	bass [voice]
b	bass [nstrument]
b	born
Bar	baritone [voice]
bc	basso continuo
bn	bassoon
BUXWV	Buxtehude-Werke-Verzeichnis [Karstädt, catalogue of Buxtehude's works]
BWV	Bach-Werke-Verzeichnis [Schmieder, catalogue of J. S. Bach's works]
c	circa [about]
c.f.	cantus firmus
cl	clarinet
collab.	in collaboration with
conc.	concerto
d	died
ded.	dedication, dedicated to
edn.	edition
facs.	facsimile
fl	flute
frag.	fragment
Gl	Gloria
glock	glockenspiel
hn	horn
hpd	harpsichord
inc.	incomplete
inst	instrument, instrumental
Jb	Jahrbuch [yearbook]
Jg.	Jahrgang [year of publication/volume]
kbd	keyboard
Ky	Kyrie
lib	libretto
Mez	mezzo-soprano
movt	movement
n.d.	no date of publication
n.p.	no place of publication
ob	oboe
obbl	obbligato
orch	orchestra, orchestral
org	organ
ov.	overture
p.	pars (1p. = *prima pars*, etc)
perf.	performance, performed by
pic	piccolo
pr.	printed
prol	prologue
pubd	published
R	photographic reprint
r	recto
rec	recorder
recit	recitative
repr.	reprinted
rev.	revision, revised by/for

RISM	Répertoire International des Sources Musicales	transcr.	transcription, transcribed by/for
		trbn	trombone
S	San, Santa, Santo [Saint]; soprano [voice]	U.	University
str	string(s)		
swv	Schütz-Werke-Verzeichnis [Bittinger, catalogue of Schutz's works]	v, vv	voice, voices
		v	verso
		va	viola
		vc	cello
sym.	symphony, symphonic	vle	violone
		vn	violin
T	tenor [voice]		
timp	timpani	z	Zimmerman catalogue [Purcell]
tpt	trumpet		

Symbols for the library sources of works, printed in *italic*, correspond to those used in *RISM*, Ser. A.

Bibliographical abbreviations

AcM	*Acta musicologica*
ADB	*Allgemeine deutsche Biographie*
AMf	*Archiv für Musikforschung*
AnMc	*Analecta musicologica*
AMw	*Archiv für Musikwissenschaft*
AMz	*Allgemeine Musik-Zeitung*
BJb	*Bach-Jahrbuch*
BMw	*Beiträge zur Musikwissenschaft*
Cw	Das Chorwerk
DAM	*Dansk aarbog for musikforskning*
DDT	Denkmäler deutscher Tonkunst
DJbM	*Deutsches Jahrbuch der Musikwissenschaft*
DTÖ	Denkmäler der Tonkunst in Österreich
EDM	Das Erbe deutscher Musik
GerberNL	*R. Gerber: Neues historisch-biographisches Lexikon der Tonkünstler*
GfMKB	*Gesellschaft für Musikforschung Kongressbericht*
HJb	*Händel-Jahrbuch*
HM	Hortus musicus
JAMS	*Journal of the American Musicological Society*
JbMP	*Jahrbuch der Musikbibliothek Peters*
Mf	*Die Musikforschung*
MGG	*Die Musik in Geschichte und Gegenwart*
ML	*Music and Letters*
MMg	*Monatshefte für Musikgeschichte*
MMR	*The Monthly Musical Record*
MQ	*The Musical Quarterly*
MR	*The Music Review*
MT	*The Musical Times*
MZ	*Muzikološki zbornik*

NA	*Note d'archivio per la storia musicale*
NM	Nagels Musikarchiv
NOHM	*The New Oxford History of Music*, ed. E. Wellesz, J. A. Westrup and G. Abraham (London, 1954–)
NRMI	*Nuova rivista musicale italiana*
NZM	*Neue Zeitschrift für Musik*
ÖMz	*Österreichische Musikzeitschrift*
PMA	*Proceedings of the Musical Association*
PRMA	*Proceedings of the Royal Musical Association*
RIM	*Rivista italiana di musicologia*
RISM	*Répertoire international des sources musicales*
SIMG	*Sammelbände der Internationalen Musik-Gesellschaft*
SJb	*Schütz-Jahrbuch*
SMA	*Studies in Music* [Australia]
SMw	*Studien zur Musikwissenschaft*
SMz	*Schweizerische Musikzeitung/Revue musicale suisse*
STMf	*Svensk tidskrift för musikforskning*
TVNM	*Tijdschrift van de Vereniging voor Nederlandse muziekgeschiedenis*
VMw	*Vierteljahrsschrift für Musikwissenschaft*
WaltherML	J. G. Walther: *Musicalisches Lexicon oder Musicalische Bibliothec*
ZIMG	*Zeitschrift der Internationalen Musik-Gesellschaft*
ZMw	*Zeitschrift für Musikwissenschaft*

Preface

This volume is one of a series of short biographies derived from *The New Grove Dictionary of Music and Musicians* (London, 1980). In its original form, the text was written in the mid-1970s, and finalized at the end of that decade. For this reprint, the texts have been re-read and modified: those for Froberger, Buxtehude and Telemann by their original authors (the second of these, in particular, has been substantially altered by the author in the light of her recent research); and that for Purcell, originally contributed by the late Sir Jack Westrup, has been revised by Nigel Fortune (with assistance from Margaret Laurie). The original discussion of Schütz's music (for which we were indebted to Kurt Gudewill) has now been replaced by a fresh text, by Colin Timms; the remainder of the Schütz entry, including the work-list (originally drawn up by Derek McCulloch) and the bibliography (originally drawn up by Stephen Baron), has been modified and brought up to date by its original author, Joshua Rifkin.

The fact that the texts of the books in the series originated as dictionary articles inevitably gives them a character somewhat different from that of books conceived as such. They are designed, first of all, to accommodate a very great deal of information in a manner that makes reference quick and easy. Their first concern is with fact rather than opinion, and this leads to a larger than usual proportion of the texts being devoted to biography than to critical discussion. The nature of a reference work gives it a particular obligation to convey received knowledge and to treat of composers' lives and works in an encyclopedic fashion, with proper acknowledgment of sources and due care to reflect different standpoints, rather than to embody imaginative or speculative writing about a composer's character or his music. It is hoped that the comprehensive work-lists and extended bibliographies, indicative of the origins of the books in a reference work, will be valuable to the reader who is eager for full and accurate reference information and who may not have ready access to *The New Grove Dictionary* or who may prefer to have it in this more compact form.

S.S.

HEINRICH SCHÜTZ

Joshua Rifkin

Colin Timms

CHAPTER ONE

Childhood and youth, 1585–1615

Heinrich – or, as he invariably wrote it, Henrich – Schütz was the greatest German composer of the 17th century and the first of international importance. Through the example of his compositions and through his teaching he played a major part in establishing the traditions of high craftsmanship and intellectual depth that marked the best of his nation's music and musical thought for more than 250 years after his death.

Schütz was born in 1585 at Köstritz (now Bad Köstritz) into a prominent bourgeois family of Franconian origin that had resided in Saxony since the mid-15th century. His birthplace belonged to the principality of Reuss and lay close to Gera, which was the capital of the region. Albrecht Schütz, his paternal grandfather, owned a local inn, 'Zum goldenen Kranich'; Christoph Schütz, Heinrich's father, served as a town clerk in Gera during the mid-1570s, then took over the inn at Köstritz on behalf of Albrecht, who had moved to Weissenfels in 1571. On 5 February 1583 Christoph, whose first wife had died and left him with three children, married Euphrosyne Bieger, daughter of the burgomaster of Gera; the couple had eight children, of whom 'Henricus' – as his name appears in the baptismal record – was the second-born and the eldest son. Two of his four younger brothers, Georg (1587–1637) and Benjamin (1596–1666), became well-known jurists. Schütz remained close to his family throughout his life;

he appears to have had a particularly warm relationship with Georg.

Schütz himself described his early years in a petition addressed to Elector Johann Georg I of Saxony on 14 January 1651; a biographical sketch that the electoral court chaplain Martin Geier appended to his funeral oration for the composer adds some further details. Both accounts contain a number of evident errors and occasionally conflict with one another; nevertheless, in the absence of extensive contemporary documentation, they remain the principal sources for information on the first three decades of Schütz's life.

According to Geier, Schütz was born on 8 October 1585 at 7 p.m.; Schütz's petition, however, gives the date as St Burkhard's Day, which falls on 14 October. The Köstritz parish records would appear to support Geier, since they indicate that the baptism took place on 9 October, old style – the system of dating used in most of Germany throughout Schütz's life. Perhaps, as Wessely (1953) suggested, Schütz intended his reference to St Burkhard's Day to mean he was born on 4 October, old style, which corresponds to 14 October, new style; but this would leave an uncommonly long gap between birth and baptism. Moreover, Schütz seems never to have used new-style datings unless writing to, from or about places that followed the Gregorian calendar – none of which applies to the petition of January 1651.

Late in the summer of 1590 Christoph Schütz and his family moved to Weissenfels, where Albrecht, who died on 28 July, had bequeathed him an inn named 'Zum güldenen Ring'. Christoph became a figure of considerable eminence in his new city, eventually serving as its burgomaster. In 1615 he purchased a second inn –

known as both 'Zur güldenen Sackpfeife' and 'Zum güldenen Esel' – which he renamed 'Zum Schützen'. According to Geier, Christoph provided his children with a thorough religious and liberal education. Heinrich quickly showed 'a singular inclination to noble music', learning 'in a short time to sing securely and very well, with particular grace'. He presumably received musical instruction from the local Kantor, Georg Weber (*c*1540–1599), as well as from the organist and sometime burgomaster Heinrich Colander, who had married the widow of Schütz's uncle Matthes.

In 1598 Landgrave Moritz ('the Learned') of Hessen-Kassel, whose varied accomplishments included considerable skill in musical composition, stayed overnight at Christoph Schütz's inn. He heard young Heinrich sing, reported Geier, and the boy's performance so pleased him that

> His Noble Grace was moved to ask the parents to allow the lad to come with him to his noble court, promising that he would be reared in all good arts and commendable virtues . . .

Encountering resistance, Moritz continued to press his case in letters, and finally, in August 1599, Christoph Schütz took his son to the landgrave's seat at Kassel.

At Moritz's court Schütz served as a choirboy and pursued his education at the Collegium Mauritianum, an academy founded by the landgrave in 1595 primarily for the children of the Hesse nobility but attended also by some of the boys in the Hofkapelle and sons of court servants. Schütz distinguished himself in all his subjects and showed a special aptitude for languages, learning Latin, Greek and French. His musical training lay in the hands of Moritz's Kapellmeister, Georg Otto, who taught at the academy. Judging from student compositions by his colleagues Friedrich Kegell and Georg Schimmelpfennig, Schütz must have received a good foundation in counterpoint; there seems no reason to suppose, however, that his instruction went much beyond the most basic level, nor that he involved himself in composition to any particular extent.

According to Schütz, 'it was never the will of my late parents that I should make a profession of music either this day or the next'; following their wishes, therefore, he

> set out, after having lost my treble voice, for the University of Marburg, in order to continue there the studies that I had already begun elsewhere in things other than music, choose a secure profession and eventually gain an honourable degree therein.

He matriculated at the university on 27 September 1608 along with Schimmelpfennig, Friedrich Kegell and Friedrich's brother Christoph, also a former choirboy at court. A cousin of Schütz's – his uncle Matthes's son, also named Heinrich – had enrolled on 18 April, and his brother Georg followed on 30 December. The relatively advanced age at which Schütz matriculated has led to speculation that the studies 'begun elsewhere' included attendance at other universities. Students at the Collegium Mauritianum, however, often remained there beyond their 20th year; on graduation most went directly to the University of Marburg, which had become the official parent institution of the collegium in 1604. More likely than not, Schütz too pursued this course. Indeed, while clearly a man of substantial erudition, he probably did not receive so extensive a formal education as most scholars have assumed. Neither he nor Geier mentioned attendance at any university other than Marburg. In a memorial poem printed with Geier's eulogy, Georg Weisse, a clergyman who seems to have known Schütz personally, wrote that the composer also studied at Frankfurt an der Oder and Jena; the matriculation registers of these universities, however, show no record of his presence. Some modern writers have claimed that Schütz studied at Leipzig, but this assertion rests simply on a failure to distinguish between the composer and his homonymous cousin.

Geier wrote that Schütz elected to study law at Marburg and quickly won distinction for his academic prowess. Nevertheless, he did not remain at the university for long. During a visit to Marburg in 1609 Landgrave Moritz advised him that since Giovanni Gabrieli, 'a widely famed but rather old musician and composer, was still alive, I should not miss the chance to hear him and learn something from him'. The landgrave often provided funds for his most gifted charges to acquire further training abroad; Christoph Kegell and the future Hesse Kapellmeister Christoph Cornet had already gone to Venice on such grants in 1604. Moritz offered Schütz a stipend of 200 thalers a year, evidently for a two-year period; Schütz accepted gratefully, even though the voyage would have countered the wishes of his parents, and left for Italy.

Soon after his arrival in Venice, Schütz later recalled,

> I perceived the gravity and difficulty of the study of composition that I had undertaken and that I had had an unsound, poor start in this; and I therefore greatly rued that I had turned away from the studies commonly pursued in German universities . . .

By the end of 1610, however, he had made such

progress that Sigismund, Margrave of Brandenburg, wrote to Moritz from Venice that

> Giovanni Gabrieli has asked me many times to write to you and urge Your Grace that you be so kind to him as to allow your servant Heinrich Schütz to remain here another year, since he is doing so well not only in composition but also in [organ] playing . . .

Weisse reported that Gabrieli even had Schütz deputize for him on occasion. Teacher and pupil evidently developed a close personal relationship. On his deathbed Gabrieli left one of his rings to Schütz; Schütz, for his part, never spoke of Gabrieli in terms other than those of highest praise, nor did he ever acknowledge anyone else as his teacher.

Gabrieli appears to have given his pupils a rigorous training grounded in traditional contrapuntal procedures but admitting some licence in the treatment of dissonance. Like most of those in Gabrieli's tutelage Schütz completed his apprenticeship – or at least its initial phase – by producing a book of five-voice madrigals (SWV1–19); he dedicated the volume to Landgrave Moritz. The foreword bears the date 1 May 1611; Schütz may have timed the collection to coincide with the end of his two-year stipend and the beginning of the renewal that Moritz evidently granted. In the petition of January 1651 – which mysteriously dates the appearance of the madrigals a year too late – he recalled with some pride that his first effort won high praise from 'the most prominent musicians in Venice of the time'.

After his third year in Venice, Schütz 'was exhorted and encouraged not only by my teacher . . . but also by . . . other leading musicians there' to continue his studies even further; he followed their advice and remained in

Italy, supported now by his parents, who had presumably become somewhat more tolerant of his musical inclinations. In August 1612, not long after the start of this fourth year in Venice, Gabrieli died. According to Geier, Schütz left Italy soon afterwards; Schütz himself, however, indicated that he did not return to Germany until 1613.

Back in his native country, Schütz resolved

> to keep to myself with the good foundations that I had now laid in music, going into hiding with them, as it were, until I had refined them somewhat further still and was then able to distinguish myself properly by bringing forth a worthy piece of work . . .

He resumed his service at Moritz's court; a register of the Hesse Kapelle from the last four months of 1613 lists him as second organist with an annual salary of 80 florins. Although he kept his pledge to publish no music, his efforts to develop his craft bore immediate fruit in a series of polychoral works – SWV36*a*, 467, 470 and 474 – all found in manuscripts copied by 1615 at the latest. His family meanwhile renewed their attempts to dissuade him from pursuing music as anything but an avocation. In the face of their 'repeated and incessant admonition' he finally decided to 'seek out once again the books that I had laid aside so long before'. According to his recollections, however, he never carried out this plan; no sooner had he made up his mind than an unexpected series of events intervened.

On 27 August 1614 the Elector Johann Georg I of Saxony wrote to Moritz that

> we understand . . . you have at present an organist by the name of Heinrich Schütz; believing as we do that you will be suspending your music because of the mourning that has set in, we would very much like to hear him . . .

Johann Georg's reference to a period of mourning presents something of a mystery, as no one of importance at the Hesse court appears to have died at this time; but the roots of his interest in Schütz prove easy to discern. Since the semi-retirement of Rogier Michael in 1613 the electoral Kapelle lacked adequate leadership; Michael Praetorius served as visiting director, but his primary responsibilities lay in Wolfenbüttel. The Saxon ruler thus had a particular need for good musicians, especially those who showed promise in composition and matters of organization. In the petition of January 1651 Schütz speculated that he had come to the elector's attention through the offices of either Christoph von Loss, the Saxon privy councillor, or Gottfried von Wolffersdorff, district administrator of Weissenfels. Loss might have had occasion to hear him when Moritz, together with his Kapelle, visited the Saxon court in 1613; family connections may have brought Schütz into contact with Wolffersdorff.

Johann Georg proposed that Schütz come to Dresden for the baptism of the elector's son August on 18 September 1614 and 'stay a while with us'. Moritz gave his consent, and Schütz left for the Saxon capital, where he presumably assisted Praetorius at the baptismal ceremonies. On 10 October Johann Georg wrote to the landgrave to announce Schütz's impending return to Kassel.

Six months later, in a letter of 25 April 1615, Johann Georg asked Moritz to lend him Schütz's services for two years, 'until we once more have at our disposal those persons' – evidently Johann Nauwach and Johann Klemm – 'whom we have sent to Italy and elsewhere to learn this art'. Moritz, though protesting that he could hardly do without Schütz for such a long period of time, acceded to the elector's request. Schütz left for Dresden at the end of August. Several writers have suggested that the months before his departure saw the beginning of a relationship between him and the landgrave's brother-in-law Count Ernst of Schaumburg-Holstein, who maintained a small Kapelle at his court in Bückeburg. For four years, starting at Easter 1615, the count's expense books allocate strikingly large payments to a musician called 'Sagittarius', whose name usually appears together with that of the cornettist and composer Johann Martin Caesar. These entries, however, can hardly refer to Schütz. While he did sometimes latinize his name as 'Henricus Sagittarius', later correspondence between Moritz and Johann Georg reveals that Schütz used the time that elapsed after his call to Dresden to put his affairs in order, which would surely have left him little or no opportunity for a visit to Bückeburg; the possibility of his having gone there on a later occasion, moreover, would seem ruled out by Schütz's busy schedule at the electoral court. In all probability the Sagittarius of Schaumburg documents was the organist Johann Sagittarius, a former colleague of Caesar's at the court of Margrave Karl of Burgau in Günzburg, and a composer represented with Caesar in the collection *RISM* 1624^{1}.

CHAPTER TWO

Early manhood, 1615–27

By the time Schütz left Kassel for Dresden in August 1615 the increasing pace of his musical activities and the favour shown him by Johann Georg must have persuaded him to abandon any thought of another career. The elector's favour, he wrote in January 1651, also helped reconcile his parents to the course that he now seemed destined to pursue. He became in effect Johann Georg's Kapellmeister. The official title still lay in the hands of Rogier Michael, and Praetorius remained, at least in principle, on call to direct the Kapelle on special occasions; but in practice, as Schütz recollected in a petition of 11 April 1651, 'I served for both of them, as I was still young'.

On 1 December 1616 Landgrave Moritz wrote to Johann Georg asking that Schütz 'rejoin us here at the earliest possible moment'. Although the composer's two-year leave had not yet expired, 'things have now befallen us such as would cost us extraordinary pains to remedy without Schütz' – probably the aged Kapellmeister Georg Otto had become incapable of performing his duties any longer. Moritz added that he also wanted Schütz as a tutor for his sons. Christoph von Loss, who received the letter, advised Johann Georg against complying with the landgrave's request. In a memorandum of 11 December Loss wrote that the electoral Kapelle could not do without someone particularly well versed in composition, knowledgeable about instruments and experienced in performance – in all of which, in my humble estimation, I know at present of no one preferable to the aforementioned Schütz . . . I fear that if we were now to dispose of him like this it would be difficult to obtain such a person in his place . . .

Loss advised that the elector seek to obtain Schütz permanently, 'so that you might henceforth not stand in danger of such a demand' as the landgrave's. Following a draft provided by his councillor, Johann Georg responded to Moritz two days later. Reminding the landgrave of the delay in Schütz's arrival at the Saxon court in 1615 and adding that a period of mourning early in 1616 had made it impossible to use his services 'for almost an entire year', the elector asked that Moritz 'be pleased to show us such kindness as to let us have the said Schütz entirely'. On 16 December Schütz – evidently not yet aware of the elector's intentions – wrote to the landgrave that

His Grace the Elector of Saxony has given me to understand that . . . I should expect no leave, doubtless because His Grace cannot dispense with my humble services in view of the coming feast days and the absence of Michael Praetorius . . . but I have no doubt that once the feast days are past it will please His Grace that I pay heed to this command from Your Grace and betake myself most quickly to Kassel . . .

Moritz replied to the elector on 24 December. Stressing once again Schütz's indispensability to him, he nevertheless agreed to let him remain in Dresden 'a while longer' until he 'might bring your Kapelle to the desired condition'; realizing, however, that this would hardly satisfy Johann Georg, Moritz asked

if it could not be so arranged that we do indeed surrender Schütz entirely to you but that he nevertheless remain in our employ and duty so that we might also make use of him now and then as the occasion arises.

Johann Georg rejected even this compromise; writing to Moritz on New Year's Day 1617, he reiterated his insistence that Schütz remain '*solely* in our employ', adding that the landgrave might borrow '*others* among our servants' should the need arise. The landgrave, whose political situation demanded that he remain on good terms with Saxony, had no choice but to capitulate; in a letter of 16 January he ceded Schütz permanently to the elector. On 17 February Johann Georg conveyed his thanks to Moritz in a letter delivered by Schütz himself, who returned to Kassel to collect his remaining effects. Schütz left Kassel for the last time late in March; on his departure, wrote Geier, the landgrave gave him a 'chain and portrait' – presumably a cameo of the sort worn by Schütz in his two authentic likenesses, although the medallions there depict Saxon rulers, not the landgrave. Following the death of Otto in November 1618, Moritz tried once again to regain Schütz from Johann Georg; the elector, replying on 25 January 1619, rejected the idea out of hand, and the post of Hesse Kapellmeister went to Christoph Cornet. Schütz nevertheless maintained amicable ties with Kassel. Not only did he continue to send his music there for at least three decades, but Moritz and his successor, Wilhelm V, seem occasionally to have entrusted young musicians to his care; a list of choirboys at the electoral court in December 1625, written in Schütz's hand, notes the presence of 'Hans Ende, sent for a year by Landgrave Wilhelm V of Kassel'.

Despite the intensity of his efforts to prise Schütz away from Landgrave Moritz, Johann Georg appears at first to have had reservations about formally naming him electoral Kapellmeister. Rogier Michael held the title until at least 30 April 1617, and a pay record of the following year reveals that the court still set aside an honorarium for Praetorius. Schütz appears in this document as 'organist and musical director'; in printed and manuscript sources covering the period from July 1617 to June 1618 he described himself as either 'musical director' or 'derzeit Kapellmeister' – a term he probably used to mean 'provisional' or 'interim' Kapellmeister. The first unqualified reference to him as Saxon Kapellmeister occurs in Johann Georg's letter to Moritz of 25 January 1619. From at least 1618 until the end of his active career at Dresden he received an annual salary of 400 florins. His duties appear to have consisted above all in the provision of music for major court ceremonies, whether primarily religious or primarily political in nature; as he remarked in his petition of January 1651, he 'most obediently served' Johann Georg at 'imperial, royal, electoral and princely assemblies' as well as at the baptisms of most of the elector's children and the weddings of all of them. He less frequently directed the music performed at ordinary religious observances; from the mid-1620s onwards this task lay increasingly in the hands of a vice-Kapellmeister.

Schütz's letters reveal that he also had the responsibilities of keeping the Kapelle adequately staffed, ensuring proper living conditions for its members and supervising the musical education of the choirboys. He took a particular interest in those among his pupils who showed creative gifts; as early as 23 September 1616 he observed in a memorandum to Christoph von Loss that a choirboy named Johannes – possibly Johann Vierdanck – 'has made a good, solid start in composition, so that something is probably to be expected of him one of these days'. During the following decades Schütz taught several notable composers, among them his cousin Heinrich Albert, Christoph Bernhard, Johann Klemm, Johann Theile and Matthias Weckmann.

Schütz's first important opportunities to present the kind of sumptuous musical display favoured by his employer came within a few months of his permanent transfer to Dresden. On 15 July 1617 the Emperor Matthias and his family came to the electoral court on a state visit. Schütz furnished the text – and no doubt the music – for a mythological ballet presented in honour of the emperor ten days later; with the possible exception of a single number, only the libretto survives. According to an ordinance for the festivities Schütz also had to make certain that 'good music. . .is not wanting. . .in church' and 'in general see to it that His Grace's Kapelle win the praise and admiration of the visitors'. At the end of October the court held an elaborate celebration for the centenary of the Reformation. A detailed account of the event by the court chaplain Matthias Hoe von Hoenegg indicates that Schütz and his musicians – 16 singers and an even larger body of instrumentalists – performed at least four concerted compositions at each of the three church services that marked the occasion; the works included some pieces that Schütz subsequently published in his *Psalmen Davids* (SWV41, 43 and 45, and perhaps 35 and 47). Not all the music from this period, however, maintains such a grand scale. The madrigal SWV Anh.1, possibly an adaptation of an Italian model, seems to date from the early Dresden years as well, as so may the Marenzio parody SWV450, at least in its first version (SWV450*a*).

As director of the largest and most important musical

establishment in Protestant Germany, Schütz inevitably found the scope of his activities widening beyond the confines of Dresden. In December 1617 he went to Gera to advise Prince Heinrich Posthumus of Reuss on the reorganization of music in the town and its schools and at the prince's court. Schütz and Heinrich Posthumus had perhaps met when the prince visited Dresden two years earlier; Schütz's brother Georg, who served as a tutor to the prince's eldest son in 1617 and possibly in 1616 too, could also have acted as an intermediary between the composer and Heinrich Posthumus. As a native of Reuss Schütz no doubt felt bound to the prince by ties of duty, and the two men may also have shared a deep mutual esteem – Heinrich enjoyed a substantial reputation both for his qualities as a ruler and for his intense devotion to music. No evidence exists, however, for the close friendship between them posited by modern scholars.

In 1618 the authorities of Magdeburg Cathedral asked Schütz, along with Praetorius and Samuel Scheidt, to oversee the reorganization of their Kapelle. In the same year Schütz wrote and published the wedding concertos swv20 and 21, the first for the marriage of the Saxon consistorial councillor Joseph Avenarius and Anna Dorothea Görlitz on 21 April, the second for the wedding of the Leipzig jurist Michael Thomas – a friend of the composer's brothers Georg and Benjamin – to Anna Schultes on 15 June. The following spring he published his first collection of sacred music, the *Psalmen Davids sampt etlichen Moteten und Concerten* swv22–47; he dedicated the volume to Johann Georg. The appearance of the psalms, which fulfilled his ambition to 'distinguish myself properly by bringing forth a worthy piece of work', coincided with the preparations for his own wedding: on 1 June 1619 he married Magdalena Wildeck, the 18-year-old daughter of an official at the electoral court. Schütz sent copies of the print along with wedding invitations to church and city councils throughout Germany, several of which responded with generous gifts; he postdated the foreword to the day of the ceremony.

On 9 August 1619 Schütz's brother Georg married Anna Gross in Leipzig; the composer no doubt attended the service, for which he wrote the concerto SWV48. Six days later he joined Praetorius, Scheidt and Johann Staden in Bayreuth at the inauguration of an organ that Margrave Christian of Brandenburg-Bayreuth had had built for the Stadtkirche there. The local organist Elias Unmüssig recalled in a poem written some years later that the four visitors all 'played splendidly' on the new instrument. The audience at the ceremonies included Heinrich Posthumus; Scheidt, who dedicated his *Pars prima concertuum sacrorum* (1622) to the prince, wrote in the foreword to the collection that he would

> always remember with what pleasure you listened to the late Michael Praetorius, Heinrich Schütz and me . . . as we sang the praises of almighty God at the assembly of princes and potentates in the illustrious court of Bayreuth.

Schütz may have travelled to Bayreuth with Prince Heinrich, who departed from Gera – some 55 kilometres south of Leipzig – on 12 August. Spitta (1894) interpreted a small payment to 'old Schütz from Weissenfels' in the records of expenses for the journey as an indication that Christoph Schütz went to Bayreuth as well; in all probability, however, the money represents a gift on the occasion of Georg Schütz's marriage.

The *Psalmen Davids* and the three wedding concertos that surrounded them inaugurated a period of steady and varied productivity for Schütz. A number of works in manuscript – swv263–4*a*, 289*a*, 326*a*, 429*a*, 430*a*, 450, 457, 459, 464, 475, 497 and Anh.*k* – seem to have originated about 1620 or not long afterwards. The psalm motet SWV51 belongs to this period as well. It appeared in a collection of 16 settings of Psalm cxvi that the Jena tax collector Burkhard Grossman published in 1623. In his foreword to the volume (*RISM* 1623[14]) Grossman wrote that he had commissioned the works four years earlier as a votive offering on his recovery from a near-fatal illness; various problems had caused him to delay publication for more than two years after he had already received all the contributions.

In 1621 Schütz composed a musical tableau for the elector's birthday, which fell on 5 March; only the libretto survives. In October of the same year he and 16 members of his Kapelle went to Breslau as part of the large retinue that Johann Georg took with him to the ceremonies proclaiming the loyalty of the Silesian estates to the Holy Roman Empire. The elector, who had negotiated the peace between the estates and Emperor Ferdinand II after the Battle of the White Mountain in 1620, served as the emperor's deputy at the formal declaration of loyalty on 3 November. Schütz wrote the *Syncharma musicum* SWV49, and possibly also the concerto SWV338, to commemorate the occasion.

In 1623 Schütz published the elegy swv52 – for which he wrote both words and music: to commemorate the interment of Johann Georg's mother, Duchess Sophia, on 28 January at Freiberg. The spring of 1623 saw the appearance of the *Historia der . . . Aufferstehung . . . Jesu Christi* SWV50, Schütz's first major publication since the *Psalmen Davids.* A volume of motets, the *Cantiones sacrae* SWV53–93, followed two years later; Schütz dedicated it to the imperial adviser Prince Johann Ulrich von Eggenberg, whom he had met on the occasion of the Emperor Matthias's visit to Dresden in 1617. Between these two larger projects came a madrigal, *Zwei wunderschöne Täublein zart*, for the wedding of the electoral courtiers Reinhart von Taube and Barbara Sibylla von Carlowitz on 10 February 1624; only Schütz's text for this piece survives. In the summer of 1625 he composed the motet SWV95 on the death of Jacob Schultes – the brother of the Anna Schultes for whose wedding he had provided music in 1618 – and the 'aria' *De vitae fugacitate* SWV94, a memorial for Magdalena Schütz's sister Anna Maria Wildeck, who died on 15 August.

Only three weeks after the death of his sister-in-law, Schütz suffered perhaps the severest personal blow of his life: on 6 September 1625 Magdalena Schütz died after a short illness. The composer and his wife had enjoyed an unusually warm and happy marriage. According to Hoe von Hoenegg's funeral oration for Magdalena, Schütz 'never knew or heard a more lovely sound or song than when he heard the voice and word of his precious wife'; she 'cared for him daily . . . when he came home from his work she was overjoyed to see him and ran with happiness to greet him'. The couple had two daughters, Anna Justina, born late in 1621 – Hoe described her as 'not yet quite four years old' at the time of her mother's death – and Euphrosina, born on 28 November 1623; Schütz, feeling unable to bring them up by himself, eventually placed them in the care of their maternal grandmother. He expressed his grief over Magdalena's loss in the continuo song SWV501, published as an appendix to Hoe's sermon. Contrary to custom, he never remarried.

In the period immediately after his wife's death, Schütz seems to have devoted himself chiefly to the composition of the so-called Becker Psalter (SWV97*a*–256*a*), a collection of simple partsongs based on the popular psalm paraphrases of the Leipzig theologian Cornelius Becker. The volume appeared early in 1628 with a dedication to the Danish-born Dowager Electress Hedwig, widow of Johann Georg's brother Christian II. In the foreword Schütz wrote that the project had grown out of occasional settings of Becker's psalms that he had made 'for the morning and evening devotions of the choirboys placed in my charge'; he had little thought of adding to these pieces, however, until

> it pleased God the almighty . . . that the sudden death of my late dear wife . . . bring to a halt such other work as I was engaged in and put this little psalter in my hands, as it were, so that I could draw greater comfort from it in my sorrow . . .

Schütz dated the foreword on the second anniversary of Magdalena's death.

In speaking of 'other work' that he had set aside Schütz may have meant a second volume of *Psalmen Davids* – as late as the 1660s he referred to the publication of 1619 as 'Part i'. He appears to have made at least a start on this undertaking at about the time the Becker Psalter went to press. The Kassel manuscripts include a series of four concerted psalms – SWV462, 466,

1. 'Heinrich Schütz, Kapellmeister to the most serene elector of Saxony, aged 42': engraving (1627/8) by August John

500 and Anh.7 – evidently copied not long after March 1627; stylistic considerations suggest further that the psalm swv476, although transmitted in a considerably later source, belongs to this group as well. Yet another psalm, swv473, survives in a manuscript copied in the late 1620s or early 1630s; but in this instance, the style of the music points to a date after 1629.

In the spring of 1627 Schütz and his Kapelle spent a month at the castle of Hartenfels at Torgau, where the elector mounted a lavish series of entertainments to celebrate the wedding of his daughter Sophia Eleonora to Landgrave Georg II of Hessen-Darmstadt. The ceremony took place on 1 April; 12 days later, in the words of a chronicler, 'the musicians enacted with music a pastoral tragi-comedy about Daphne'. The laconic reporter had in fact witnessed the performance of the first opera created in Germany. Martin Opitz adapted the libretto from the *Dafne* written by Rinuccini for Peri more than 30 years earlier; Schütz's music, like that of all his stage works, does not survive. In a more modest contribution to the festivities he wrote a little duet, swv96, to conclude a set of German villanellas written by his colleague Johann Nauwach and dedicated to the landgrave and Sophia Eleonora on their wedding day (*RISM* 1627[9]). A similar work, the canzonetta swv441, survives in a partially autograph manuscript dating from about this time or not long afterwards and thus presumably originated in the middle or late 1620s as well.

In August 1627 Johann Georg visited Heinrich Posthumus at the castle of Osterstein, near Gera, and it is possible that Schütz accompanied him. The presence of a libretto for *Dafne* in the castle library has led some writers to suggest, without apparent foundation, that Schütz directed a performance of at least part of his opera for the prince. In the autumn Schütz went with Johann Georg to the electoral assembly held at Mühlhausen from 4 October to 5 November. He submitted a memorandum listing a group of six singers and 12 instrumentalists whom he hoped to take with him; as a contingency measure he added a reduced list of performers. The elector apparently let him have the larger ensemble, since the concerto SWV465, written for the assembly, exceeds the instrumental forces detailed in the second list.

CHAPTER THREE

Middle age, 1628–45

Although Saxony did not take part in the Thirty Years War for more than a decade after it began in 1618, economic pressures began to affect the electoral court in the late 1620s. On Palm Sunday 1628 the singers and instrumentalists of the Kapelle submitted a petition, written for them by Schütz, asking Johann Georg for back wages; 'those who were not in Mühlhausen and Torgau', the document emphasized, 'have been given barely one month's pay in the space of a year'. In a memorandum of 14 July Schütz himself had to ask for some expenses owed him for a year; from this time on the theme of financial worry became increasingly persistent in his correspondence. Hampered by the deteriorating conditions at court, he resolved, in Geier's words, 'to journey abroad once again'. On 22 April 1628 he wrote to the elector asking permission to pay another visit to Italy; the letter makes it clear that he had broached the subject on previous occasions. Johann Georg eventually granted the request, and on 11 August, according to Geier, Schütz left Dresden. On 3 November, new style, he wrote from Venice announcing his arrival after a long and difficult journey. Unexpected delays had made the voyage far more costly than he had predicted, and he asked Johann Georg for additional money to help defray future expenses.

Since his first visit to Italy, Schütz wrote in his

letter, 'everything has changed, and the music in use at princely banquets, comedies, ballets and other such productions has markedly improved'. He directed his energies to absorbing the new developments. In an elegy printed by Geier, the Dresden court poet David Schirmer wrote that Schütz enjoyed the aid of 'the noble Monteverdi', who 'guided him with joy and happily showed him the long-sought path'; Schütz later paid tribute to Monteverdi with the concerto swv356, an adaptation of the older master's *Armato il cor* and *Zefiro torna e di soavi accenti.* Monteverdi may have advised him on dramatic monody in particular. Writing on 6 February 1633 to Friedrich Lebzelter, the Saxon emissary in Hamburg, Schütz recalled that

> during my recent journey to Italy I engaged myself in a singular manner of composition, namely how a comedy of diverse voices can be translated into declamatory style and be brought to the stage and enacted in song – things that to the best of my knowledge . . . are still completely unknown in Germany . . .

His words do not make clear whether he actually composed an opera in Venice or merely studied the methods of the newest theatrical pieces; they do, however, suggest that *Dafne* had not used a true recitative style.

On 29 June 1629, new style, Schütz reported sending a consignment of music and instruments back to Germany and having engaged the Mantuan violinist Francesco Castelli for the electoral court. A passage in the diaries of Philipp Hainhofer, a political agent and art dealer from Augsburg who visited Dresden the following autumn, indicates that the purchase of instruments took Schütz as far as Lombardy – probably to Cremona, where the court had five violins on order. On 24 August, new style, the composer requested an advance against his salary to help repay some debts that he had incurred in preparing for his imminent departure from Italy. During these final weeks in Venice he published his *Symphoniae sacrae* SWV257–76; in the preface, dated 'XIV. Calend. Sept.' (19 August), he described the collection as the fruits of his encounter with the 'fresh devices' used by the newer Italian composers 'to tickle the ears of today'. He dedicated the volume to the elector's musically inclined eldest son, who also bore the name Johann Georg.

Schütz returned to Germany with Castelli and Caspar Kittel, a former choirboy at Dresden whom the elector had sent to Venice in 1624. On 26 October 1629 Hainhofer, arriving home from his visit to Saxony, 'found before me the three musicians who had come from Italy and had already been waiting for me in Augsburg for eight days'. After staying with Hainhofer another two days, Schütz and his party continued on their way, proceeding first to Leipzig – where Schütz, as he recalled in a letter of the following 30 April, received the money advanced him by the elector – and then to Dresden. A receipt signed by the instrumentalist Wilhelm Günther indicates that the musicians reached court by 20 November 1629 and that Schütz brought three new cornetts and four 'cornettini' with him from Venice.

In the months following his return from Italy, Schütz had to provide music for two major celebrations at court: the marriage of the elector's daughter Maria Elisabeth to Duke Friedrich III of Holstein-Gottorf on 21 February 1630, and the centenary of the Augsburg Confession on 5–7 June. Between these events he visited Leipzig, perhaps to buy music at the Easter fair; before leaving he asked for another advance on his salary, explaining that he had unexpectedly had to use the last one to cover transport expenses and thus remained in debt to several merchants. A pay record of the period shows that apart from the advances he had received virtually no salary since 1628.

In January 1631 Schütz published the motet SWV277 as a memorial to his friend Johann Hermann Schein, who had died in Leipzig the preceding 19 November; the printer's foreword indicates that Schein had requested the work when Schütz visited him on his deathbed. The two composers had probably known each other since the years 1613–15, when Schein worked for Gottfried von Wolffersdorff; Schein also had ties with Georg and Benjamin Schütz. The autumn of 1631 brought further bereavement to Schütz with the deaths of his father and father-in-law in early October. The autumn also saw Saxony enter the Thirty Years War for the first time. At an assembly held in Leipzig from 10 February to 3 April 1631 – to which Schütz and his Kapelle accompanied Johann Georg – the German electors and their allies had resolved to steer a middle course between the Catholic League and the anti-imperial opposition spearheaded by King Gustavus II Adolphus of Sweden. But on 11 September Johann Georg formed an alliance with Gustavus Adolphus, and Saxon troops joined the Swedish army at the Battle of Breitenfeld in November. The expense of Saxony's military effort spelt the end of any hope of improvement in the Kapelle. Although 13 singers and a slightly larger group of instrumentalists remained in 1632, their numbers soon declined drastically, and musical activities at the court ground almost to a halt.

On 6 February 1633 Schütz received a letter from Friedrich Lebzelter informing him that Crown Prince Christian of Denmark had asked if he would go to Copenhagen to direct the music at the prince's forthcoming wedding to Johann Georg's daughter Magdalena Sibylla. Responding the same day, Schütz told Lebzelter that the invitation represented a welcome chance to escape a situation in which 'I am of less than no use'. For almost nine months, he wrote, he had sought permission to spend a year in Lower Saxony – which in the language of the time encompassed not only the present-day territory of that name but also Mecklenburg, Bremen and Holstein; Schütz may have particularly wished to visit Hamburg, whose rich musical activity had remained unimpaired by the war. Although none of his requests for leave had met with success, the composer felt that Johann Georg might give more consideration to Prince Christian's wishes. Schütz offered to supply the Danish court with works 'not only of my invention as the poorest but also by the most prominent composers in Europe' and in particular to present Christian – whose fondness for music he knew from the prince's visits to Dresden – with a dramatic composition in the style that he had learnt in Venice.

Three days later Schütz renewed his petition for leave; he volunteered to undertake the proposed journey at his own expense, asking only that 'if in the meantime some payment should be made to the musicians my share be accorded me and not withheld'. On 1 March Prince Christian sent Schütz a passport and addressed a formal request to Johann Georg for his services; at the same time Christian wrote to his aunt, the Electress Hedwig, seeking her support for his plan. Several months appear to have elapsed, however, before Schütz left for Denmark. In July Prince Christian wrote to urge that he come as quickly as possible; but even this does not seem to have had any immediate result. Spitta (1894) suggested that Schütz might have spent the summer months in Güstrow at the court of Duke Johann Albrecht of Mecklenburg-Güstrow, a cousin of whose commissioned a reprinting of the Becker Psalter in 1640; the published accounts of the duke's Kapelle, however, show no trace of Schütz's name. In all probability the composer remained at the electoral court to prepare and direct the music for thanksgiving services held on 6–7 September to commemorate the victory at Breitenfeld two years before. On 18 November Lebzelter wrote to Christian from Hamburg that Schütz and his entourage had arrived there two months earlier and would proceed to Denmark as soon as the prince reached Haderslev, where the royal family maintained one of its residences. A register of the electoral Kapelle from April 1634 reveals that Matthias Weckmann and the instrumentalist Daniel Hämmerlein accompanied Schütz on his journey. While Hämmerlein may have continued the rest of the way to Denmark, Weckmann would appear to have remained in Hamburg; his pupil Johann Kortkamp later reported that Schütz took Weckmann there to study with Jacob Praetorius for a period of three years, and this could hardly refer to any other occasion. Schütz presumably left Hamburg shortly after Lebzelter wrote to Christian; he stopped to visit the prince at Haderslev and then, on 6 December, set off for Copenhagen bearing a letter of introduction to the king's privy councillor Ditlev Reventlow. On his arrival in the capital a few days later he received the title of Kapellmeister to King Christian IV, at an annual salary of 800 reichsthalers, starting on 10 December.

The wedding festivities for the crown prince and Magdalena Sibylla lasted from 3 to 18 October 1634. Schütz appears to have brought a number of musicians from Germany for the occasion to augment the already sizable Danish *kapel*. The organist Michael Cracowit, who had worked in Denmark the previous year, wrote on leaving a post at Danzig that Schütz had invited him to return to his former position; the English flautist John Price came from Dresden and Heinrich Albert, employed at the time at Königsberg, composed an aria for the entry of the Saxon princess into Copenhagen on 30 September. A chronicle published by the Copenhagen bookseller J. J. Holst in Latin (1635), Danish (1637) and German (1648) versions reveals that the entertainments during the celebration included a ballet and two dramatic pieces with music and dancing: a *Comoedia de raptu Orithiae*, performed on 8 October, and a *Comoedia de Harpyriarum profligatione*, given on 12 October. Although Holst does not mention him by name, Schütz no doubt wrote the music to both comedies, especially as a notice in the court records states that the king conferred directly with him about 'the monsters that will be employed' in them. Neither work survives, nor, with a single exception – the canzonetta swv278, sung as part of a tableau in a procession on 13 October – does anything else composed by Schütz for the wedding. Schütz remained in Denmark until May 1635. Perhaps, as Brodde (1972) suggested, news of his mother's death at the start of February prompted his ultimate departure; pressure from Johann Georg, however, who surely wanted his Kapellmeister back from an over-extended leave, seems a more probable cause. On 4 May King Christian wrote a letter of thanks for Schütz to take to the elector in which he asked for the composer's early return 'so that he may bring to completion the work that he has begun with our *kapel*'. As a parting gift the king made Schütz a present of 200 reichsthalers and a gold chain with a portrait – perhaps the one shown in paintings of the composer. Schütz received his final payment on 10 May. After leaving Copenhagen he proceeded south to the palace at Nykøping, where Prince Christian gave him letters to the elector and the Electress Hedwig, and a passport, dated 25 May, to Dresden 'and from there back into this kingdom'. On 14 June Johann Georg wrote to King Christian announcing Schütz's return. Ten days later the court held a festive service to celebrate the Peace of Prague, which created a unified German front – including Saxony – against the Swedes and seemed to promise an early end to the fighting; Schütz no doubt directed the music.

Directly after his return to Dresden, Schütz appears to have assembled a collection of his most recent works for his former employers at Kassel. On 30 March Wilhelm V had written to the composer recommending a musician for the Danish *kapel* and taken the opportunity to ask if Schütz – who had 'previously been in the habit of honouring our Kapelle at all times with your new compositions and pieces but not done so for a while' – would 'send at the earliest possible moment those pieces which you have recently composed and our Kapelle still does not have, and also [would] not object to providing us with whatever else you compose in the future'. Manuscripts still at Kassel as well as an inventory of music at the Hesse court from 1638 indicate that Schütz sent several works later published in the *Kleine geistliche Concerte* (SWV287*a*, 293, 296*a*, 298, 300, 301*a*, 302*a*, 304*a*, 316*a*, 317, 325, 331*a*), the second book of *Symphoniae sacrae* (SWV341*a*, 348*a*, 349, 352*a*, 361*a*) and the *Geistliche Chor-Music* (SWV455), along with a handful of compositions never printed (SWV449 and 460, and a lost setting of *Christ lag in Todesbanden*). One of the pieces in the group, the German *Nunc dimittis* SWV352*a*, bears a dedication to Christoph Cornet, 'ever to serve him with greatest affection'; Cornet, who died of the plague shortly before 2 August 1635, had no doubt commissioned the work as his funeral music.

On 3 December 1635, barely four months after Cornet's death, Heinrich Posthumus died. Schütz wrote the largest and most important of all his funeral compositions, the *Musicalische Exequien* SWV279–81, for the prince's interment on 4 February 1636. In keeping with a frequent practice of the time, Heinrich Posthumus had arranged many details of his own funeral, even leaving instructions about the text and character of the music he wished performed. It does not appear, however, that he actually asked Schütz to compose the *Exequien* for him; the printed order of service for the burial states explicitly that the commission came from Heinrich's widow and sons.

In the autumn of 1636 Schütz published his first collection of music for seven years, the *Erster Theil kleiner geistlichen Concerten* SWV282–305; he dedicated the volume to Heinrich von Friesen, head of the appellate court at Dresden. According to Schütz's preface, the war had forestalled the appearance of 'some of the musical works that I have composed' – perhaps a reference to the series of psalms begun in the late 1620s – 'which I have . . . had to withhold for want of publishers'. The concertos represented an interim measure, taken 'so that the talent God granted me . . . might not lie wholly fallow but rather create and bring forth something, however small'.

By the time the *Kleine geistliche Concerte* appeared, the hopes raised by the Peace of Prague had vanished. France, which had tacitly supported Sweden, entered openly into the war, initiating a new phase of the hostilities that brought catastrophe on a scale previously unimagined. Schütz attempted to return to Denmark. As early as September 1635 Prince Christian reported to one of his father's ministers that the composer had written to offer his services. In a petition for leave from 1 February 1637 Schütz noted that he had left his 'best pieces of music' in Denmark and had a stipend from the crown prince as well as other prospects awaiting him there; he hoped Johann Georg would allow him at least to retrieve the music. Perhaps in anticipation of a prolonged absence from Dresden, Schütz took steps at this time to safeguard interests important to himself and his children. Seeking to extend the protection that his compositions enjoyed in Saxony to the neighbouring territories of the Holy Roman Empire, he petitioned Ferdinand III for a copyright privilege; on 3 April the emperor granted his request and issued a patent forbidding the unauthorized reprint of 'those works of music that he has composed, partly in Latin, partly in German, both sacred and secular, and in part without text', for a period of five years. As Wessely (1953–4) and Steude (1967) suggested, Schütz probably sought the copyright in response to the pirating of the concertos SWV39 and 291 in *RISM* 1638[5], a collection edited in the imperial free city of Nordhausen and apparently published, despite its date, early in 1637. On 26 April Schütz and his brothers Johann and Benjamin instituted proceedings to enrol their daughters as beneficiaries of a family trust administered by the city fathers of Chemnitz. Hard pressed financially because of the war, the town failed to take appropriate action on the matter, and the trust eventually became a source of recurring trouble for Schütz and his relations.

The journey to Denmark evidently never materialized: no mention of it occurs either in Danish records or in Geier's biography. Matthias Weckmann appears to have carried out Schütz's mission in his stead; on 31 July 1637 the elector issued a passport authorizing Weckmann, 'whom we have for some time previous been keeping in Hamburg for further study and practice in music, to travel to Holstein and Denmark, there pick up various items pertaining to his art and belonging to us, and then make his way with them back here to us'. A letter of Schütz's dated 8 November 1638 suggests that he had to abandon going to Denmark because of obligations arising from the death of his brother Georg, which seems to have occurred in the late spring or early summer of 1637 – the last evidence of Georg's activity comes from a poem written in memory of a relation who died on 17 April, while the series of events recounted in the letter would appear to exclude a date later than the end of July. Schütz became responsible for the education of Georg's children, one of whom, whose school had fallen into disarray because of the war, he brought to live with him and installed as a treble in the electoral

Kapelle; at the time of writing the boy had already sung in the ensemble 'for more than a year'. The composer no doubt remained in Dresden during this entire period; a report in Geier of a visit to 'Brunswick and Lüneburg' in 1638 probably refers to Schütz's service the following year at the court of Georg of Calenberg, one of the many dukes in the proliferous house of Brunswick-Lüneburg.

The long series of personal losses that Schütz suffered throughout the 1630s culminated with the death of his elder daughter, Anna Justina, in the early months of 1638 – Geier gives the year, and a letter of Schütz's written 26 May 1641 indicates that the child died before 10 July. In late autumn Schütz and his Kapelle provided music for the wedding of Prince Johann Georg and Princess Magdalena Sybilla of Brandenburg. Town pipers and trumpeters from Dresden and elsewhere joined the depleted corps of electoral musicians for the festivities, which lasted from 13 to 20 November and concluded with a five-act opera-ballet on the Orpheus legend. Only the libretto of this work, by the Wittenberg poet August Buchner, survives; the title-page describes the music as 'composed by the electoral Kapellmeister Heinrich Schütz in the Italian manner'.

In his letter of 8 November 1638 Schütz wrote that he would shortly have to leave Dresden because of pressing family business in Weissenfels and probably remain away for the winter. On 12 January 1639 he and his immediate relations completed the sale of a vineyard that had belonged to Christoph Schütz. Towards the end of spring, after his return to Dresden, the composer published his second volume of *Kleine geistliche Concerte* (SWV306–37); he dedicated the collection to

Christian IV's youngest son, Prince Frederik. In the preface, dated 2 June 1639, Schütz apologized for offering the prince 'so small and simple a piece of work', noting once again how the war made it impossible to bring out the 'other and (without boasting) better works that I have at hand'. Within a few months of the concertos' appearance he took another extended leave from the electoral court, going to Hanover and Hildesheim to serve as Kapellmeister to Georg of Calenberg. The duke's account books show payments to Schütz, at a salary of 500 thalers per year, for a period of 18 months starting at Michaelmas 1639; the earliest reference to his absence from Dresden occurs in an electoral memorandum of 23 November. The origins of Schütz's ties with Georg of Calenberg remain obscure; perhaps the two men had come into contact earlier in the decade when the composer had spoken of visiting Lower Saxony. On the way to Georg's court Schütz apparently stopped in Halle, where the elector's son August – the same August whose baptism had first brought him to Dresden – resided as administrator of the archbishopric of Magdeburg. Johannes Zahn, organist of the city's Marktkirche, sought Schütz's aid in a salary dispute with the church fathers; 'when I gave the electoral Saxon Kapellmeister Heinrich Schütz to understand of my hardship and duties a short while ago', Zahn wrote in a petition, 'he was astonished at my duties and the meagre recompense, and thus spoke warmly on my behalf to the head of the council'. On 24 December Georg of Calenberg's treasury recorded a payment of 8 thalers to a burgher in Hanover for 'what Kapellmeister Schütz . . . consumed in his house'. Schütz had probably not stayed there very long; Duke Georg, despite having established

Hanover as his official residence, actually lived in Hildesheim, some 30 kilometres away. An entry in the autograph album of a young theology student, with the characteristic inscription 'Ut Sol inter planetas, Ita MUSICA inter Artes liberales in medio radiat', attests Schütz's presence at Hildesheim on 29 January 1640. From July to September of that year Georg met with Amalia Elisabeth, Landgravine of Hessen-Kassel, in Göttingen to form a military alliance; if, as seems likely, Schütz accompanied the duke on the journey, he would have had an opportunity for renewed contact with the court at which he spent the greater part of his youth.

Schütz stayed less than a full 18 months with Georg of Calenberg; his correspondence shows that he returned to Dresden by the first week of 1641. He found the electoral Kapelle in a state of almost total collapse. In November 1639 Johann Georg had made an effort to strengthen the ensemble by appropriating the services of some young musicians employed by his eldest son; but even with this measure, the group numbered barely ten members, and salaries, despite frequent pleas for assistance, continued to go unpaid. Writing to the elector on 7 March 1641, Schütz likened the situation to that of a patient in his death throes. Mindful that the war prevented an immediate restoration of the Kapelle to full strength, he urged that Johann Georg at least ensure its eventual rebirth by making provision for the training of eight boys as singers and instrumentalists; nothing, however, seems to have come of the proposal. At the close of the letter Schütz revealed that he himself had just recovered from a grave illness; fear of impending death might account for the uncommonly agitated tone of a letter of 17 February concerning the overdue payment of 1000 florins borrowed from a trust that he had established for his daughter. Moser (1936) suggested that Schütz wrote the concerto swv346–7, published in the second book of *Symphoniae sacrae*, as a token of gratitude for his recovery; the idea would seem specially plausible in view of the fact that the first part exists in a manuscript version copied in the early or mid-1640s (swv346*a*).

Schütz signed his letter of 7 March 1641 with the remark 'written and left behind in Dresden'; he evidently went to Weissenfels, perhaps to visit his sister Justina, the only member of his family who still lived there. On 26 May he wrote from Weissenfels to ask the elector's help in recovering the money owed to his daughter's trust; the elector, whose eldest son had incurred a large debt with the very parties who had borrowed the 1000 florins from Schütz, failed to take any action on the matter. On 14 September, in Dresden once again, Schütz prepared orders assigning – or, more likely, reassigning – Weckmann and three other musicians to the service of Prince Johann Georg.

Schütz spent most of the years 1642–4 in Denmark. Payments to him as chief Kapellmeister, once again at an annual salary of 800 reichsthalers, began on 3 May 1642. Whether he actually arrived at the Danish court by then, however, seems open to question. Only eight days earlier he wrote from Dresden to Ferdinand III seeking renewal of the copyright privilege that the emperor had granted him in 1637; on 15 July he made arrangements for adding a small complement of choirboys to Prince Johann Georg's fledgling Kapelle. An undated memorandum of the period suggests further that he remained in Dresden to lead the music at the baptism of the prince's first daughter, which took place in late September. Schütz would doubtless have reached Denmark, however, by the end of October, since Geier and others report that he directed the music at the double wedding of King Christian's twin daughters in November. The funeral sermons for Andreas Gleich and Clemens Thieme indicate that both musicians accompanied Schütz to Copenhagen. Moser (1936) wrote that Weckmann, Philipp Stolle and Friedrich Werner went as well; but none of these three appears in Danish records until 1643.

Schütz received his last payment from the Danish court on 30 April 1644. After leaving Copenhagen he went again to the Brunswick-Lüneburg territories. Until the spring of 1645 he seems to have lived mostly in the city of Brunswick, possibly in the home of a merchant named Stephan Daniel, whom later correspondence describes as a friend. His known artistic activities during this period all centred around the nearby court of Wolfenbüttel, the residence of Duke August the younger of Brunswick-Lüneburg; Schütz may have come to August's notice through Georg of Calenberg, a cousin of the duke's. Musical affairs at court appear to have lain chiefly in the hands of August's wife, Sophie (or Sophia) Elisabeth, a princess from the ducal house of Mecklenburg-Güstrow and a composer of considerable talent. A letter of Schütz's, written from Brunswick on 22 October 1644, reveals that he advised the duchess in her creative efforts – acknowledging the receipt of some arias that she had sent him, he remarked that 'Your Grace has made notable gains as a result of my modest instruction' – and that she aided him in negotiating the sale of a small organ belonging to a church in Hamburg, where he had presumably stopped on the way from Denmark. The letter further hints at a more ambitious collaboration between the composer and Sophie Elisabeth. Following a visit to Hildesheim on personal business, Schütz wrote, he planned to meet the duchess at Wolfenbüttel 'to discuss and consider whatever is necessary to the completion of the musical work that we have at hand'. The preparations no doubt concerned the *Theatralische neue Vorstellung von der Maria Magdalena*, a spectacle presented at the end of the year. The court poet Justus Georg Schottelius later published two songs from the work with the remark that Schütz had provided the settings. Sophie Elisabeth presumably contributed to the piece as well, as Schütz's reference to her arias could indicate.

On 23 February 1645 Schütz stood godfather to the third child of Delphin Strungk, the organist at St Mary's, Brunswick; on 17 March, after a short absence from Brunswick, he wrote to tell Sophie Elisabeth that a musician whom he had recommended to her and her husband had unexpectedly left the area. In the weeks that followed Schütz probably went to Wolfenbüttel to attend a festive birthday celebration for Duke August held there on 10 April; a poem of his printed five years later in a volume of panegyrics to the duke may come from this occasion. Schütz left the region of Brunswick and Wolfenbüttel soon afterwards; on 21 May he wrote from Leipzig to inform Johann Georg of his imminent return to Dresden. In Leipzig, Schütz evidently made the acquaintance of the young Johann Rosenmüller, to whose *Paduanen . . . mit drey Stimmen*, published in the autumn of 1645, he contributed a gratulatory poem 'sent from Dresden' following his arrival there.

CHAPTER FOUR

Old age, 1645–56

In 1645, when he was almost 60 years of age and with over 30 years of service to the Saxon court, Schütz felt that the time had come to withdraw from active duty as Kapellmeister. 'Since the electoral Kapelle has gone completely to ruin in these parlous times, and I in the meanwhile have grown old', he wrote in his letter of 21 May 1645, 'it is now my only wish that I might henceforth live free from all regular obligations'. Schütz asked for a pension of 200 thalers a year along with the right to retain his title and direct the Kapelle on special occasions, 'particularly when foreign rulers or emissaries are present'. In a memorandum of 28 September, written as he prepared to leave Dresden for a journey to Leipzig and Weissenfels, he added that he hoped to set up a home with his sister at Weissenfels; he requested permission to remain there until Easter and to return whenever possible in the future. Schütz no doubt harboured the thought of settling permanently in the city of his childhood, where, with 'more peaceful circumstances and greater freedom', he could 'complete the various musical works that I have begun'. Johann Georg granted his wishes only in part: he appears to have let him go to Weissenfels almost every autumn or winter for the next decade, but he did not let him retire.

The journey referred to in Schütz's memorandum of 23 September 1645 lasted, as planned, well into the following year. On 9 March 1646 the Dresden court organist Johann Klemm – who appears not infrequently to have served the composer as a secretary and copyist – noted in a letter to Johann Hermann Schein's son Johann Samuel that Schütz had 'recently been in Leipzig but [is] now once again staying in Weissenfels'. Schütz presumably kept his promise to return to court at Easter. He remained only briefly. In a memorandum submitted on 30 July he revealed his intention of going back to Weissenfels, and he left Dresden within a few weeks. He evidently went first to Calbe an der Saale, where August of Saxony, the administrator of Magdeburg, had a summer residence; August's betrothal, to Princess Anna Maria of Mecklenburg-Schwerin, on 6 September may have provided the occasion for the visit. On 7 September Schütz wrote from Calbe to an otherwise unknown correspondent named Christian Schirmer, who had solicited his opinion on the celebrated dispute between Paul Siefert and Marco Scacchi. In 1643 Scacchi had published a treatise entitled *Cribrum musicum ad triticum Syferticum* in which he subjected some compositions of Siefert's to a withering critique for their contrapuntal deficiencies; Siefert, who had provoked the attack by claiming that Italians no longer knew how to write correct counterpoint, responded in 1645 with a polemic entitled *Anticribratio musica ad avenam Schachianam*. Schütz declined to render an immediate judgment on the matter since, as he informed Schirmer, he had not had the chance fully to digest the relevant documents. Nevertheless, he wrote,

I for my part would indeed have wished that Master Siefert had not instigated this affair, since in my estimation, as far as I have been able to gather from a quick and casual perusal, Master Scacchi . . . is an extremely well-grounded musician; and thus I will have no choice, it seems to me, but to agree with him in many things, especially when, after three months' travel, I have returned to Weissenfels, my new home and dwelling, and have examined these writings more carefully.

In a second letter, evidently written early in 1648, Schütz affirmed his support for Scacchi, observing that the ideas set forth in the *Cribrum musicum* recalled the precepts according to which 'I too in my youth was drilled and instructed by my teacher Giovanni Gabrieli of blessed memory'; he further urged the completion of a counterpoint treatise promised by Scacchi, 'with which he will surely create something of great value, above all to our German nation'.

On 30 October 1646, writing from an unstated location to the organist Martin Knabe at Weissenfels, Schütz indicated that he would reach Weissenfels in about three weeks. Midway through the winter he made a short visit to Weimar to attend a birthday celebration for Eleonora Dorothea, the wife of Duke Wilhelm of Saxe-Weimar. The duke noted in his journal that Schütz arrived at court on 7 February; five days later 'Herr Schütz and Magister Dufft' – the poet Christian Timotheus Dufft, who had come from Gotha for the occasion – 'had a most excellent song of thanks performed by the Kapelle'. The composition, SWV368, appeared in print soon afterwards. Schütz left Weimar on 15 February. His correspondence reveals that he returned to Weissenfels and stayed there at least until Palm Sunday – 11 April in 1647 – after which he went back to Dresden, where he arrived on 26 April.

Shortly after his return to Dresden, Schütz brought out his first collection of music for eight years, the *Symphoniarum sacrarum secunda pars*, SWV341–67; he dedicated the volume to Prince Christian of Denmark. In the prefatory material – the date of which, 1 May 1647, may represent the third anniversary of his departure from Copenhagen – Schütz wrote that he had in fact completed the *symphoniae* 'some years ago' and presented them in manuscript to Christian while serving at the Danish court. The continuing disruptions caused by the Thirty Years War, as well as the knowledge that few German performers had a sufficient understanding of music composed in the 'modern Italian manner', had made him reluctant to publish them; only when they began to circulate widely in faulty copies did he change his mind. Apart from the *Symphoniae sacrae*, few compositions of Schütz's occur in sources of the early and mid-1640s. The Breslau organist Ambrosius Profe included the concerto SWV338 and the dialogue SWV339 in the collection *RISM* 1641[3]; five years later the concerto SWV340 appeared in another volume assembled by Profe, *RISM* 1646[4]. Manuscripts from this period in Kassel contain the dialogues SWV444 and 477, and the concerto SWV456. A related group of sources in the same collection includes the madrigals SWV438 and 442, the concerto SWV469 and a handful of works subsequently published in the third book of *Symphoniae sacrae* (SWV398*a*, 401*a*, 406*a*, 416*a*, 418*a*); but these copies probably originated closer to the end of the decade. The *Magnificat* SWV468, a work found outside Kassel in a source of uncertain date, so closely resembles the larger pieces in the Kassel manuscripts that its composition must surely have fallen in roughly the same period as theirs.

August 1647 brought one of the few truly happy events of Schütz's later years: the engagement of his daughter Euphrosina to Christoph Pincker, a jurist in Leipzig. In December Schütz wrote to Wilhelm of Saxe-Weimar and August of Saxony inviting them to send representatives to the wedding. August had celebrated his own marriage only a short time before, on 23 November. Schütz presumably supplied music for the occasion; neither his correspondence nor any other documentation, however, indicates that he attended the ceremony, which took place in Schwerin. A covering note added to the invitation for Duke Wilhelm, on the other hand, reveals that the composer made a short visit to Leipzig at the start of the new year in connection with his preparations for Euphrosina's marriage. The wedding took place in Dresden on 25 January; two days later, the musicians of the electoral Kapelle presented the bride and groom with a ballet whose libretto, signed by the court singer Johann Georg Hofkontz, praised Schütz as 'the Orpheus of our time'.

Between 15 and 28 January 1648 Georg of Calenberg's brother Friedrich, the Duke of Celle, ordered a payment to Schütz of 20 thalers. The sum probably represents a combined wedding gift and honorarium for the second book of *Symphoniae sacrae*, a copy of which appears in a later inventory of music belonging to the Celle Kantorei. On the other hand, two local historians, possibly drawing on documents no longer extant, have stated that Schütz served as Kapellmeister to Friedrich's court for some years after 1648 (Schuster, 1905; Cassel, 1930). If so, he would no doubt have functioned only in an advisory capacity; there seems no reason to imagine that he actually visited Celle at any time.

In his wedding invitation to Wilhelm of Saxe-Weimar

Schütz expressed the hope of getting permission to go to Weissenfels as soon after the ceremony as possible – 'to reactivate', he added in the covering note, 'my musical veins, which have lately become dried out as never before, and set them flowing again'. In the event, however, he seems to have remained in Dresden; a memorandum concerning the appointment of a young musician shows him at court on 29 March. He presumably devoted the greater part of his time to overseeing the publication of his *Geistliche Chor-Music* SWV369–97, a set of motets that – to judge by his remarks to Duke Wilhelm – he must have completed well before the end of the preceding year; the volume appeared with a foreword dated 21 April. Written in a carefully regulated contrapuntal style and, according to the foreword, with the aim of encouraging young composers to crack the 'hard nut' of traditional polyphony before they attempted the concerted idiom, the *Geistliche Chor-Music* appears to represent a practical response on Schütz's part to the Scacchi–Siefert controversy; several remarks in the foreword echo both Scacchi's *Cribrum musicum* and the composer's own recent commentary on the dispute. Schütz underlined the pedagogic intent of the volume by dedicating it to the civic fathers of Leipzig as a present for the choir of the Thomaskirche, already famous as one of the most important training institutions in Germany.

On 17 July 1648 Johann Georg wrote to his 'very dear cousin and kinsman' Duke Wilhelm of Saxe-Weimar that 'our most esteemed dear cousin's delegated Kapellmeister Heinrich Schütz has dutifully come to see us concerning his departure and in connection therewith given us to know that he is of a mind to attend most obediently upon Your Excellency'; the title given Schütz suggests that the elector had formally placed him at the disposal of Duke Wilhelm, whose Kapelle evidently lacked a permanent director at that time. Schütz himself presumably took the letter to Weimar. Thiele ('Heinrich Schütz und Weimar', 1954) proposed a connection between this visit and a possible homecoming celebration for Wilhelm's eldest son, who had returned on 8 July from a long voyage; the duke's journal, however, sheds no further light on the matter. Schütz no doubt took advantage of his presence in Weimar to supply the court with copies of his newest works. An inventory of the ducal Kapelle of 1662 lists a substantial number of compositions by him, several of which no longer exist; the lost works include a pair of German sacred madrigals 'that he left with us the last time he came here'.

On 12 December 1648 Schütz bought a house in Halle; the reason for the purchase remains unknown. A letter from his son-in-law to the city fathers of Chemnitz – against whom Schütz had taken action over some money owed his daughter from the legacy that they administered – reveals that the composer had returned to Weissenfels by the beginning of February 1649. He remained there until well into the spring. On 16 May of that year Hofkontz wrote in a memorandum to the elector that Schütz 'has now once again sat for a year in Weissenfels and concerned himself with the Kapelle hardly or not at all'; although Hofkontz's peevish tone no doubt stemmed at least in part from personal animosity – Schütz had prevented his receiving a promised appointment as vice-Kapellmeister – his complaint about the neglect of the Kapelle does not seem wholly unjustified. Whether or not Hofkontz's remarks had any repercussions, Schütz's correspondence shows that he returned to Dresden within little over a month and stayed there without interruption for two years or slightly longer.

During Schütz's absence from Dresden the Peace of Westphalia, signed on 24 October 1648, at last brought the Thirty Years War to a close. But the peace had little immediate effect on musical conditions at the electoral court. Although both Johann Georg and his eldest son increased the size of their Kapellen – by the spring of 1651 the elector had 19 singers and instrumentalists in his employ, the prince 18 – the court remained heavily in debt, and the musicians continued to go without pay. Ceremonies commemorating the end of the war did not take place until 22 July 1650, following the departure of the occupying Swedish forces. An ordinance of the day's events reveals that the music performed included the early psalm SWV45; manuscript and stylistic evidence suggests further that Schütz composed the psalm SWV461 for the occasion. From 14 November to 11 December the court held a series of festivities for the marriage of Johann Georg's youngest sons, Christian and Moritz, to the princesses Christiana and Sophia Hedwig of Holstein-Glücksburg. At the nuptial service on 19 November, a chronicler reported, 'some pieces composed for this princely wedding by the Kapellmeister Heinrich Schütz were performed'. Some writers have suggested that Schütz also furnished the music for the ballet *Paris und Helena*, to a libretto by David Schirmer, which Prince Johann Georg had presented two weeks later; but no solid evidence for his authorship seems to exist.

In the brief period between the peace celebration and the wedding Schütz brought out his *Symphoniarum sacrarum tertia pars* SWV398–418; the foreword bears the date Michaelmas 1650. Schütz seems to have regarded this volume – the last collection of new works that he published – as the culmination of his long career in Dresden. He dedicated it to Johann Georg, to whom he had dedicated his first publication as Saxon Kapellmeister; he accompanied its presentation to the elector with the often quoted petition of 14 January 1651 (see fig.2), in which, after recounting the circumstances of his life up to his arrival in Dresden and reviewing what he had achieved there, he made a renewed plea for release from his duties. Given the toll, he wrote, of his lifelong

> study, travel, writing and other constant work (which has been inescapably called for by my demanding profession and office, the difficulty and gravity of which, in my estimation, only very few . . . can truly comprehend . . .)

and given his 'old age and failing eyesight and strength', he felt that he could no longer provide suitable leadership for the Kapelle; 'if, moreover,' he added, 'I am not to endanger and even utterly destroy my health, I must, on the advice of my physicians, refrain as far as possible from sustained study, writing and reflection'. He also feared that he might suffer the same fate as an elderly Kantor of his acquaintance

> who wrote to me a while ago complaining bitterly that the young councillors of his town were very dissatisfied with his old style of music and were thus eager to be rid of him, and therefore told him straight to his face in the town hall that after 30 years neither a tailor nor a Kantor was still of any use in the world . . . and while I would not expect this from Your Electoral Highness's sons . . . such a thing could befall me at

2. First page of Schütz's petition (dated 14 January 1651) to Elector Johann Georg I of Saxony the hands of others, even perhaps from some of the up-and-coming young musicians . . .

As he had done six years earlier, he asked the elector to

free me from my regular obligations (so that I may gather together and complete what remains of the musical works that I began in my youth and have them printed for my remembrance)

and requested a pension and the right to keep his title. He reported that the castrato Giovanni Andrea Bontempi, a singer in Prince Johann Georg's Kapelle and an aspiring composer, had offered to conduct for him; Schütz warmly supported the proposal and hoped that the elector would acquiesce in it.

Johann Georg evidently returned Schütz's petition unread. On 29 February the composer wrote to the elector's private secretary, Christian Reichbrodt, submitting the document a second time; 'since I fear', he remarked,

that it might be rather on the long side for you to present to our most gracious lord and for him to listen to, I would ask most dutifully that you do not hurry unduly to present it but rather wait for the proper time, even if it should take another quarter year . . .

Neither this attempt, however, nor a reminder contained in a note to Reichbrodt of 11 April appears to have had any effect. Meanwhile the lot of the Kapelle grew worse. On 14 August Schütz and Hofkontz – who would seem to have made peace with one another – appealed to the elector's son Christian, the administrator of the Kapelle, to intercede with Johann Georg on behalf of the musicians, who had received no money in years and 'live in such misery as to move a stone in the earth to pity'. Five days later Schütz reported to Reichbrodt that

there is such unbearably great distress and lamentation among the elector's entire musical company that I for my part . . . would, God knows, rather be a Kantor or organist in some little town than stay any longer in such circumstances . . .

Things had got so bad for one of the singers, Schütz had heard, 'that he is living like a sow in a pigsty, has no bedstead, sleeps on straw [and] has already pawned his coat and doublet'. He himself had lent the musicians whatever he could but no longer had the means to do so. Hopeful nevertheless of some improvement, he suggested a number of measures that he thought might help; he even volunteered to defer his plans for retirement if the elector 'would show me the kindness of assigning me a young, qualified substitute' – Schütz now proposed Christoph Bernhard, Bontempi evidently having become director of Prince Johann Georg's Kapelle – to conduct in his place.

Not long after his letter to Reichbrodt Schütz left Dresden for a short visit to Weissenfels. City records there show that he purchased a house before the end of the year; two letters of Bernhard's indicate that Schütz returned to Dresden between 17 November 1651 and 24 January 1652. On 4 February 1652 Schütz wrote to inform Reichbrodt that four members of the Kapelle had resolved to leave the elector's service. On 28 May, writing again to Reichbrodt after three weeks of inactivity due to an attack of erysipelas, he reported that the Kapelle now stood to lose the singer about whom he had told him the previous year:

For want of money the bass recently pawned his clothes again and then stalked about his house just like a beast in the forest; now he is bestirring himself once more and has had his wife tell me that he must and will be gone. . . . But it is dreadful, truly dreadful, that such an exquisite voice should be lost to the Kapelle. To be sure, nothing else about his character is worth particularly much, and his tongue wants its daily wash in the wine-barrel; but a throat that wide needs more moisture than many a narrow one, and even if the good fellow ever really got his meagre salary it would not be enough for great banquets; and if anyone wants to see how this fellow can really behave and maintain his house, then he should, I believe, give him his pittance when it is due – but so long as this does not happen, one can hardly call him a spendthrift . . .

The continued lack of response to his petition for retirement finally moved Schütz to seek the aid of the court marshal, Heinrich von Taube, to whom he wrote on 26 June 1652 explaining yet again his reasons for wanting to withdraw from active service and adding that 'because of the . . . present wretched state of the electoral musicians the desire to work with them has been totally soured within me'. This attempt had no more effect than any of the previous ones. At the close of his note Schütz remarked that he planned to visit Halle and Weissenfels in the near future 'to pull together what few assets I still have'. By this time, it would appear, he had given up any idea of living out his final days in Weissenfels. In a letter of 16 June 1653 he spoke explicitly of having to 'spend my remaining time in Dresden (since I now see . . . it will not be granted me to live elsewhere)'; and as early as April 1651 he had told Reichbrodt that he would agree to 'live and die here in Dresden' if Johann Georg came to his aid in the litigation against the Chemnitz authorities – which the elector evidently did.

Schütz presumably did not begin his journey to Halle and Weissenfels until after 11 October 1652, when the elector's daughter Magdalena Sibylla, the widow of the Danish crown prince, married Duke Friedrich Wilhelm of Saxe-Altenburg. He had written the continuo song SWV434 for the couple's engagement a year earlier, and he no doubt supplied music for the wedding as well. The festivities scheduled for the occasion included a ballet–opera, *Der triumphierende Amor*, with a libretto by Schirmer; a death in the electoral family two weeks before the wedding, however, forced the cancellation of the performance. Apart from SWV434 only one work of Schütz's appears to survive from this period: the elegy SWV419, written in memory of Anna Margaretha Brehme, the wife of the court librarian, who died on 21 September 1652.

In Halle, which he reached before the end of December, Schütz sold the house that he had purchased four years earlier. Whether he went on to Weissenfels as planned appears uncertain. A receipt from 21 April 1653 indicates that he returned to Dresden fairly early in the new year, and he may have had to go there directly from Halle. He did go to Weissenfels, however, later in the spring, and attempted to negotiate the sale of some property that he owned nearby. The prospective buyer suggested paying for the land by transferring to Schütz a debt owed the buyer's wife by a young prince from the house of Reuss; on 16 June the composer wrote to Heinrich II of Reuss-Gera – the eldest son of Heinrich Posthumus, and the prince's uncle and guardian – asking approval for the arrangement. Schütz reported at the close of the letter that he would soon have to go back to Dresden; Prince Johann Georg wanted him to audition three musicians newly arrived from Italy and also to participate in celebrations planned for the feast of St John the Baptist on 24 June – the name day of the elector, the prince and the latter's eldest son. Ernst Geller, the director of entertainments at court, prepared an arcadian drama for the occasion. An account of the performance by the official diarist notes that the Kapelle took part but provides no details about the music itself.

On 11 August Schütz wrote a prefatory letter for the treatise *Von den Madrigalen* by the poet and jurist Caspar Ziegler, a stepbrother of Benjamin Schütz's wife. Calling the madrigal 'that poetic genre most perfectly suited to the creation of an artfully wrought composition', Schütz observed that Ziegler's demonstration of how German writers could adopt its flexible verse forms for use in their native idiom answered a need long felt by musicians; he himself, for example, had 'scraped together a little volume of sundry poems' – presumably verse of the rigidly strophic type practised in Germany – and 'I know best what trouble it cost me before I could shape it into music even approximately in the Italian manner'. If, as these words could suggest, Schütz actually assembled a collection of secular compositions, he never published it, and no further trace of it survives.

Ziegler's treatise must have provided a rare moment of satisfaction in a time of mounting troubles. With the electoral Kapelle severely depleted and its remaining members living in penury, even the music for regular Sunday services evidently fell into disarray; a letter of 21 August from Schütz to Reichbrodt, Taube and Jacob Weller – the court chaplain since the death of Hoe von Hoenegg eight years before – reveals that the increasingly powerful Prince Johann Georg attempted to restore order by having Schütz and Bontempi, each presumably with his own Kapelle, take charge of the performances in alternate weeks. The action came as an affront to Schütz. As he pointed out in his letter, the direction of music on ordinary Sundays traditionally lay not in his hands but in those of the vice-Kapellmeister; and notwithstanding his apparent high regard for Bontempi, it added insult to injury that 'an old and, I hope, not undeserving servant' should have to appear before the public on an equal footing with 'a man three times younger than I and castrated to boot'. Whether this protest had any effect remains unknown. On the same day Schütz addressed a new petition for retirement to the elector; once again he appears not even to have received an answer.

To make matters worse for Schütz, Prince Johann Georg, who already had a number of Italian musicians in his service, apparently placed the three who had recently arrived in the electoral Kapelle; the action provoked considerable religious and nationalistic resentment, and rumours began to circulate holding Schütz – who clearly had never concealed his admiration for Italian music – responsible for it. On 23 August Schütz wrote to seek the prince's help in clearing himself of the charge. 'I can but grievously deplore', he remarked, 'the fact that towards the end of my life, and after the most humble, laborious and well-intentioned services that I have rendered for so long at this electoral court, the planets and elements would now seem all to conspire and rise up against me'; as he stated in his letter to Reichbrodt, Taube and Weller two days earlier, 'I would sooner wish my death than remain any longer in such vexed circumstances'.

On 21 September 1653, no doubt in preparation for his usual autumn visit to Weissenfels, Schütz submitted a list of the 'foremost motives and reasons' moving him to press for his retirement. Advancing age, he noted, had now affected even his creative powers. While he could still 'work out my compositions as soundly and as well as ever before, it all takes much longer and is more difficult'; he feared that the continuing obligation to 'invent new pieces for the two separate collegia musica that we now have here' would leave him unable to carry out his own musical plans. As he had often indicated before, he wished to 'complete' his works; in addition, he wrote, he hoped to create a setting of Luther's prose psalter written 'in such a way that the congregation in the church may easily be able to learn these melodies and sing with them' — a project that seems never to have come to fruition.

On 14 April 1654 Schütz sent a brief note from Weissenfels to Heinrich II of Reuss-Gera – whose response to his previous letter seems to have gone astray – inquiring again about the payment for his land; even with the message of assurance from Heinrich that soon followed, however, it appears that the sale ultimately fell through. The composer returned to Dresden before the end of spring. Conditions in the electoral Kapelle remained unchanged; although a letter of Schütz's from 19 June suggests the existence of a plan to give the musicians at least a portion of their long-overdue salaries – for which they had again petitioned earlier in the year – an appeal submitted to Johann Georg by Hofkontz and the organist Christoph Kittel on 30 November makes it clear that the idea never became a reality. Three days later Weller remarked in a note to the elector that it had become all but impossible to have so much as the Lord's Prayer sung in the palace church.

During the winter, which he presumably spent in Weissenfels, Schütz suffered the loss of his daughter

Euphrosina, who died in Leipzig on 11 January 1655; her funeral oration reports that he visited her on her deathbed. Poems of condolence, printed with the sermon, came from Bontempi, Hofkontz, Bernhard and Ziegler among others. Schütz returned to Dresden by 29 May, when he addressed yet another – and, as fate eventually decreed, his last – petition for retirement to the elector. The spring, summer and autumn saw an exchange of letters with Sophie Elisabeth of Brunswick-Lüneburg concerning some new appointments to the ducal Kapelle at Wolfenbüttel. After the departure or death of their Kapellmeister Stephanus Körner not long before, the duchess and her husband evidently decided to entrust the supervision of their musical establishment to Schütz. The composer arranged for Johann Jacob Löwe von Eisenach to assume leadership of the Kapelle, helped procure singers and instrumentalists, and provided choirboys whom he had trained. He would also seem once again to have advised Sophie Elisabeth on her compositions; a postscript to a letter of his from 24 July requests payment for a 'copyist who wrote out Your Grace's little psalter for me in Weissenfels'. The work in question probably corresponds, at least in part, to Sophie Elisabeth's *Christ-Fürstliches Davids-Harpfen-Spiel*, published in 1667. In the same postscript Schütz asked for written confirmation of his own position as absentee Kapellmeister and for payment of the salary promised him, 'particularly since at this time . . . I am especially needy because of the costs for the funeral of my last daughter'. The duke responded with a contract providing for an annual recompense of 150 reichsthalers – a larger amount than previously discussed; Schütz signed the document on 23 August. In a letter of

27 November, however, he informed the duchess that he had still not seen anything of his money, and future payments also fell into arrears on more than one occasion. Nevertheless, the relationship between Schütz and the Wolfenbüttel court remained a warm one and lasted at least until Duke August's death in 1666.

Schütz's activities in Dresden, meanwhile, appear to have continued unabated. He no doubt directed the electoral Kapelle at special services held on 24 and 25 September for the centenary of the Peace of Augsburg; and the court diaries – which only at this period begin to provide detailed information about the music sung in the palace church – record several performances under his direction in the spring and early summer of 1656: on 6 April he presented his *Resurrection History* swv50, and he led the music at Whitsun (25 May), Trinity Sunday (1 June), and the feasts of St John the Baptist (24 June) and the Visitation (2 July). On the last three occasions the Kapelle had to repeat the service in the elector's private chambers; Johann Georg, almost exactly the same age as Schütz, had fallen gravely ill and could no longer attend church. Throughout the summer his condition worsened, and on 8 October he died. The interment took place in Freiberg on 4 February 1657. As a final act of duty to the man whom he had served for more than 40 years – and whose passing, despite everything, must have affected him deeply – Schütz composed the twin settings of the German *Nunc dimittis* swv432–3.

CHAPTER FIVE

Last years, 1657–72

Schütz's tribulations in Dresden finally came to an end in 1656, with the death of the elector Johann Georg I. The new elector, Johann Georg II, combined his own Kapelle with that of his father and the entire ensemble was placed under the direction of Bontempi and the recently engaged Vincenzo Albrici. Schütz, now freed from daily responsibilities, received the title of chief – or, in some documents, senior – Kapellmeister and a pension most likely equal to half his former salary. He continued to write new works for major occasions and no doubt helped shape musical policy at court; it does not appear, however, that he conducted the Kapelle again. Within a few months of the old elector's funeral Schütz sold his house in Dresden. The only traces of his whereabouts until the end of the decade – a prefatory letter, dated 21 September 1657, to the *Aelbianische Musen-Lust* of his younger colleague Constantin Christian Dedekind, and an album entry signed 24 August 1659 – place him at Weissenfels; a document of 1658, however, indicates that Johann Georg II expected him to visit court three or four times a year and provided him with a pair of horses for each journey. Schütz probably spent most of his time during this period preparing a revised and expanded version of his Becker Psalter, which he brought out in 1661 (SWV97–256). In the preface to the volume he reported that the elector, who wished to promulgate the Psalter throughout Saxony, had ordered the new edition shortly after coming to power; for his own part, Schütz wrote, he 'would rather have devoted the little time of my life that remains to the revision and completion of various other, more artful inventions that I have undertaken'. The composer evidently finished work on the project well before its publication; attacks on Becker's texts from certain literary circles made him delay sending the volume to press until at last a number of 'eminent and knowledgeable people' persuaded him to do so.

In 1657 Christoph Kittel published a group of Schütz's smaller choral works under the title *Zwölff geistliche Gesänge* (SWV420–31); although, as the edition stated, these products of the composer's 'spare time' appeared with his approval and – as Steude (1982–3) discovered – he himself furnished the paper for the volume, Schütz does not seem to have taken a direct hand in the preparation of the music. He probably had nothing whatever to do with a second publication of the same year, a setting of Psalm xxiii for eight voices – now lost, but perhaps identical with SWV33 or 398*a* – that a Strasbourg printer announced in the catalogues of the Frankfurt and Leipzig book fairs.

In the spring of 1660 Schütz made a visit to Wolfenbüttel, possibly to attend celebrations for the birthday of Duke August on 10 April; he would appear in any event to have arrived by 5 May, the date of a note from Löwe von Eisenach asking him to intercede with the duke in support of a request for leave. Payment accounts reveal that Schütz received 225 thalers in overdue salary and 50 thalers for travel expenses on 6 June, no doubt in preparation for his departure. In the weeks immediately following he probably returned to Weissenfels for a short time, then went to Dresden to oversee the publication of the revised Becker Psalter. On 10 April 1661, Duke August's 82nd birthday, he sent two copies of the new print to Wolfenbüttel, one each for the duke and the duchess; he had spent 'more than eight months in Dresden', he wrote in the accompanying letter, guiding the volume through the press. These months seem also to have witnessed the first performance of the *Historia der . . . Geburth . . . Jesu Christi* SWV435. The court diaries describe the music at Christmas Vespers in 1660 as 'the birth of Christ in recitative style'; this can hardly refer to anything but Schütz's composition, the earliest known German setting of the nativity story to have the evangelist's words sung in recitative instead of the traditional unaccompanied chant.

Near the end of his letter to Duke August Schütz expressed the hope of returning to Weissenfels as soon as possible; he appears to have gone as planned and stayed in Weissenfels for about 18 months. His absence from the electoral court coincided with a major renovation of the palace church. Performances of sacred music did not wholly cease during this period; on 15 June 1662 the court diarist noted the presentation of a motet by Schütz, *Aquae tuae, Domine*, which does not survive. According to the same chronicler, Schütz wrote a new setting of Psalm c, 'with trumpets', for the festive reopening of the church on 28 September 1662; this work, too, does not survive. Although the composer evidently did not attend the performance, which Vincenzo Albrici conducted, he went to Dresden shortly afterwards for the wedding of the elector's

3. Heinrich Schütz in his 70s: portrait by Christoph Spetner (1617–99)

daughter Erdmuth Sophia to Margrave Ernst Christian of Brandenburg-Bayreuth; a note in the court diaries records payments to Schütz and other members of the Kapelle for ceremonial robes. The festivities lasted from 18 October to 13 November and reached their musical highpoint with the performance of Bontempi's opera *Il Paride* on 3 November. On his way to Dresden Schütz stopped at Leipzig, where, on 7 October, he wrote a prefatory epigram for the *Geistliche Arien, Dialogen und Concerten* of Werner Fabricius.

On 23 November the court diarist reported a second performance of *Aquae tuae, Domine*. Schütz no doubt attended, as he seems to have remained in Dresden after the wedding of the elector's daughter. Towards the end of March 1663 he addressed an appeal – no longer extant but referred to in later correspondence – to Johann Georg for 500 thalers owed him from tax revenues, evidently as his salary for the previous two-and-a-half years; although the elector quickly ordered the money to be paid, the head of the treasury failed to do so and eventually left his post with the matter still unsettled. On 21 May the composer signed a receipt for 300 thalers from August of Brunswick-Lüneburg at the spa of Teplitz, some 50 kilometres south of Dresden; as Pirro (1913) suggested, Schütz had no doubt gone to Teplitz with Johann Georg's wife, who went there every May for at least four weeks. Letters from the following months indicate that he returned to Dresden by the beginning of summer and stayed there until about the end of autumn. While at court, both before and after his excursion to Teplitz, Schütz apparently devoted much of his energy to organizing a Kapelle for the elector's brother Moritz, Duke of Saxe-Zeitz. Documents from Zeitz, where Moritz had established a new residence early in the year, reveal that Schütz arranged the appointments of Löwe von Eisenach as Kapellmeister and Clemens Thieme as chief instrumentalist, and himself received the title of 'Kapellmeister von Haus aus'; letters from Schütz of 14 July and 29 September show that he also procured instruments for the new ensemble, supervised the training of three choirboys whom the duke sent to Dresden in mid-April and even suggested architectural modifications to improve the acoustics of the palace church then under reconstruction.

On 10 January 1664 Schütz wrote from Leipzig to August of Brunswick-Lüneburg announcing that he had sent a number of his printed works – 'as many as I am able to muster for now' – to Wolfenbüttel for the duke's library, one of the largest and most famous in Europe. The consignment fulfilled a request that August had evidently made some months earlier; apologizing for his 'neglect and delay' in providing the music, Schütz explained that he had not had access to it until 'three weeks ago, following my return from Dresden to Weissenfels'. In addition to the prints, Schütz had hoped to send copies of pieces still in manuscript, which he described as 'better worked out than those already mentioned'; this would have happened, he wrote,

> and I would have proceeded with their publication, if I had not lacked the means to have it done, and if, as was originally my intention, I could have used for this purpose the retainer or allowance that Your Noble Grace has most generously allotted me – which, however, because of my meagre regular income, I have mostly had to take and use for my scanty provisions and succour.

Schütz never did see his backlog of earlier works into

print, nor did he publish any of the few compositions that he wrote in the remaining years of his life. His failure to do so may not have stemmed from financial exigencies alone. When a partial edition of the *Christmas History* swv435, containing only the evangelist's recitatives, appeared later in 1664 – an edition that the composer seems to have neither instigated nor supervised – the preface stated that Schütz had withheld the work's more richly scored concerted movements from publication 'since he has observed that his inventions would hardly attain their proper effect anywhere but in well-appointed princely Kapellen'; similar considerations might have stood in the way of issuing other works for large forces. Schütz may also have feared that his music would no longer attract much attention in a world dominated by a younger generation of composers; perhaps he felt that the fate of the old musician whose story he had recounted to Johann Georg I 13 years before had now overtaken him.

On 1 May 1664 Moritz of Saxe-Zeitz celebrated the opening of his rebuilt palace church with a festive consecration service. Schütz evidently attended the ceremony; a letter of Clemens Thieme's and a set of statutes for the ducal Kapelle both record his presence in Zeitz – which lay only 20 kilometres from Weissenfels – less than a week later. It does not seem likely, however, that he wrote any music for the occasion, as Werner (1922) contended; this assignment would surely have fallen to Löwe von Eisenach. On concluding his visit Schütz no doubt returned directly to Weissenfels. The autumn and winter months presumably saw him at work on the initial version of his *St John Passion* (swv481*a*); according to court diaries, the electoral Kapelle sang the piece for the first time on Good Friday, 24 March 1665. Schütz probably did not hear the performance. A note from Löwe von Eisenach to Duke Moritz indicates that he made a brief visit to Zeitz in early or mid-April, which he could scarcely have done if he had spent Holy Week in Dresden; and a copy of the Passion sent to August of Brunswick-Lüneburg on the occasion of the duke's birthday bears the inscription 'Weissenfels, 10 April 1665'.

On 10 July 1665 Schütz stood godfather at the baptism of his former colleague David Pohle's son Augustus in Halle. In the autumn he may have gone to Dresden to attend a special service held on 15 October for the birthday of the elector's wife; the court diary reports that Schütz wrote a new setting of Psalm c – identical, judging from its description, with swv493 – for the event. The diary further records the performance of an otherwise unknown concerto, *Renunciate Johannis quae audistis*, on 16 December. Schütz's letters definitely place him in Dresden the following spring. On 1 and 3 May 1666 he wrote to Johann Georg inquiring about the 500 thalers still owed him from three years earlier; the outcome of the matter remains unknown. A month before, the court diaries reveal, the second and third of Schütz's Passions – those according to St Matthew, swv479, and St Luke, swv480 – received their first performances, the former on 1 April, the second Sunday before Easter, the latter a week afterwards on Palm Sunday. On Good Friday, 13 April, the choir repeated the *St John Passion* given the previous year; Schütz probably created the revised version of the work (swv481) for this presentation.

Few external events mark the final years of Schütz's life. On 1 January 1667 the court diarist reported a performance of Psalm cl in 'Kapellmeister Schütz's new composition with trumpets and timpani'. A second reference to a 'new' setting of Psalm cl, this time without mention of trumpets and drums, occurs in the programme of a service held on 22 July 1668 to celebrate a recently concluded peace treaty; whether this in fact indicates a new composition or merely denotes a repeat presentation of the work heard the previous year remains uncertain since no music survives from either occasion. During this period the composer evidently continued to live at Weissenfels. According to Mattheson (1725), Johann Theile, the last of Schütz's many pupils, went there to learn 'the true fundamentals of composition' from the aged master after studying at the University of Leipzig; Theile matriculated at the university in 1666 and remained in Leipzig at least until the publication of his *Weltliche Arien und Canzonetten* early the following year. He may have gained access to Schütz through the latter's son-in-law Christoph Pincker, whose name appears among the dedicatees of Theile's arias.

Towards the end of the 1660s a dramatic increase in the salaries of the electoral Kapelle brought Schütz's retirement pay to 800 thalers per year – twice what he had earned in active service under Johann Georg I, and two-thirds of the amount given to each of the four Italian Kapellmeister who actually led the ensemble. In the last year of the decade the elector presented him with a gilded cup 'in gracious remembrance', possibly to honour the composer on reaching the age of 84, one of the duodecimal *Stufenjahre* regarded by German tradition as the major dividing-points in life. At about this time or not long afterwards, Schütz, no doubt mindful of the pledge that he had made to die in Dresden, gave up his home at Weissenfels. A contract of 14 January 1672 concerning the disposition of some family property speaks of him as already having left the city, and all further evidence of his whereabouts from 1670 until his death points to Dresden. As the account of his funeral in the court diaries reveals, Schütz settled in rented quarters near the electoral palace. Increasingly weak and hard of hearing, he could, in Geier's words,

> but rarely go out or attend the preaching of God's word, but had mostly to stay at home, where he passed the greater part of his time reading Holy Scripture and other books of learned theologians . . . and maintained himself in abstemious and temperate fashion . . .

He began to make preparations for his end. A poem by Dedekind, dated 1 September 1670 and written at Schütz's behest, commemorates the completion of his tomb in Dresden's Frauenkirche. Mattheson (1740) reported that Schütz also wrote to Christoph Bernhard in Hamburg asking for a five-voice setting, 'in the Palestrina style of counterpoint', of Psalm cxix.54, the words chosen as the motto for his funeral sermon; the composition, which does not survive, reached Schütz in 1670. 'My son,' he told Bernhard in his letter of thanks, 'you have done me a great kindness in sending the motet for which I asked. I do not know how to improve a single note in it'.

Psalm cxix, the source of Schütz's funeral text, served in its entirety as the basis for his last composition, a monumental eight-voice work in 11 *partes* (SWV482–92). Dedekind recollected in the preface to his *König Davids göldnes Kleinod* of 1674 that Schütz had written the psalm 'shortly before his blessed end' and had expressly designated it as his swan-song; the composer often suggested, Dedekind added, 'that I might sometime wish to write instrumental parts for it', but 'modesty bade me not be so bold as to presume to such a thing'. In 1671 Schütz had a set of partbooks containing Psalm cxix and an 'appendix' of two smaller pieces – Psalm c SWV493, and the German *Magnificat* SWV494 – copied and fitted with printed title and index pages; a note on the organ part in the composer's shaky hand reveals that he presented the manuscript to Johann Georg II in the hope that the elector

would grant this admittedly meagre work the kindness of . . . letting it be tried out and sung in Your Majesty's court chapel, in the two choir lofts erected across from each other above the altar, by eight good voices with two small organs.

For some time before his death, according to Geier, Schütz

had been subject to occasional severe attacks of apoplexy, which, however, he always withstood with the help of useful medicaments. On . . . 6 November [1672], however, he arose, hale and hearty to be sure, and got dressed, but, after 9 o'clock, as he was looking for something in his chamber, was overtaken by a sudden weakness and apoplexy, which forced him to sink to the floor and remain there helpless; and although, when his people came to him, helped him up and quickly set him on a bed in the sitting room, he recovered somewhat and even spoke coherently, the apoplexy had struck him so hard that, after saying audibly that he placed everything in God's gracious will, he was no longer capable of speech; and even when his physician was quickly summoned to him and applied all his diligence to helping him with choice medicaments and to strengthening his being, still, little was to be done for him. His father confessor was also summoned and read and recited sundry prayers and proverbs to him, at which he gave to understand several times by nods of his head and with his hands that he had his Jesus in his heart, whereupon the father confessor gave him the final blessing; and then he lay there as if sleeping, quietly, until finally his breath and pulse grew weaker and faded away, and, as it struck four, midst the prayers and songs of those around him, he passed on softly and blissfully, without a single tremor.

Schütz's funeral took place on 17 November 1672, eleven days after he died; according to the court diaries, Johann Georg II arranged for the ceremony and sent a personal delegate to represent him. The funeral party evidently gathered at the home of the composer, where the coffin still lay. Johann Ernst Herzog, a young pastor attached to the court, gave the farewell address, and the Kapelle members in attendance sang one of Schütz's works. The procession then made its way to the nearby Frauenkirche, where Geier delivered the funeral oration; before and after the sermon, the diarist reported, the German members of the chapel performed 'four pieces, the first of which was composed by the former vice-Kapellmeister Christoph Bernhard, the other three, however, by the late Kapellmeister himself for voices and instruments'. Schütz was buried in a portico just outside the chancel. Although the rebuilding of the church in 1721 destroyed his grave, a description of it survives in a collection of epitaphs and inscriptions gathered by the sexton Johann Gottfried Michaelis in 1714. The body lay beneath a square tablet of black marble that carried the words 'Heinricus Schützius Seculi sui Musicus excellentissimus Electoralis Capellae Magister MDCLXXII'. On a wall close by stood a relief showing an open book inscribed with the Horatian motto 'Vitabit libitinam' (*Odes* 3.xxx.7) and resting on a skull, with a pair of alabastar trumpets in the background and a garland of rue enclosing the whole; the same emblem, in slightly varied form, appears in the portrait of Schütz that Christian Romstet engraved for the print of Geier's sermon. Just under the relief a square brass plaque presented a Latin epitaph:

Heinrich Schütz, Christian Asaph, a joy for foreigners, a light for Germany, immortal adornment to the chapel of the most serene Saxon electors Johann Georg I and II, over which he presided for 57 years. That which in him was mortal he laid down beneath this monument, erected through the munificence of the elector, in the 87th year of his life and in the 1672nd year of our reckoning.

CHAPTER SIX

Portraits

Two authentic portraits of Schütz have come to light. The earlier, an engraving by the Dresden artist August John (1603– after 1678), shows the composer at the age of 42 in resplendent court dress with a medallion bearing the image of Johann Georg I (see (fig.1, p.20). The second portrait, executed in oils by the Leipzig painter Christoph Spetner (1617–99) and housed today in the Bibliothek der Karl-Marx-Universität, Leipzig, portrays a considerably older Schütz in his Kapellmeister's robe, with a rolled sheet of music in his hand further indicative of his station; this work provided the model for Romstet's engraving in the funeral sermon. An inscription associating Spetner with Stedten, near Querfurt, implies that the painting originated before October 1654, when the artist became a citizen of Leipzig; but Möller (1984) has argued that the medallion worn by Schütz shows the profile of Johann Georg II, which would point to a date of 1657 or later (see fig.3, p.62).

Maerker (1937–8) and Benesch (1963) proposed identifying Schütz as the subject of Rembrandt's *Portrait of a Musician* of 1633 (Corcoran Gallery of Art, Washington, DC); but Sass (1971) disposed of this possibility on iconographic grounds – grounds amply confirmed by the more recent discovery of the John portrait – and biographical evidence eliminates virtually any chance of a meeting between artist and composer at the time in question. The common identification of Schütz as Kapellmeister in the engraving of the electoral Kapelle done by David Conrad for Bernhard's *Geistreiches Gesangbuch* of 1676 seems more plausible: the figure not only shows a marked facial resemblance to Schütz but also wears a chain and medallion like that shown in the authentic portraits. Nevertheless, the engraving can hardly count as, or even derive from, an authentic rendering of a scene in which Schütz participated, since it shows the palace church as it looked after the renovations of 1662, by which time he seems no longer to have conducted the Kapelle. If Conrad had in fact wished to portray Schütz, he might simply have copied his features from Romstet.

A further likeness of Schütz may appear in an illustrated scroll depicting the funeral of Duke August of Saxony on 4 February 1616. The procession includes a group of eight boys and four men labelled, collectively, 'choirboys and Kapellmeister' (repr. in Schnoor, 1948); the men could represent Schütz, Rogier Michael, Praetorius and the Kapelle administrator. The lack of clearly individuated features, however, as well as the absence of comparative material, renders any attempt at a more precise identification purely speculative.

In 1935 the Prussian State Library – now the Deutsche Staatsbibliothek, Berlin – acquired an anonymous miniature in oils labelled 'Heinricus Sagittarius' and dated 1670. The picture quickly became well-known, widely reproduced and much commented on, one writer even went so far as to say that it conveys the 'force of [Schütz's] personality with an immediacy rare in musical portraiture' (Rifkin, *The New Grove*). Recent investigations by art historians, however, have exposed it as a modern forgery.

CHAPTER SEVEN

Introduction to the works

I Sources

Some 500 of Schütz's works have survived. The major portion appears in 14 prints, containing large-scale individual works or collections of smaller ones, that Schütz enumerated in a handwritten catalogue sent to August of Brunswick-Lüneburg in 1664 (see Haase, 1973; Walter, 1973); he had previously published a similar list, covering the first ten volumes in the series, as an appendix to the *Symphoniarum sacrarum secunda pars* SWV341–67. Both lists assign opus numbers to all the prints, though only four of them (opp.10–13) actually bear such designations on their title-pages. With the exception of the *Zwölff geistliche Gesänge* SWV420–31, which he authorized but did not edit, Schütz himself oversaw the production of each volume, and he even supplied the paper – with a watermark containing the family crest and his personal monogram – for several of them. In a number of instances he continued to correct and revise even after publication; the ducal library at Wolfenbüttel (Herzog August Bibliothek) preserves nine prints in copies given by Schütz to Duke August, and several of these show significant additions and alterations in the composer's hand.

In addition to the pieces contained in the central series of prints, two major works survive in partial printed form. The recitative portions of the *Historia, der*

... *Geburth* ... *Jesu Christi* SWV435 appeared in an edition no doubt sanctioned, but not prepared, by the composer. The sole extant copy, now in the Staatsbibliothek Preussischer Kulturbesitz, Berlin, once belonged to Rudolph August of Brunswick-Lüneburg, the eldest son of August the younger and Sophie Elisabeth, and might have come to the Wolfenbüttel court as a gift from Schütz; several manuscript revisions found in the volume could thus have the composer's authority. In 1671 Schütz had the title-pages and a table of contents printed for the *Königs und Propheten Davids hundert und neunzehender Psalm* SWV482–94, but the work itself, like most of SWV435, remained in manuscript.

A handful of smaller occasional compositions (SWV20, 21, 48, 49, 52, 94, 95, 277, 278, 432–3, 453, 464) appeared as individual prints, only some of these under Schütz's auspices. Very few works occur in contemporary collections: one each in *RISM* 1623^{14}, 1627^{9} and 1646^{4}, and two in 1641^{3}; the funeral sermon *RISM* 1652^{6a} contains one further composition, as does the poetic collection *David Schirmers* ... *poetische Rauten-Gepüsche* (Dresden, 1663). Schütz demonstrably or presumably supplied the music to the editors of all these volumes. On the other hand, the printing of excerpts from SWV50 in *RISM* 1637^{3} and of SWV39 and 291 in the companion volume *RISM* 1638^{5} clearly did not have his sanction; SWV39 and the material from SWV50 appear in arrangements verging on falsifications, while SWV291, identical in every respect with the version published in the first book of *Kleine geistliche Concerte*, seems to represent a simple act of piracy.

As Schütz indicated in several prefaces and letters, war and economic hardship forced him to leave a substantial part of his output unpublished. At least half of these works no longer survive. Fires in the 18th century destroyed the older portions of the Dresden Kapelle library, which Mattheson (1740) described as preserving a 'great number' of compositions by Schütz, and the library of the Danish royal *kapel*, which presumably contained all the music Schütz wrote during his years in Copenhagen. Important repositories in Lüneburg and Weimar, known through their inventories, exist no longer. The most significant concentration of works in manuscript still extant is in the Gesamthochschul-Bibliothek, Kassel; these sources, which are the remains of a larger collection belonging to the Hesse Hofkapelle, originated almost without exception either in Kassel – no doubt on the basis of copies provided by Schütz – or in Dresden. Several of the manuscripts from Dresden have titles or other textual entries in Schütz's hand; the sources for SWV441, 470, 474 and Anh.1 contain autograph musical material as well (see fig.4). The manuscript collection of the royal Swedish Kapellmeister Gustaf Düben which is in the Universitetsbibliotek, Uppsala, includes a fair number of works by Schütz, some in copies made locally, others in copies of probable Dresden origin. Before World War II a third important group of Schütz sources existed in the Staats- und Universitätsbibliothek, Königsberg; these apparently came from the collection of the Naumburg Kantor Andreas Unger. The war also accounted for the destruction of a smaller body of manuscripts, mostly of secondary significance, in the Stadtbibliothek, Danzig. Finally, a number of Schütz's works occur in collective and other secondary manuscripts written near Dresden and held there today in the Sächsische Landesbibliothek.

II The music

Even allowing for lost works known only from library catalogues and other sources, Schütz's output consists almost entirely of sacred vocal music. Apart from the symphonies in his vocal works, no instrumental music by him survives – not even for organ, though his skill as an organist was often acclaimed. The reason for this is that the composition of instrumental music was not one of his responsibilities as electoral Kapellmeister. His secular output, also, reflects his status: it is dominated by his lost theatrical works, while his few surviving German songs, madrigals and concertos, some of an occasional nature, suggest that he was not often required to set a small-scale secular German text. His op.1, of course, is a book of secular Italian madrigals; nevertheless, the bulk of his output is sacred.

These sacred works are extremely varied in nature. There are two histories (on the Resurrection and the Nativity, SWV50, 435), three Passions (SWV479–81), a German Requiem (SWV279–81) and a setting of the Seven Last Words (SWV478). But the majority of his works are settings in motet or concertato style of relatively short texts, mainly from the Bible. Schütz occasionally rearranged or omitted biblical words but normally set them with little alteration. His earlier printed collections are dominated by texts from the Old Testament (especially the Psalms and the Song of Songs) and short devotional poems, all of which invite a subjective response. His later publications, on the other hand, include more New Testament texts, which are less amenable to such treatment.

This change of emphasis does not seem to relate to liturgical requirements, for most of Schütz's works do not appear to have been intended for regular liturgical use. None of his collections is arranged liturgically, and at most a handful of individual pieces bear rubrics assigning them to specific occasions in the church year. Items from the *Psalmen Davids* swv22–47 were sung at Dresden as introits and at the beginning of Vespers on major feasts and special occasions, but they and the three books of *Symphoniae sacrae* swv257–76, 341–67 and 398–418, in particular, are ambitious works: large-scale settings, with instruments, which would have been too elaborate for frequent performance. Some of his large-scale compositions were probably political works, and it is clear from the texts of the *Cantiones sacrae* swv53–93 that these motets were intended for private devotions. The Becker Psalter swv97–256 and the *Zwölff geistliche Gesänge* swv420–31, on the other hand, are functional collections intended for liturgical or domestic use and consisting of modest harmonizations of simple melodies for SATB and continuo. The Becker Psalter presents the psalms in Biblical order, but the revised edition of 1661 includes a register grouping them according to the liturgical calendar. Schütz's settings became a standard feature of services in the Dresden court chapel, where his histories and Passions were also presented regularly, sometimes in place of lessons. He wrote a comparatively small number of works based strictly on chorale melodies, and these few (e.g. swv41, 467) belong mostly to his early period; his other settings of hymn texts make at most occasional reference to the associated melody (swv94) or are completely free in their thematic substance (swv316).

Only two of Schütz's collections consist entirely of Latin works, the *Cantiones sacrae* (1625) and the first book of *Symphoniae sacrae* swv257–76 (1629); most of his works are settings of German texts, and it is from these that the character of his music largely derives. He was clearly very responsive to the sounds, rhythms and meaning of German words, all of which are vividly reflected in the nature of his melodic invention. At the same time he possessed to a remarkable degree the ability to conceive a large-scale musical design and to use his material to construct a convincing musical argument. The forms of his works derive largely from this twofold approach to the setting of texts: his music not only reflects and underlines the meaning of individual phrases and words but also deepens and reinforces the significance of the text as a whole. These qualities are also to be found in his Latin works, and even in his Italian madrigals.

The following survey of Schütz's output is based almost entirely on his major published collections and large-scale compositions. As the work-list shows, however, he also composed, throughout his life, a considerable number of isolated pieces, many of which remained in manuscript and some of which are by no means small-scale; among these is a group of psalm settings dating from the 1620s which may have been intended to form part of a second volume of *Psalmen Davids*. Furthermore, some of the pieces in his collections were written many years before they were published or are revisions of earlier works. For all of these reasons, the following account is regrettably incomplete and over-simplified. Nevertheless, a study of the published collections and large-scale works will provide a broad framework against which the isolated printed and manuscript works, which are the subject of continuing musicological research, will eventually be able to be viewed.

CHAPTER EIGHT

Early works, 1611–28

Schütz's *Primo libro de madrigali* SWV1–19 (1611) represents the culmination of his studies with Giovanni Gabrieli and exhibits an astonishingly complete mastery of the techniques associated with the Italian madrigal of the late Renaissance. The texts are nearly all from Guarini and Marino, while the settings (with one exception) are for five unaccompanied voices and reveal a lively interest in aspects of texture, harmony and rhythm. The voices are often treated in twos or threes, in the manner of Wert and Monteverdi, and the wide range of note-values employed (from *longa* to semiquaver) recalls the rhythmic contrasts of Gesualdo. Among the harmonic irregularities favoured by Schütz are chromatic progressions and false relations, the simultaneous sounding of a dissonance and its resolution, and the simultaneous use of two dissonances – particularly in the form of slow passing notes in parallel thirds or sixths against a held chord in two or three other parts. In their straining after effect, these settings might be described as 'mannerist'; yet the effects are prompted entirely by the words and are always harnessed to a larger rhetorical design. The final madrigal is dedicated to Schütz's patron, Landgrave Moritz of Hessen-Kassel, and through its scoring for two antiphonal choirs forms a tribute to his teacher, Gabrieli. All in all, the collection displays a remarkable insight into the madrigal and constitutes a most impressive op.1.

Early works

During the eight years that preceded the publication of op.2, Schütz moved to Dresden and composed his first surviving polychoral sacred works. Most of these were included in his *Psalmen Davids* (1619), a collection of 'German psalms in the Italian style, to which I had been assiduously introduced by . . . Herr Johan Gabriel'. The 26 items are apparently arranged in two halves: the first consists entirely of psalms, the last of which (*Ich danke dem Herrn* swv34) ends with a doxology based on Gabrieli's *Lieto godea*; the second half, on the other hand, comprises a variety of pieces – seven complete psalms (swv35–8, 43–5), two 'concerte' (swv39, 47) and a 'moteto' (swv42) set to psalm texts, a 'canzon' (*Nun lob, mein Seel, den Herren* swv41) based on a chorale paraphrase of Psalm 103, a 'moteto' (*Ist nicht Ephraim mein teurer Sohn* swv40) and a 'concert' (*Zion spricht* swv46) set to texts from Jeremiah and Isaiah respectively.

It is from the early 17th century that the rise of German psalm settings dates; previously, Latin had prevailed, even in Protestant circles. Among the earliest composers of German settings were Calvisius, Praetorius and Franck, but Schütz surpassed all his predecessors in the scale and richness of his settings. These are scored for one or two 'cori favoriti', comprising the best singers and/or instrumentalists, in most cases reinforced by one or two 'Capellen', additional (and usually optional) choirs of voices and/or instruments providing extra richness of sonority. Detailed instructions on the composition and layout of the choirs, and on the use of instruments, are given in the preface. Schütz creates a magnificent variety of colours and textures: sometimes his forces are divided into choirs of equal ranges (SATB, SATB), sometimes into high and low (SSAT, ATBB);

4. Autograph MS from the Cantus I part of Schütz's madrigal 'Vier Hirtinnen, gleich jung, gleich schon' SWV *Anh.1, with the text completed in another hand* some choirs are entirely vocal, others call for instruments such as the violin, cornetto, trombone or bassoon. With the aid of such forces he explores all the possibilities of antiphonal composition, including a more-or-less literal echo throughout *Jauchzet dem Herren* SWV36. The textures are predominantly homophonic and often employ quite rapid chordal declamation or psalmodic recitation. Other signs of Italian influence include soloistic writing for the 'favoriti', both singly and in combination, and the occasional use of triple metre. These elements are held together by strong harmonic designs and sometimes by means of a repetition scheme. In short, if the *Madrigali* demonstrated his mastery of late Renaissance polyphony, the *Psalmen Davids* display Schütz's command of the textural and stylistic possibilities of the polychoral medium.

A further new medium, the German *historia*, was conquered in his next publication, the *Resurrection History* (1623). Although it is modelled on the *Auferstehungshistorie* (1568) of Antonio Scandello, one of his predecessors at Dresden, it represents an advance for Schütz and for the genre. The text is derived from all four gospels. The words of the Evangelist are set to the so-called 'Easter tone', a recitation formula resembling psalm-recitation which had been common in Germany since the mid-16th century. The part is written largely in plainsong notation and accompanied by four viols which may however be replaced, if necessary, by an organ or lute. For most of the time, the viols play sustained chords beneath the recitation, but at cadences the vocal part is written in mensural notation and accompanied by changes of harmony in the viols. If Schütz thus takes pains to control the cadences, he also gives the viols the freedom to improvise *passaggi* during their held notes, as was common in contemporary Italian *falso bordone*.

Although the Evangelist's recitation is marvellously sensitive to the words and includes occasional expressive flourishes, Schütz's powers of characterization are more apparent in the 'concerted' sections which set the utterances of the various characters. Some of these are set in the traditional manner as duets – Mary Magdalene for two sopranos and Jesus an alto and a tenor – but others are more realistically scored: the three Marys as three sopranos and Cleophas as a tenor. All are accompanied by continuo. What impresses in these concerted sections is the passionate intensity of the music, which often recalls the compelling quality of the duets in Monteverdi's Seventh Book of Madrigals (1619). All the resources of the medium – imitative and homophonic textures and rapid exchanges, over static and moving basses; sequential repetition; rhythmic dislocation; chromatic harmonies – are deployed in order to convey the meaning of the words as powerfully as possible, and the effect of these sections is enhanced by contrast with the restraint of the Evangelist's part. The narrative and dramatic substance of the work is prefaced, punctuated and rounded off by three brief but perfectly judged choruses which provide a satisfying textural and tonal framework for the whole.

Schütz's next publication explores the possibilities of the polyphonic and concertato motet and again represents an advance in his development. The *Cantiones sacrae* (1625) consists of 35 motet movements, many of which belong together in pairs or larger groups, and five Grace settings. The motet texts are taken mainly from the *Precationes* of Andreas Musculus, a Latin prayer book – first published in 1553 – that became popular among Protestants as well as Catholics and was reprinted many times; a few come from the Bible and other sources. The texts are extremely penitential in mood and intensely personal in feeling, and these qualities are vividly reflected in Schütz's settings, which were composed over an extended period and display a variety of styles. Several of the motets, including those on biblical texts (e.g. swv55, 61–2), are set in a highly developed polyphonic style reminiscent of that of his madrigals, while some employ a more modern concertato texture in which the continuo is indispensable (e.g. swv70, 84, 85–7). Every piece, whatever its style, is infused with an extraordinary intensity of feeling deriving from Schütz's response to the words. In addition to the methods listed above, he uses a wider range of rhythmic techniques, a higher level of dissonance and chromaticism and bolder melodic gestures. His responses are occasionally so extreme as to threaten the coherence of the texture or structure, but unity is assured by a variety of means, including augmentation and diminution (see ex.1) and motivic relationships (ex.2). The prayers of Musculus were originally intended for private use, and Schütz's settings – the most subjective works he ever wrote – were probably meant as sacred chamber music in an aristocratic house, possibly that of the dedicatee, who was a convert to Catholicism.

The settings of the Graces are much more restrained and stand apart from the body of the collection. Chordal textures of the kind he uses in the simplest of these may also be seen in the Becker Psalter (1628). The psalm paraphrases of Cornelius Becker had been published in

Ex.1 *Deus misereatur nostri* SWV55

original a minor 3rd higher

1602 and quickly become popular among Lutherans. Schütz began to compose his settings in the early 1620s, for the devotions of the choirboys in his charge, and took up the task with renewed vigour after the death of

his wife in 1625. His publication of 1628 consists of 103 settings, most of them original (though twelve employ chorale melodies); all of them were revised or completely replaced in the third edition (1661), which also included settings of the psalms he had not tackled before. They are simple harmonizations for SATB, with figured bass for organ and with just enough melisma and syncopation to temper the severity of the chordal textures. These are admittedly modest works, yet like the larger works of his earlier years they display the refinement and judgment that are among the hallmarks of his artistic maturity.

CHAPTER NINE

Middle-period works, 1629–50

The middle of Schütz's career is dominated by the sacred symphony and concerto. The period begins with his second visit to Italy (1628–9) and the publication of his first book of *Symphoniae sacrae* and ends with the appearance of his third book in 1650. Between these dates he issued two books of *Kleine geistliche Concerte* and other works. While each of the major publications of his earlier period explored a different style or genre, most of his middle-period collections present a sizable repertory of sacred music in the contemporary Italian concertato idiom. Instruments no longer merely double or elaborate vocal lines but become increasingly differentiated from them in style; and the continuo is now an essential component of the musical texture.

The first book of *Symphoniae sacrae* SWV257–76 was published in Venice and was the first sustained demonstration of his mastery of the few-voice concertato. The settings of the Latin texts are in three to six parts, plus continuo, each 'part' being for a solo voice or instrument. The scorings vary from one voice plus two instruments (and continuo) to one plus four (e.g. the celebrated *Fili mi, Absalon* SWV269, for bass voice and four trombones), three plus three and four plus two; violins, recorders and bassoons are among the other instruments required. The vocal and instrumental combinations are reminiscent of those in the *Psalmen*

Davids, but in only one of the pieces in the *Symphoniae sacrae* (*Veni, dilecte mi* SWV274) are the forces divided into antiphonal choirs.

The influence of Italian music is very apparent, as much in the clear articulation of structures as in the nature of the ornamentation. The articulation is achieved by a variety of means, including alternation of duple and triple metres (e.g. SWV258), contrast between recitative and aria styles (SWV260) and repetition of words and music in the manner of a refrain. In *O quam tu pulchra es* SWV265, for tenor, baritone, two violins and continuo, the opening words, after their initial appearance, recur six times, like an *idée fixe*, and on the last occasion are set to a rising sequence in which their voluptuousness is most graphically portrayed (ex.3, pp.92–3). This sequence is also brought back at the end of the second part of the piece, *Veni de Libano* SWV266, transformed into duple metre and with richer harmony (ex.4, pp.94–5). It thus provides a satisfying conclusion to a work which recalls Grandi's setting of the text but which far surpasses it in sensuousness as well as technique.

The structures are also articulated by the use of the instruments. Generally speaking, these share the vocal material. Sometimes they anticipate the vocal sections with a short 'sinfonia', sometimes they follow them, and elsewhere they combine with the voices in a variety of contrapuntal and homophonic textures typical of the concertato of the period. The writing for two high instruments is reminiscent of the trio sonata textures of Turini and Castello, just as the writing for pairs of voices recalls the duets of Monteverdi and his contemporaries.

A striking contrast with the *Symphoniae sacrae* is created by Schütz's next large-scale composition, his *Musicalische Exequien* SWV279–81 (1636). This German Requiem, which was commissioned for the funeral of Prince Heinrich Posthumus of Reuss by his widow and children, consists of three items based on words from

Ex.3 *O quam tu pulchra es* SWV265

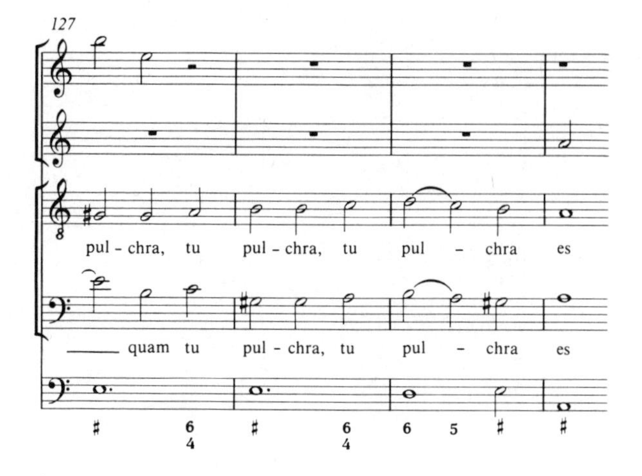

the Bible and from hymns – a 'Concerto in the form of a German burial Mass', a motet and a setting of the *Nunc dimittis*. The concerto is scored for six solo voices and continuo, but an optional choir ('capella') may be used to double the soloists at specified points (i.e. when they all sing together). In the first section of the concerto ('Nacket bin ich') these 'tutti' passages deliver the words of the Kyrie eleison. The second section ('Also hat Gott die Welt geliebt') corresponds to the Gloria of the Mass; here some of the 'capella' passages incorporate chorale melodies, and some are repeated, imparting a musical structure to this more extended section.

The motet *Herr, wenn ich nur Dich habe* is scored for two antiphonal choirs (SATB) with or without organ; the texture is predominantly chordal and the music moves with great solemnity. The words of the *Nunc dimittis* ('Herr, nun lässest du') are set for a five-part choir (AATTB) placed near the organ; a second choir (SSB), which represents two seraphim and a 'Beata

Ex.4 *Veni de Libano* SWV 266

anima cum seraphinis' and should be situated some distance away, sings the words 'Selig sind die Toten' and thus provides a heavenly gloss on the basic text. The effect is hauntingly beautiful, yet achieved by the simplest of means. Indeed, the whole of the Requiem is

characterized by its simplicity of texture and harmony. This restraint bears witness to Schütz's sensitivity and helped him create an atmosphere of nobility and serenity – but it did not prevent him from composing ingenious and highly appropriate designs.

Shortly before the funeral of Heinrich Posthumus, Schütz had published the first part of his *Kleine geistliche Concerte* swv282–305; the second part, swv306–37, followed in 1639. The concertos reflect the privations of the Thirty Years' War. As he said in his preface, music was in decline and in some places discontinued altogether: hence the (almost total) absence of instruments – the most obvious respect in which the *Concerte* differ from the *Symphoniae sacrae*. In addition, about half of the texts are rather gloomy. Some are taken from hymns, religious poems or St Augustine, but the majority are short biblical sayings ('Sprüche'). The few-voice concertato with continuo only is the last distinctive medium that Schütz was to cultivate, and his *Concerte* are his first publication of small-scale settings of German texts. His 'Spruch' settings represent the culmination of the development of the genre in the early 17th century and were the first by a German composer trained in Italy.

There are 56 concertos – nine for solo voice, 21 for two voices, nine for three, eleven for four and six for five. Almost every conceivable vocal combination is represented, but pairs of equal voices predominate in the duets – and in the trios, where they are joined by a third voice below. The limitation of the forces means that the nature and structure of the music are more than usually dependent on the words. This is particularly true of the solos, where the greatest vocal freedom is to be found. The normal style is a mixture of recitative and arioso – *Eile mich, Gott* swv282 is even marked 'in stylo oratorio'. In the best of the solos (e.g. *Was hast du verwirket* swv307) the nature of the thematic material is largely determined by the text, the meaning of which is then enriched or intensified by the way in which the material is subsequently employed. Melodic ideas are repeated and varied – extended, augmented or syncopated, for example – in the voice, and imitated or anticipated in the bass. The texture is bound together by such contrapuntal means, which also contribute to a unity of affect. Refrains and contrasts of metre help to articulate the structures, but the musical coherence of these works is largely motivic and organic.

The same is broadly true of the duets, in which the greater range of textural possibilities compensates for a slight loss of freedom. They do include a little recitative (e.g. swv312) but normally proceed in 'aria' style. The voices generally engage in imitative counterpoint or rapid exchanges above an independent bass, but sometimes the bass has the same material and occasionally the texture is homophonic. As in Renaissance polyphony, the balance between counterpoint and homophony is one of the rhetorical resources of these works. The influence of Italian music may be seen in some of the ornamentation (e.g. the dotted rhythmic patterns in swv313) and in the 'walking bass' of *Habe deine Lust* swv311, which is by far the longest and most intricately organized of the duets and ends with an ostinato section.

The concertos for three, four and five voices are less amenable to generalization. Among them are a number of chorale settings, in which the various lines and verses of the hymns are distinguished by changes of texture and, sometimes, of metre. In *Wir glauben all an einen Gott* swv303 and *Ich ruf zu dir* swv326 the chorale melody provides material for a series of largely imitative sections; in *Ich hab mein Sach Gott heimgestellt* swv305

– a later version of swv94 (1625) – and *Allein Gott in der Höh* swv327, the bass that accompanies the first verse of the chorale is used as the basis for a set of strophic variations; and in *Nun komm der Heiden Heiland* swv301 the beginning of the chorale melody occasionally appears in long notes in the bass in the manner of a cantus firmus. By comparison with the solos and duets, the trios may seem rhythmically rather square; but swv325 incorporates appealing sequential passages and swv322 makes use of lively repartee and of some ornate material in its closing section. The luxurious *Veni, sancte Spiritus* swv328, for pairs of sopranos and tenors, recalls the Monteverdi of the *Vespers*; the voice of God is vividly evoked in the florid passages for bass voice at the start of *Die Stimm des Herren* swv331, and the characters of Mary and the Angel are effectively distinguished by contrasting rhythms in the dialogue *Sei gegrüsset* swv333/334, the only concerto to call for instruments. All in all, the *Kleine geistliche Concerte* present the rhetoric of Schütz's music in its most concentrated form and show that the principles of monody could be combined with those of counterpoint to create an extraordinarily powerful kind of musical declamation and expression.

By 1647, the date of his next published collection, Schütz was well over 60 and anxious to retire from his post as Kapellmeister at Dresden. His second book of *Symphoniae sacrae* swv341–67, as he explained, had been ready for some years – several pieces in it survive in earlier versions dating from the middle of the previous decade – but had not been published because the contemporary Italian style had remained largely unknown in Germany. These symphonies for one, two and three

5. A page of the Bassus 1 part of Schütz's 'Nun komm, der Heiden Heiland' SWV*301, from 'Erster Theil kleiner geistlichen Concerten' (1636), with a correction and the addition of Latin text in the composer's hand*

voices, two obbligato violins 'or other like instruments' and continuo are even more Italianate than his earlier Italianate works. Triple metre is far more prevalent, and aria sections are thus more clearly distinguished from recitative. The standardization of the instrumentation represents a significant advance over the first book of *Symphoniae sacrae*, in which the number and disposition of instruments had been quite variable. The writing for violins is more idiomatic than hitherto, both in the separate 'sinfonia' sections and in the arias, where they engage in imitative interplay with the voice(s). As well as taking a share of the florid ornamentation, the instruments also contribute to echo effects (swv344) and have passages in *stile concitato* (swv350); and in common time they occasionally combine with a voice in homophonic passages, sometimes marked 'tarde' or 'submisse', which suggest a kind of accompanied recitative (ex.5). *Es steh Gott auf* swv356 is based on the duets *Armato il cor* and *Zefiro torna* from Monteverdi's *Scherzi musicali* (1632), and *Der Herr ist mein Licht* swv359 is faintly reminiscent of *Chiome d'oro* from his Seventh Book of Madrigals – of which Schütz made a contrafactum (swv440) at around this time. *Von Gott will ich nicht lassen* swv366 is a carefully planned setting of nine verses of the eponymous chorale, and the collection ends with a textural tour de force, *Freuet euch des Herren* swv367, in which the violins have double-stops marked 'tremolant' and a continuo realization is partly written out.

The following year witnessed the publication of Schütz's *Geistliche Chor-Music* swv369–97, a collection of 29 German motets in five to seven parts with continuo; a handful of them call for instruments, but most are

Ex.5 *Was betrübst du dich, meine Seele?* SWV353

purely choral. The collection may appear to be a reaction against the style of the *Symphoniae sacrae*, but it was actually published as a practical contribution to a current musical debate and therefore stands outside the main stream of his development at the time. In his dedication Schütz stressed the importance to composers of mastering the traditional style of counterpoint before attempting the concertato idiom, and his motets were intended to demonstrate the continued vitality of vocal polyphony. Nearly half the texts were taken from the New Testament. His use of modality and rhythmic flexibility helped him to avoid the comparative regularity of more modern styles and to achieve his usual high level of text expression and interpretation: epitomised on the one hand by the composure of *Selig sind die Toten* swv391 and on the other by the exuberance of *Die Himmel erzählen* swv386, a later version of swv455 (1635). This was Schütz's second and last book of motets: it is not his most spectacular publication, but it displays his consummate mastery and dignity and is undoubtedly the most important German motet collection of the 17th century.

His third book of *Symphoniae sacrae* swv398–418 (1650) complements the second in being scored for three to six voices, with instruments. All the symphonies require two violins, apart from *Wo der Herr* swv400, which calls for violin and cornetto. Most are also scored for an optional *complementum* of four voices or instruments, some calling for a bassoon. There are some exceptional combinations, such as the *complementum* of two instruments in *Siehe, wie fein* swv412, the optional choirs in *Saul, Saul* swv415 and *Komm, heiliger Geist* swv417 and the *complementum* of four viols in *Herr, wie lang* swv416. Most of the pieces begin with a sinfonia. The 'solo' sections are very Italianate indeed, with frequent use of triple metre and of semiquaver passages in common time, and the emphasis on duets, for voices and instruments, is still extremely strong. The sections for the *complementum* are largely homophonic, the simple but powerful chord progressions being sonorously scored, and are generally used to introduce a new portion of text, state a refrain, create antiphony or reach an effective conclusion. Throughout the collection the music displays the highest level of technical assurance and expressive power. Every piece is to some extent unique, but among the features that stand out are the vivid characterization of the dramatic dialogue *Mein Sohn* swv401, the forceful interventions of the chorus in *Es ging ein Sämann aus* swv408 and *Seid barmherzig* swv409, the use of the word 'Vater' as an ostinato in *Vater unser* swv411, the overpowering, almost terrifying combination of polychorality and *stile concitato* in *Saul, Saul*, the recitative for the six voices in *Herr, wie lang* and the treatment of the chorale melodies in *Komm, heiliger Geist*. In the variety and resourcefulness of its scoring the collection is strongly reminiscent of the *Psalmen Davids*, but it far surpasses it in richness and maturity of musical language. This third book of *Symphoniae sacrae* was Schütz's last collection of concertato settings for voices and instruments and it represents the climax of his development of the mid-17th century Italian style. Spitta (1894) said that it laid the foundation for all German sacred music for the next 100 years.

CHAPTER TEN

Late works, 1651–72

The final period in Schütz's creative career is dominated by his three Passions and by two oratorical works. The first collection of this period, the *Zwölff geistliche Gesänge* SWV420–31 (1657), is comparatively unimportant: the pieces were composed in his spare time for domestic use or church choirs, not for the electoral chapel, and were assembled for publication by Christoph Kittel, the Dresden court organist, who also appended his own setting of *O süsser Jesu Christ* (*cf* SWV427). The collection opens with four Mass movements (SWV420–23) and includes German versions of some of the Graces that had appeared in the *Cantiones sacrae*. The settings are for SATB with optional continuo and employ an extremely restrained style of polyphony which gives little indication as to when they were written. The style is not dissimilar to that of the Becker Psalter, of which the revised and enlarged third edition appeared in 1661.

The *Christmas History* SWV435 appears to have been performed at Dresden in 1660. The Evangelist's part ('Chor des Evangelisten'), consisting of recitative for tenor and continuo, was published there in 1664, but all the other parts ('Chor der Concerten in die Orgel') remained in manuscript, as Schütz felt that they could be performed only by a few well-appointed princely chapels (he invited choirmasters to buy manuscript copies of these sections or to compose their own settings). Apart from the opening chorus, of which only the figured bass is extant, and the second trombone part of one of the 'Concerte', the work has fortunately survived complete. The text is compiled from *Matthew* and *Luke* and the Evangelist's recitative is in a thoroughly contemporary style. The narrative is carefully punctuated by rests, the music is frequently illustrative or expressive of the text, and a line is sometimes repeated to round off a section. The 'Concerte', or 'Intermedia', on the other hand, are short concerted movements for various combinations of solo voices and instruments, setting the words of the different characters in the story. The forces chosen are invariably appropriate to the personages concerned and strongly reinforce the considerable degree of characterization that Schütz achieves in the music itself. A framework is provided by introductory and concluding choruses. Partly because the story is so familiar, the *Christmas History* is undoubtedly one of Schütz's most approachable works. But this is also because the music is so consistently attractive. The tonality is predominantly major, and the invention is unfailingly fresh and imaginative. It is impossible to think of a contemporary choirmaster who could have composed the Intermedia in so appealing a manner.

The composition of Schütz's three late Passions was anticipated some years earlier by that of the *Seven Last Words* swv478, the origins and date of which are unknown. The last utterances of Christ on the cross are connected by brief passages of narrative from the gospels and framed by the first and last verses of the chorale 'Da Jesus an dem Kreuze stund'. The chorale verses are set in motet style for SATTB; the words of Christ are sung by the second tenor, accompanied by two instruments (possibly violins) and continuo, in a highly flexible and expressive recitative-like style; the robbers are an alto and a bass, and the Evangelist's part is taken at different times by a soprano, alto, first tenor and SATB quartet. The entire piece could be performed by five singers, two violins and an organ, were it not for the two brief five-part symphonies sandwiched between the chorales and the body of the work. The nature and modest scale of the text doubtless account for the limited forces and restrained musical language employed, but these restrictions intensify the moving quality of a work that combines Passion traditions with the style of the oratorio.

Schütz's Passions according to St Matthew, St Luke and St John (swv479–81) are the last great examples of the 'dramatic' Passion, in which the words of the Evangelist (tenor) and individual protagonists are set to traditional recitation formulae while those of the other characters and the crowd are given to a chorus. In accordance with liturgical practice, the Passions are entirely unaccompanied. The recitation departs further from the traditional formulae than does that in the *Resurrection History*: the setting is mainly syllabic, but melismas and leaps are used to convey pictorial or emotional words – the effect of which is heightened by the general restraint – and the differences between the various voices and melodic styles provide the means for musical characterization.

The Passions begin and end with motet-like choruses, but most of the choral sections are relatively terse settings of the words of the various groups of characters in the story: the disciples, priests, crowd and so on. These choruses are often vividly realistic and demonstrate Schütz's great skill in portraying dramatic situations. The short, highly characterized rhythmic phrases are generally treated imitatively, and the transparent textures that result allow the words to come through with great force as well as clarity. In the *St John Passion* the insistence of the 'Crucify' choruses is intensified toward the end by 'presto' markings and the chorus in which the soldiers cast lots for Christ's raiment ends with a passage reminiscent of a hocket. The choruses in the *St Matthew Passion* employ a wider range of vocal combinations and achieve a greater variety of char-

Ex.6 *St Matthew Passion* SWV479
(a) The disciples of Jesus

acterization: the two false witnesses are pictured in a canon at the second, and the disciples' interpolation 'Herr, bin ichs' (ex.6*a*) and the crowd's 'Barrabam' (ex.6*b*) possess a terrifying urgency.

Although these Passions are deliberately restricted in their means, they are certainly not restricted in their effect: every interval in the recitation is significant, every entry in the choruses makes a mark. In these respects they represent a distillation of the essential elements of

Late works

Schütz's art – complete command of melody and of vocal counterpoint and full use of the range and power of expression that they could achieve.

CHAPTER ELEVEN

Achievement and reputation

Schütz inevitably appears to some to be a limited composer: his music is mostly sacred, and only those familiar with German are likely to understand it fully. But this limitedness is partly illusory – since his sacred works exploit an enormous range of forms, styles and scoring – and partly offset by his depth: the quality of his insight and inspiration. It is difficult to study a work by Schütz without becoming drawn into his thought processes – the way in which he viewed and responded to the text, planned the structure, composed and used the material and ultimately enriched the text by his setting. His clarity and strength of conception and perfection of execution are everywhere apparent; his intelligence and musicianship are clear on every page. And he was also practical, as his alternative scorings and careful directions for performance show.

He was described on his tombstone as 'Heinricus Schützius Seculi sui Musicus excellentissimus Electoris Capellae Magister', and during his life he enjoyed a reputation in Protestant Germany unmatched by that of any composer. Poets as eminent as Christoph Kaldenbach and Johann Rist wrote verses in his praise, and Bontempi honoured him with the dedication of a composition treatise (*Nova quatuor vocibus componendi methodus*, 1660). As early as 1620 the composer Michael Altenburg mentioned Schütz alongside Michael Praetorius as one of the outstanding musicians of Germany; ten years later the poet Paul Fleming wrote that 'when Schütz's songs resound, Saxony rejoices'. In 1657 a candidate for the post of Thomaskantor at Leipzig referred to Schütz as 'Parentem nostrae Musicae modernae', and in 1667 Gabriel Reuschel, a brother of the composer Michael M. Reuschel, apostrophized Schütz as 'thou prince of German singers'. As late as 1690 Wolfgang Caspar Printz recalled that Schütz was regarded in the middle of the century as 'the very best German composer'.

He was beyond doubt the greatest German musician of the 17th century. Apart from composing works of the very highest quality, he also guided German music through a difficult historical period. He provided a smooth transition from the late Renaissance to the mid-Baroque and brought together the style of traditional counterpoint and that of Italian monody. He spread the most recent Italian developments in German-speaking lands, at a time when they were riven by war, and did more than anybody else to bring the style of German sacred music up to date. In these respects he provided the essential foundation on which the masterpieces of German sacred music of the late Baroque were to be built.

WORKS

Editions: *H. Schütz: Sämmtliche Werke*, ed. P. Spitta and others (Leipzig, 1885–94, 1909, 1927/*R*1968–73) [G]
H. Schütz: Neue Ausgabe sämtlicher Werke, ed. W. Bittinger, W. Breig, W. Ehmann and others (Kassel, 1955–) [N]
H. Schütz: Sämtliche Werke, ed. G. Graulich and others (Stuttgart, 1971–) [S; references without vol. nos. indicate separately pubd single edns.]

Catalogue: *Schütz-Werke-Verzeichnis (SWV): Kleine Ausgabe*, ed. W. Bittinger (Kassel, 1960); suppl. in W. Breig (1979); complete, ed. Breig (in preparation) [SWV; A = Anhang]

Unless otherwise stated, entries in parentheses are earlier versions of the preceding work, with the same scoring. Opus nos. in square brackets come from the list of publications appended to Symphoniarum sacrarum secunda pars, and from the handwritten catalogue of works sent by Schütz to August of Brunswick-Lüneburg in 1664 (see chapter 7, p.75). Indications of scoring retain the nomenclature of the source in ambiguous cases. This applies particularly to designations of subensembles: 'capella' in the sources means (a) doubling ensemble, (b) optional ensemble included intermittently, or (c) ensemble always used as a whole; 'complementum' corresponds to (b), 'ripieno' and 'tutti' to (a). (The marking 'tutti' is retained when it seems to prescribe the addition of doubling voices.) Voices left undesignated in the sources, or labelled according to function rather than type (e.g. 'quinta pars'), are assigned on the basis of their notated range (with parts in Mez clef normally given to S, parts in Bar clef listed as B), unless the source includes directions for transposition, in which case the transposed range determines the labelling. 'Alternating' reflects the layout within a single part and does not necessarily indicate mode of performance. Lost parts are given in square brackets. Only principal sources are given for works in MS; all MS nos. for *D-B*, *Dlb*, *DS* and *Z* have prefix Ms.Mus., all for *Kl* have prefix 2° Ms.Mus., all for *W* have prefix Cod.Guelf., all for *S-Uu* have prefix Vok. mus. i hdskr. Dates and places of provenance of MSS are given in square brackets; place of provenance is present location of source unless otherwise stated.

cap. – *capella*
compl. – *complementum*
rip. – *ripieno*

* – *inst part (or part of uncertain scoring) fully or partly texted*
\+ – *doubling part or parts*
† – *Schütz's personal copy survives in D-W*
†† – *only known or complete copy lost since 1945*

Numbers in right-hand margins denote references in the text.

SURVEY OF WORKS AND SOURCES — 75–81

(printed works with opus numbers)

Il primo libro de madrigali, [op.1] (Venice, 1611)†

Psalmen Davids sampt etlichen Moteten und Concerten, [op.2] (Dresden, 1619)

Historia der frölichen und siegreichen Aufferstehung unsers einigen Erlösers und Seligmachers Jesu Christi, [op.3] (Dresden, 1623)

Cantiones sacrae, [op.4] (Freiberg, 1625)†

Psalmen Davids, hiebevorn in teutzsche Reimen gebracht, durch D. Cornelium Beckern, und an jetzo mit ein hundert und drey eigenen Melodeyen . . . gestellet, [op.5] (Freiberg, 1628) [rev. and enlarged 3rd edn. as [op.14], 1661]

Symphoniae sacrae, [op.6] (Venice, 1629)†

Musicalische Exequien . . . dess . . . Herrn Heinrichen dess Jüngern und Eltisten Reussen, [op.7] (Dresden, 1636)††

Erster Theil kleiner geistlichen Concerten, [op.8] (Leipzig, 1636)†

Anderer Theil kleiner geistlichen Concerten, [op.9] (Dresden, 1639)†

Symphoniarum sacrarum secunda pars, op.10 (Dresden, 1647)† — 75

Musicalia ad chorum sacrum, das ist: Geistliche Chor-Music . . . erster Theil, op.11 (Dresden, 1648)† — 75

Symphoniarum sacrarum tertia pars, op.12 (Dresden, 1650)† — 75

Zwölff geistliche Gesänge, op.13 (Dresden, 1657)† — 75

Psalmen Davids . . . jetzund . . . auffs neue übersehen, auch . . . vermehret, [op.14] (Dresden, 1661)† [rev. and enlarged edn. of [op.5], 1628] — 89

(other printed works) — 76

See SWV20, 21, 48, 49, 51, 52, 94–6, 277, 278, 338–40, 368, 419, 432–3, 434, 435 (recit only), 453, 464, 482–94 (title-page and index only), 501 (*works in manuscript*) 77–8

In *D-B*; *BĨB*; *Dlb*; *DS*; *FBo*; *Kl*; *KMs*; *LEm*; *W*; *WF*; *WRh*; *Z*; formerly Preussische Staatsbibliothek, Berlin; formerly Singakademie, Berlin; formerly Dreikönigskirche, Dresden; formerly Marienkirche, Helmstedt; Stefan Zweig's private collection, London; formerly Stadtbibliothek, Breslau; *S-Uu*; formerly Staats- und Universitätsbibliothek, Königsberg

---

SWV

Il primo libro de madrigali, [op.1] (Venice, 1611)†; G ix, N xxii, Si — 6, 78, 82, 85

1 O primavera (B. Guarini), 2 S, A, T, B
2 2p.: O dolcezze amarissime (Guarini), 2 S, A, T, B
3 Selve beate (Guarini), 2 S, A, T, B
4 Alma afflitta, che fai (G. B. Marino), S, A, 2 T, B
5 Così morir debb'io (Guarini), S, A, 2 T, B
6 D'orrida selce alpina (A. Aligieri), 2 S, A, T, B
7 Ride la primavera (Marino), 3 S, A, B
8 Fuggi, fuggi, o mio core (Marino), 3 S, A, B
9 Feritevi, ferite (Marino), 2 S, A, T, B
10 Fiamma ch'allacia (A. Gatti), 2 S, A, T, B
11 Quella damma son io (Guarini), 2 S, A, T, B
12 Mi saluta costei (Marino), 3 S, A, B
13 Io moro (Marino), 3 S, A, B
14 Sospir che del bel petto (Marino), 3 S, A, B
15 Dunque à Dio (Guarini), 3 S, A, B
16 Tornate, o cari baci (Marino), 2 S, A, T, B
17 Di marmo siete voi (Marino), 3 S, A, B
18 Giunto è pur, Lidia (Marino), 3 S, A, B
19 Vasto mar, nel cui seno (Dialogo) (?Schütz), I: S, A, T, B; II: S, A, T, B

Die Wort Jesus Syrach . . . auff hochzeitlichen Ehrentag des . . . Herrn Josephi Avenarii (Dresden, 1618) — 15, 17, 76

20 Wohl dem, der ein tugendsam Weib hat, I: T, *3 cornetts; II: S, A, T, B; bc (org), for wedding of Joseph Avenarius and Anna Dorothea Görlitz, Dresden, 21 April 1618; G xiv

Concert mit 11 Stimmen: auff des . . . Herrn Michael Thomae . . . hochzeitlichen EhrenTag (Dresden, 1618) — 15, 17, 76

21 Haus und Güter erbet man von Eltern, I: T, *3 trbn/bn; II: T, *3 cornetts/vn; III: 2 S, B; bc (+ lutes, hpd), for wedding of Michael Thomas and Anna Schultes, Leipzig, 15 June 1618; G xiv

Psalmen Davids sampt etlichen Moteten und Concerten, [op.2] (Dresden, 1619) — 15–16, 17, 18, 19, 79, 83, 85, 90, 103

22 Der Herr sprach zu meinem Herren (Psalmus 110), I: S, A, T, B; II: S, A, T, B; cap. a *5 ad lib; bc; G ii, N xxiii
23 Warum toben die Heiden (Psalmus 2), I: S, A, T, B; II: S, A, T, B; 2 cap. a *4 ad lib; bc; G ii, N xxiii
24 Ach Herr, straf mich nicht (Psalmus 6), I: S, A, T, B; II: S, A, T, B; bc; G ii, N xxiii
25 Aus der Tiefe (Psalmus 130), I: S, A, T, B; II: S, A, T, B; bc; G ii, N xxiii
26 Ich freu mich des (Psalmus 122), I: S, A, T, B; II: S, A, T, B; 2 cap. a *4 ad lib; bc; G ii, N xxiii
27 Herr, unser Herrscher (Psalmus 8), I: 2 S, A, T; II: A, T, 2 B; cap. a *5 ad lib; bc; G ii, N xxiii
28 Wohl dem, der nicht wandelt (Psalmus 1), I: 2 S, A, B; II: A, 2 T, B; bc; G ii, N xxiii, S
29 Wie lieblich sind deine Wohnunge (Psalmus 84), I: 2 S, A, B; II: 2 T, 2 B; bc; G ii, N xxiii, S
30 Wohl dem, der den Herren fürchtet (Psalmus 128), I: 2 S, A, T; II: A, T, 2 B; bc; G ii, N xxiii, S
31 Ich hebe meine Augen auf (Psalmus 121), I: S, A, T, B (+ cap. ad lib); II: S, A, T, B (+ cap.); bc; G ii, N xxiv, S
32 Danket dem Herren, denn er ist freundlich (Psalmus 136), I: 3 S, T; II: A, 2 T, B; 2 cap. a *4 ad lib; bc; G ii, N xxiv
33 Der Herr ist mein Hirt (Psalmus 23), I: 2 S, A, T (+ cap. ad lib); II: S, A, T, B (+ cap.); bc; G ii, N xxiv, S — 60
34 Ich danke dem Herrn (Psalmus 111), I: S, A, T, B; II: S, A, T, B; 2 cap. a *4 ad lib; bc (doxology 'Imitatione sopra: Lieto godea . . . di Gio.[vanni] Gab.[rieli]'); G ii, N xxiv — 83
35 Singet dem Herrn ein neues Lied (Psalmus 98), I: S, A, T, B; II; S, A, T, B; bc (? = Singet dem Herrn ein newes Lied, Ps.98, perf. Dresden, 31 Oct 1617, according to Hoe von Hoenegg (16!8)); G iii, N xxiv, S — 14, 83

SWV		
36	Jauchzet dem Herren, alle Welt (Psalmus 100), I: S, A, T, B, bc; II: S, A, T, B, bc; G iii, N xxiv	83, 85
(36*a*	Jauchzet dem Herren, alle Welt (Ps. 100), I: S, A, T, B; II: S, A, T, B, III: S, A, T, B, *D-Kl* 49*r* [1614–15] (bc in source added 1616–17, unauthentic); N xxviii)	7
37	An den Wassern zu Babel (Psalmus 137), I: S, A, T, B; II: S, A, T, B; bc; G iii, N xxiv	83
38	Alleluja, lobet den Herren (Psalmus 150), I: S, A, T, B; II: S, A, T, B; 2 cap. (I: *cornett/vn (alternating with cornett, vn), *2 cornetts/vn, *trbn/bn (alternating with bn, bn/trbn); II: *cornett/rec (alternating with cornett, rec), *2 trbn, *trbn/bn (alternating with bn, trbn, bn/trbn)) ad lib; bc; G iii	83
39	Lobe den Herren, meine Seele (Concert), I: S, A, T, B (+ cap. ad lib); II: S, A, T, B (+ cap.); bc (org); G iii	33, 76, 83
40	Ist nicht Ephraim mein teurer Sohn (Moteto), I: 2 cornetts/S, cornett/T, *cornett/trbn; II: A, *3 trbn; 2 cap. a *4 ad lib; bc; G iii	83
41	Nun lob, mein Seel, den Herren (Canzon), I: S, A, T, B (+cap. ad lib if inst cap. not used); II: S, A, T, B (+ cap. ad lib if inst cap. not used); 2 cap. (I: 2 vn, 2 viole, vle; II: 4 cornetts, trbn) ad lib if vocal cap. not used; bc (= Nun lob mein Seel den Herrn, auff 4 Chor, perf. Dresden, 1 Nov 1617, according to Hoe von Hoenegg (1618)); G iii	14, 79, 83
42	Die mit Tränen säen (Moteto), I: S, T, *3 trbn; II: S, T, *3 trbn; bc; G iii	83
43	Nicht uns, Herr (Psalmus 115), I: T, *3 cornetts; II (cap.): S, A, T, B; III: A, *3 trbn; bc (= Nicht uns Herr . . . Psalm 115 mit drey Choren, perf. Dresden, 2 Nov 1617, according to Hoe von Hoenegg (1618)); G iii	14, 83
44	Wohl dem, der den Herren fürchtet (Psalmus 128), I: T, *4 cornetts; II: A, *vn, *3 trbn; cap. I: S, A, T, B; cap. II a *4 ad lib; bc; G iii	83
45	Danket dem Herren, denn er ist freundlich (Psalmus 136), I: 2 S, A, T; II: A, *3 trbn; cap. a *5; tpts and timp; bc (= 136. Psalm mit Trommeten und Heerpauken, perf. Dresden, 2 Nov 1617, according to Hoe von Hoenegg (1618)); G iii	14, 47, 83
46	Zion spricht, der Herr hat mich verlassen (Concert), I: S, T, *3 cornetts, *bn; II: S, T, *4 trbn; 2 cap. a *4 ad lib; bc; G iii	83
SWV		
47	Jauchzet dem Herren, alle Welt (Concert), I: A, T, *2 fl (alternating with *2 cornetts), *bn; II: S, T; III: S, *vn, *3 viole; cap. a *5 ad lib; bc (+ lutes) (? = Concert mit 4 Choren Jubilate Deo, perf. Dresden, 2 Nov 1617, according to Hoe von Hoenegg (1618)); G iii, S	14, 83
	Der 133. Psalm . . . auff die hochzeitliche Ehrenfrewde Herrn Georgii Schützen (Leipzig, 1619)††	16, 17, 76
48	Siehe, wie fein und lieblich ists, 2 S, A, T, B, mute cornett/vn, vn/fl, vle/bn, bc (org), for wedding of Georg Schütz and Anna Gross, Leipzig, 9 Aug 1619 (earlier version of no.412); G xiv	
	Syncharma musicum (Breslau, 1621)††	17, 76
49	En novus Elysiis, I: T, 3 cornetts; II: T, 3 bn; III (coro aggiunto): 3 S, B; bc (org), for declaration of loyalty of the Silesian estates, Breslau, 3 Nov 1621 (copy formerly in Stadtbibliothek, Breslau, incl. 2nd text, Wo Gott nicht selbst bei uns wäre, added in later hand); G xv, N xxxviii	
	Historia der frölichen und siegreichen Aufferstehung unsers einigen Erlösers und Seligmachers Jesu Christi, [op.3] (Dresden, 1623)	18, 58, 76, 78 79, 85–6, 106
50	Die Auferstehung unsres Herren Jesu Christi, 3 S, 2 A, 3 T, 2 B, 4 viols, bc (swv50 Anm.2–3, in *D-Kl* 49*v*, 53*x* [1623–7], unauthentic); G i, N iii	
	in *RISM* 1623[14]:	17, 76
51	Das ist mir lieb, 2 S, A, T, B; G xii, N xxviii, S	17
	Kläglicher Abschied von der churfürstlichen Grufft zu Freybergk (Freiberg, 1623)	17–18, 76
52	Grimmige Gruft, so hast du dann (Schütz), S, bc, for burial of Duchess Sophia of Saxony, Freiberg, 28 Jan 1623; G xviii, N xxxvii, facs. edn. (n.p., 1922)	
	Cantiones sacrae, [op.4] (Freiberg, 1625)†	18, 79, 80, 86–7, 104
53	O bone, o dulcis, o benigne Jesu, 2 S, A, B, bc (org); G iv, N viii	
54	2p.: Et ne despicias, 2 S, A, B, bc (org); G iv, N viii	
55	Deus misereatur nostri, 2 S, A, B, bc (org); G iv, N viii	87, *88*
56	Quid commisisti, o dulcissime puer, S, A, T, B, bc (org); G iv, N viii	

57 2p.: Ego sum tui plaga doloris, S, A, T, B, bc (org); G iv, N viii
58 3p.: Ego enim inique egi, S, A, T, B, bc (org); G iv, N viii
59 4p.: Quo, nate Dei, S, A, T, B, bc (org); G iv, N viii
60 5p.: Calicem salutaris accipiam, S, A, T, B, bc (org); G iv, N viii
61 Verba mea auribus percipe, S, A, T, B, bc (org); G iv, N viii — 86
62 2p.: Quoniam ad te clamabo, S, A, T, B, bc (org); G iv, N viii — 86
63 Ego dormio, S, A, T, B, bc (org); G iv, N viii
64 2p.: Vulnerasti cor meum, S, A, T, B, bc (org); G iv, N viii
65 Heu mihi, Domine, S, A, T, B, bc (org); G iv, N viii
66 In te, Domine, speravi, 3 S, T, bc (org); G iv, N viii
67 Dulcissime et benignissime Christe, 3 S, T, bc (org); G iv, N viii
68 Sicut Moses serpentem in deserto exaltavit, S, A, T, B, bc (org); G iv, N viii
69 Spes mea, Christe Deus, S, A, T, B, bc (org); G iv, N viii
70 Turbabor, sed non perturbabor, S, A, T, B, bc (org); G iv, N viii — 86
71 Ad Dominum cum tribularer clamavi, S, A, T, B, bc (org); G iv, N viii
72 2p.: Quid detur tibi, S, A, T, B, bc (org); G iv, N viii
73 Aspice, Pater, piissimum filium, S, A, T, B, bc (org); G iv, N ix
74 2p.: Nonne hic est, S, A, T, B, bc (org); G iv, N ix
75 3p.: Reduc, Domine Deus meus, S, A, T, B, bc (org); G iv, N ix
76 Supereminet omnem scientiam, 2 S, A, B, bc (org); G iv, N ix
77 2p.: Pro hoc magno mysterio pietatis, 2 S, A, B, bc (org); G iv, N ix
78 Domine, non est exaltatum cor meum, S, A, T, B, bc (org); G iv, N ix
79 2p.: Si non humiliter sentiebam, S, A, T, B, bc (org); G iv, N ix
80 3p.: Speret Israel in Domino, S, A, T, B, bc (org); G iv, N ix
81 Cantate Domino canticum novum, 2 S, A, B, bc (org); G iv, N ix, S
82 Inter brachia Salvatoris mei, S, A, T, B, bc (org); G iv, N ix
83 Veni, rogo, in cor meum, 3 S, T, bc (org); G iv, N ix
84 Ecce advocatus meus apud te, S, A, T, B, bc (org) (see also no.304); G iv, N ix — 86
85 Domine, ne in furore tuo, S, A, T, B, bc (org); G iv, N ix — 86, *89*
86 2p.: Quoniam non est in morte, A, T, B, bc (org); G iv, N ix — 86
87 3p.: Discedite a me omnes qui operamini, S, A, T, B, bc (org); G iv, N ix — 86

88 Oculi omnium in te sperant, 3 S, T, bc (org) (on no.429*a*); G iv, N ix
89 2p.: Pater noster, qui es in coelis, 3 S, T, bc (org) (as no.88; see also no.92); G iv, N ix
90 3p.: Domine Deus, pater coelestis, 3 S, T, bc (org) (as no.88); G iv, N ix
91 Confitemini Domino, 3 S, T, bc (org) (on no.430*a*); G iv, N ix
92 2p.: Pater noster, qui es in coelis, 3 S, T, bc (org) (as no.91; text and music = no.89); G iv, N ix
93 3p.: Gratias agimus tibi, 3 S, T, bc (org) (as no.91); G iv, N ix

De vitae fugacitate: aria . . . bey Occasion des . . . Todesfalles der . . . Jungfrawen Anna Marien Wildeckin (Freiberg, 1625) — 18, 76, 80
94 Ich hab mein Sach Gott heimgestellt, 2 S, A, T, B, bc, on death of Anna Maria Wildeck, Dresden, 15 Aug 1625 (earlier version of no.305); G xii, N xxxi — 98

Ultima verba psalmi 23 . . . super . . . obitu . . . Jacobi Schultes (Leipzig, 1625) — 18, 76
95 Gutes und Barmherzigkeit werden mir folgen, S, [S], A, T, [T, B], bc (org), on death of Jacob Schultes, Leipzig, 19 July 1625; G xviii, N xxxi

in *RISM* 1627⁹: — 76
96 Glück zu dem Helikon (?M. Opitz), A, T, bc; G xv, N xxxvii — 21

Psalmen Davids, hiebevorn in teutzsche Reimen gebracht, durch D. Cornelium Beckern, und an jetzo mit ein hundert und drey eigenen Melodeyen . . . gestellet, [op.5] (Freiberg, 1628, 2/1640) [all versions with *a* nos.]; rev. and enlarged 3rd edn., as Psalmen Davids . . . jetzund . . . auffs neue übersehen, auch . . . vermehret, [op.14] (Dresden, 1661)† [nos.97–256] — 19, 28, 59–60, 61, 79, 87–9, 104
97 Wer nicht sitzt im Gottlosen Rat, S, A, T, B, bc (org); G xvi, N vi
(97*a* Wer nicht sitzt im Gottlosen Rat, 2 S, A, B)
98 Was haben doch die Leut im Sinn, S, A, T, B, bc (org); G xvi, N vi
(98*a* Was haben doch die Leut im Sinn, S, A, T, B)
99 Ach wie gross ist der Feinde Rott, S, A, T, B, bc (org); G xvi, N vi

SWV
(99*a* Ach wie gross ist der Feinde Rott, S, A, T, B)
100 Erhör mich, wenn ich ruf zu dir, S, A, T, B, bc (org); G xvi, N vi
(100*a* Erhör mich, wenn ich ruf zu dir, S, A, T, B)
101 Herr hör, was ich will bitten dich, S, A, T, B, bc (org); G xvi, N vi
(101*a* Herr hör, was ich will bitten dich, 2 S, A, B)
102 Ach Herr, mein Gott, S, A, T, B, bc (org); G xvi
(102*a* Ach Herr, mein Gott, S, A, T, B; N vi)
103 Auf dich trau ich, S, A, T, B, bc (org); G xvi
(103*a* Auf dich trau ich, S, A, T, B; N vi)
104 Mit Dank wir sollen loben, S, A, T, B, bc (org); G xvi
(104*a* Mit Dank wir sollen loben, S, A, T, B; N vi)
105 Mit fröhlichem Gemüte, S, A, T, B, bc (org); G xvi, N vi
(105*a* Mit fröhlichem Gemüte, 2 S, A, B)
106 Wie meinst du's doch, S, A, T, B, bc (org); G xvi, N vi
(106*a* Wie meinst du's doch, S, A, T, B)
107 Ich trau auf Gott, S, A, T, B, bc (org); G xvi, N vi
(107*a* Ich trau auf Gott, S, A, T, B)
108 Ach Gott vom Himmel, sieh darein, S, A, T, B, bc (org); G xvi, N vi
(108*a* Ach Gott vom Himmel, sieh darein, S, A, T, B)
109 Ach Herr, wie lang willt du, S, A, T, B, bc (org); G xvi, N vi
(109*a* Ach Herr, wie lang willt du, S, A, T, B)
110 Es spricht der Unweisen Mund wohl, S, A, T, B, bc (org); G xvi, N vi
(110*a* Es spricht der Unweisen Mund wohl, S, A, T, B)
111 Wer wird, Herr, in der Hütten dein, S, A, T, B, bc (org); G xvi, N vi
(111*a* Wer wird, Herr, in der Hütten dein, 2 S, A, B; N vi)
112 Bewahr mich, Gott, ich trau auf dich, S, A, T, B, bc (org); G xvi, N vi
(112*a* Bewahr mich, Gott, ich trau auf dich, S, A, T, B)
113 Herr Gott, erhör die Grechtigkeit, S, A, T, B, bc (org); G xvi
(113*a* Herr Gott, erhör die Grechtigkeit, 2 S, A, B; N vi)
114 Ich lieb dich, Herr, von Herzen sehr, S, A, T, B, bc (org); G xvi, N vi
(114*a* Ich lieb dich, Herr, von Herzen sehr, 2 S, A, B)
115 Die Himmel, Herr, preisen sehr, S, A, T, B, bc (org); G xvi
(115*a* Die Himmel, Herr, preisen sehr, 2 S, A, B; N vi)
116 Der Herr erhör dich in der Not, S, A, T, B, bc (org); G xvi, N vi
(116*a* Der Herr erhör dich in der Not, 2 S, A, T)

SWV
117 Hoch freuet sich der König, S, A, T, B, bc (org); G xvi
(117*a* Hoch freuet sich der König, 2 S, A, B; N vi)
118 Mein Gott, mein Gott, S, A, T, B, bc (org); G xvi, N vi
(118*a* Mein Gott, mein Gott, S, A, T, B)
119 2p. of no.118: Ich will verkündgen in der Gmein, S, A, T, B, bc (org); G xvi, N vi
(119*a* 2p. of no.118*a*: Ich will verkündgen in der Gmein, S, A, T, B)
120 Der Herr ist mein getreuer Hirt, S, A, T, B, bc (org); G xvi, N vi
(120*a* Der Herr ist mein getreuer Hirt, S, A, T, B)
121 Die Erd und was sich auf ihr regt, S, A, T, B, bc (org); G xvi
(121*a* Die Erd und was sich auf ihr regt, S, A, T, B; N vi)
122 Nach dir verlangt mich, S, A, T, B, bc (org); G xvi, N vi
(122*a* Nach dir verlangt mich, S, A, T, B)
123 Herr, schaff mir recht, S, A, T, B, bc (org); G xvi, N vi
(123*a* Herr, schaff mir recht, 2 S, A, B)
124 Mein Licht und Heil ist Gott der Herr, S, A, T, B, bc (org); G xvi, N vi
(124*a* Mein Licht und Heil ist Gott der Herr, 2 S, A, B)
125 Ich ruf zu dir, Herr Gott, S, A, T, B, bc (org); G xvi, N vi
(125*a* Ich ruf zu dir, Herr Gott, 3 S, T)
126 Bringt Ehr und Preis dem Herren, S, A, T, B, bc (org); G xvi
(126*a* Bringt Ehr und Preis dem Herren, S, A, T, B; N vi)
127 Ich preis dich, Herr, S, A, T, B, bc (org); G xvi, N vi
(127*a* Ich preis dich, Herr, 2 S, A, B)
128 In dich hab ich gehoffet, Herr, S, A, T, B, bc (org); G xvi, N vi
(128*a* In dich hab ich gehoffet, Herr, S, A, T, B)
129 Der Mensch für Gott wohl selig ist, S, A, T, B, bc (org); G xvi, N vi
(129*a* Der Mensch für Gott wohl selig ist, S, A, T, B)
130 Freut euch des Herrn, ihr Christen all, S, A, T, B, bc (org); G xvi, N vi
(130*a* Freut euch des Herrn, ihr Christen all, S, A, T, B)
131 Ich will bei meinem Leben, S, A, T, B, bc (org); G xvi, N vi
(131*a* Ich will bei meinem Leben, S, A, T, B)
132 Herr, hader mit den Hadern mein, S, A, T, B, bc (org); G xvi, N vi
(132*a* Herr, hader mit den Hadern mein, S, A, T, B)
133 Ich sags von Grund meins Herzen frei, S, A, T, B, bc (org); G xvi, N vi
(133*a* Ich sags von Grund meins Herzen frei, S, A, T, B)

134 Erzürn dich nicht so sehre, S, A, T, B, bc (org); G xvi, N vi
(134*a* Erzürn dich nicht so sehre, S, A, T, B)
135 Herr, straf mich nicht in deinem Zorn, S, A, T, B, bc (org); G xvi, N vi
(135*a* Herr, straf mich nicht in deinem Zorn, 2 S, A, B)
136 In meinem Herzen hab ich mir, S, A, T, B, bc (org); G xvi
(136*a* In meinem Herzen hab ich mir, S, A, T, B; N vi)
137 Ich harrete des Herren, S, A, T, B, bc (org); G xvi, N vi
(137*a* Ich harrete des Herren, S, A, T, B)
138 Wohl mag der sein ein selig Mann, S, A, T, B, bc (org); G xvi, N vi
(138*a* Wohl mag der sein ein selig Mann, S, A, T, B)
139 Gleichwie ein Hirsch eilt mit Begier, S, A, T, B, bc (org); G xvi, N vi
(139*a* Gleichwie ein Hirsch eilt mit Begier, S, A, T, B)
140 Gott, führ mein Sach, S, A, T, B, bc (org); G xvi, N vi
(140*a* Gott, führ mein Sach, S, A, T, B)
141 Wir haben, Herr, mit Fleiss gehört, S, A, T, B, bc (org); G xvi, N vi
(141*a* Wir haben, Herr, mit Fleiss gehört, S, A, T, B)
142 Mein Herz dichtet ein Lied mit Fleiss, S, A, T, B, bc (org); G xvi, N vi
(142*a* Mein Herz dichtet ein Lied mit Fleiss, S, A, T, B)
143 Ein feste Burg ist unser Gott, S, A, T, B, bc (org); G xvi, N vi
(143*a* Ein feste Burg ist unser Gott, S, A, T, B; N vi)
144 Frohlockt mit Freud, S, A, T, B, bc (org); G xvi, N vi
(144*a* Frohlockt mit Freud, S, A, T, B)
145 Gross ist der Herr, S, A, T, B, bc (org); G xvi, N vi
(145*a* Gross ist der Herr, S, A, T, B)
146 Hört zu, ihr Völker in gemein, S, A, T, B, bc (org); G xvi, N vi
(146*a* Hört zu, all Völker in gemein, S, A, T, B)
147 Gott, unser Herr, S, A, T, B, bc (org); G xvi, N vi
(147*a* Gott, unser Herr, 2 S, A, B)
148 Erbarm dich mein, o Herre Gott, S, A, T, B, bc (org); G xvi, N vi
(148*a* Erbarm dich mein, o Herre Gott, S, A, T, B)
149 Was trotzst denn du, Tyrann, S, A, T, B, bc (org); G xvi, N vi
150 Es spricht der Unweisen Mund wohl, S, A, T, B, bc (org) (text and music = no.110); G xvi, N vi
151 Hilf mir, Gott, S, A, T, B, bc (org); G xvi, N vi
152 Erhör mein Gbet, S, A, T, B, bc (org); G xvi, N vi
(152*a* Erhör mein Gbet, S, A, T, B)
153 Herr Gott, erzeig mir Hülf und Gnad, S, A, T, B, bc (org); G xvi, N vi
154 Sei mir gnädig, o Gott, S, A, T, B, bc (org); G xvi, N vi
(154*a* Sei mir gnädig, o Gott, S, A, T, B)
155 Wie nun, ihr Herren, seid ihr stumm, S, A, T, B, bc (org); G xvi, N vi
156 Hilf, Herre Gott, errette mich, S, A, T, B, bc (org); G xvi, N vi
157 Ach Gott, der du vor dieser Zeit, S, A, T, B, bc (org); G xvi, N vi
158 Gott, mein Geschrei erhöre, S, A, T, B, bc (org); G xvi, N vi
(158*a* Gott, mein Geschrei erhöre, 2 S, A, B)
159 Mein Seel ist still in meinem Gott, S, A, T, B, bc (org); G xvi, N vi
160 O Gott, du mein getreuer Gott, S, A, T, B, bc (org); G xvi
(160*a* O Gott, du mein getreuer Gott, 2 S, A, B; N vi)
161 Erhör mein Stimm, Herr, S, A, T, B, bc (org); G xvi, N vi
(161*a* Erhör mein Stimm, Herr, S, A, T, B)
162 Gott, man lobt dich in der Still, S, A, T, B, bc (org); G xvi, N vi
163 Jauchzet Gott, alle Lande sehr, S, A, T, B, bc (org); G xvi, N vi
164 Es wollt uns Gott genädig sein, S, A, T, B, bc (org); G xvi, N vi
(164*a* Es wollt uns Gott genädig sein, S, A, T, B)
165 Es steh Gott auf, S, A, T, B, bc (org); G xvi, N, vi
(165*a* Es steh Gott auf, S, A, T, B)
166 Gott hilf mir, denn das Wasser dringt, S, A, T, B, bc (org); G xvi, N vi
(166*a* Gott hilf mir, denn das Wasser dringt, S, A, T, B)
167 Eil, Herr, mein Gott, zu retten mich, S, A, T, B, bc (org); G xvi
(167*a* Eil, Herr, mein Gott, zu retten mich, 2 S, A, B; N vi)
168 Auf dich, Herr, trau ich allezeit, S, A, T, B, bc (org); G xvi, N vi
169 Gott, gib dem König auserkorn, S, A, T, B, bc (org); G xvi, N vi
170 Dennoch hat Israel zum Trost, S, A, T, B, bc (org); G xvi, N vi
(170*a* Dennoch hat Israel zum Trost, S, A, T, B)
171 Warum verstösst du uns so gar, S, A, T, B, bc (org); G xvi, N vi
172 Aus unsers Herzen Grunde, S, A, T, B, bc (org); G xvi, N, vi
173 In Juda ist der Herr bekannt, S, A, T, B, bc (org); G xvi, N vi
174 Ich ruf zu Gott mit meiner Stimm, S, A, T, B, bc (org); G xvi, N vi
175 Hör, mein Volk, mein Gesetz und Weis, S, A, T, B, bc (org); G xvi

SWV

(175*a* Hör, mein Volk, mein Gesetz und Weis, 2 S, A, B; N vi)
176 Ach Herr, es ist der Heiden Heer, S, A, T, B, bc (org); G xvi, N vi
(176*a* Ach Herr, es ist der Heiden Heer, S, A, T, B)
177 Du Hirt Israel, höre uns, S, A, T, B, bc (org); G xvi, N vi
178 Singet mit Freuden unserm Gott, S, A, T, B, bc (org); G xvi, N vi
179 Merkt auf, die ihr an Gottes Statt, S, A, T, B, bc (org); G xvi, N vi
180 Gott, schweig du nicht so ganz und gar, S, A, T, B, bc (org); G xvi, N vi
(180*a* Gott, schweig du nicht so ganz und gar, S, A, T, B)
181 Wie sehr lieblich und schöne, S, A, T, B, bc (org); G xvi
(181*a* Wie sehr lieblich und schöne, 2 S, A, B; N vi)
182 Herr, der du vormals gnädig warst, S, A, T, B, bc (org); G xvi, N vi
183 Herr, neig zu mir dein gnädigs Ohr, S, A, T, B, bc (org); G xvi, N vi
184 Fest ist gegründet Gottes Stadt, S, A, T, B, bc (org); G xvi, N vi
185 Herr Gott, mein Heiland, Nacht und Tag, S, A, T, B, bc (org); G xvi, N vi
186 Ich will von Gnade singen, S, A, T, B, bc (org); G xvi, N vi
(186*a* Ich will von Gnade singen, S, A, T, B)
187 2p. of no.186: Ach Gott, warum verstösst du nun, S, A, T, B, bc (org); G xvi, N vi
(187*a* 2p. of no.186*a*: Ach Gott, warum verstösst du nun, S, A, T, B)
188 Herr Gott Vater im höchsten Thron, S, A, T, B, bc (org); G xvi, N vi
(188*a* Herr Gott Vater im höchsten Thron, S, A, T, B)
189 Wer sich des Höchsten Schirm vertraut, S, A, T, B, bc (org); G xvi, N vi
(189*a* Wer sich des Höchsten Schirm vertraut, S, A, T, B)
190 Es ist fürwahr ein köstlich Ding, S, A, T, B, bc (org); G xvi, N vi
(190*a* Es ist fürwahr ein köstlich Ding, S, A, T, B; N vi)
191 Der Herr ist König herrlich schön, S, A, T, B, bc (org); G xvi, N vi
192 Herr Gott, dem alle Rach heimfällt, S, A, T, B, bc (org); G xvi, N vi
(192*a* Herr Gott, dem alle Rach heimfällt, 3 S, T)
193 Kommt herzu, lasst uns fröhlich sein, S, A, T, B, bc (org); G xvi, N vi

SWV

194 Singet dem Herrn ein neues Lied, S, A, T, B, bc (org); G xvi, N vi
195 Der Herr ist König überall, S, A, T, B, bc (org); G xvi, N vi
196 Singet dem Herrn ein neues Lied, S, A, T, B, bc (org); G xvi, N vi
(196*a* Singet dem Herrn ein neues Lied, S, A, T, B)
197 Der Herr ist Köng und residiert, S, A, T, B, bc (org); G xvi, N vi
198 Jauchzet dem Herren alle Welt, S, A, T, B, bc (org); G xvi, N vi
(198*a* Jauchzet dem Herren alle Welt, S, A, T, B)
199 Von Gnad und Recht soll singen, S, A, T, B, bc (org); G xvi, N vi
200 Hör mein Gebet und lass zu dir, S, A, T, B, bc (org); G xvi, N vi
201 Nun lob, mein Seel, den Herren, S, A, T, B, bc (org); G xvi
(201*a* Nun lob, mein Seel, den Herren, 2 S, A, B; N vi)
202 Herr, dich lob die Seele mein, S, A, T, B, bc (org); G xvi, N vi
(202*a* Herr, dich lob die Seele mein, S, A, T, B)
203 Danket dem Herren, lobt ihn frei, S, A, T, B, bc (org); G xvi, N vi
204 Danket dem Herrn, erzeigt ihm Ehr, S, A, T, B, bc (org); G xvi, N vi
205 Danket dem Herren, unserm Gott, S, A, T, B, bc (org); G xvi, N vi
206 Mit rechtem Ernst, S, A, T, B, bc (org); G xvi, N vi
207 Herr Gott, des ich mich rühmte viel, S, A, T, B, bc (org); G xvi, N vi
(207*a* Herr Gott, des ich mich rühmte viel, S, A, T, B)
208 Der Herr sprach zu meim Herren, S, A, T, B, bc (org); G xvi
(208*a* Der Herr sprach zu meim Herren, S, A, T, B; N vi)
209 Ich will von Herzen danken Gott dem Herrn, S, A, T, B, bc (org); G xvi, N vi
(209*a* Ich will von Herzen danken Gott dem Herrn, S, A, T, B)
210 Der ist fürwahr ein selig Mann, S, A, T, B, bc (org); G xvi, N vi
(210*a* Der ist fürwahr ein selig Mann, S, A, T, B)
211 Lobet, ihr Knecht, den Herren, S, A, T, B, bc (org); G xvi, N vi
212 Als das Volk Israel auszog, S, A, T, B, bc (org); G xvi, N vi
213 Nicht uns, nicht uns, Herr, lieber Gott, S, A, T, B, bc (org); G xvi, N vi
(213*a* Nicht uns, nicht uns, Herr, lieber Gott, S, A, T, B)
214 Meim Herzen ists eine grosse Freud, S, A, T, B, bc (org); G xvi, N vi

(214*a* Meim Herzen ists eine grosse Freud, S, A, T, B)
215 Lobt Gott mit Schall, S, A, T, B, bc (org); G xvi, N vi
(215*a* Lobt Gott mit Schall, S, A, T, B)
216 Lasst uns Gott, unserm Herren, S, A, T, B, bc (org); G xvi, N vi
217 Wohl denen, die da leben, S, A, T, B, bc (org); G xvi, N vi
(217*a* Wohl denen, die da leben, S, A, T, B)
218 2p. of no.217: Tu wohl, Herr, deinem Knechte, S, A, T, B, bc (org); G xvi, N vi
219 3p.: Lass mir Gnad widerfahren, S, A, T, B, bc (org); G xvi, N vi
220 4p.: Du tust viel Guts beweisen, S, A, T, B, bc (org); G xvi, N vi
221 5p.: Dein Wort, Herr, nicht vergehet, S, A, T, B, bc (org); G xvi, N vi
222 6p.: Ich hass die Flattergeister, S, A, T, B, bc (org); G xvi, N vi
223 7p.: Dir gbührt allein die Ehre, S, A, T, B, bc (org); G xvi, N vi
224 8p.: Fürsten sind meine Feinde, S, A, T, B, bc (org); G xvi, N vi
225 Ich ruf zu dir, mein Herr und Gott, S, A, T, B, bc (org); G xvi, N vi
226 Ich heb mein Augen sehnlich auf, S, A, T, B, bc (org); G xvi
(226*a* Ich heb mein Augen sehnlich auf, S, A, T, B; N vi)
227 Es ist ein Freud dem Herzen mein, S, A, T, B, bc (org); G xvi, N vi
(227*a* Es ist ein Freud dem Herzen mein, S, A, T, B)
228 Ich heb mein Augen auf zu dir, S, A, T, B, bc (org); G xvi, N vi
229 Wär Gott nicht mit uns diese Zeit, S, A, T, B, bc (org); G xvi, N vi
(229*a* Wär Gott nicht mit uns diese Zeit, S, A, T, B)
230 Die nur vertraulich stellen, S, A, T, B, bc (org); G xvi, N vi
231 Wenn Gott einmal erlösen wird, S, A, T, B, bc (org); G xvi, N vi
232 Wo Gott zum Haus nicht gibt sein Gunst, S, A, T, B, bc (org); G xvi, N vi
(232*a* Wo Gott zum Haus nicht gibt sein Gunst, S, A, T, B)
233 Wohl dem, der in Furcht Gottes steht, S, A, T, B, bc (org); G xvi, N vi
234 Die Feind haben mich oft gedrängt, S, A, T, B, bc (org); G xvi, N vi
(234*a* Die Feind haben mich oft gedrängt, S, A, T, B)
235 Aus tiefer Not schrei ich zu dir, S, A, T, B, bc (org); G xvi, N vi
(235*a* Aus tiefer Not schrei ich zu dir, S, A, T, B; N vi)
236 Herr, mein Gemüt und Sinn du weisst, S, A, T, B, bc (org); G xvi, N vi
(236*a* Herr, mein Gemüt und Sinn du weisst, S, A, T, B)
237 In Gnaden, Herr, wollst eindenk sein, S, A, T, B, bc (org); G xvi, N vi
238 Wie ists so fein, lieblich und schön, S, A, T, B, bc (org); G xvi, N vi
(238*a* Wie ists so fein, lieblich und schön, S, A, T, B)
239 Den Herren lobt mit Freuden, S, A, T, B, bc (org); G xvi, N vi
240 Lobt Gott von Herzengrunde, S, A, T, B, bc (org); G xvi, N vi
(240*a* Lobt Gott von Herzengrunde, S, A, T, B)
241 Danket dem Herren, gebt ihm Ehr, S, A, T, B, bc (org); G xvi, N vi
(241*a* Danket dem Herren, gebt ihm Ehr, 2 S, A, B)
242 An Wasserflüssen Babylon, S, A, T, B, bc (org); G xvi, N vi
(242*a* An Wasserflüssen Babylon, S, A, T, B)
243 Aus meines Herzen Grunde, S, A, T, B, bc (org); G xvi, N vi
(243*a* Aus meines Herzen Grunde, S, A, T, B)
244 Herr, du erforschst mein Sinne, S, A, T, B, bc (org); G xvi, N vi
245 Von bösen Menschen rette mich, S, A, T, B, bc (org); G xvi, N vi
246 Herr, mein Gott, wenn ich ruf zu dir, S, A, T, B, bc (org); G xvi, N vi
(246*a* Herr, mein Gott, wenn ich ruf zu dir, S, A, T, B)
247 Ich schrei zu meinem lieben Gott, S, A, T, B, bc (org); G xvi, N vi
248 Herr, mein Gebet erhör in Gnad, S, A, T, B, bc (org); G xvi, N vi
249 Gelobet sei der Herr, mein Hort, S, A, T, B, bc (org); G xvi, N vi
250 Ich will sehr hoch erhöhen dich, S, A, T, B, bc (org); G xvi, N vi
251 Mein Seel soll loben Gott den Herrn, S, A, T, B, bc (org); G xvi, N vi
(251*a* Mein Seel soll loben Gott den Herrn, S, A, T, B)
252 Zu Lob und Ehr mit Freuden singt, S, A, T, B, bc (org); G xvi
(252*a* Zu Lob und Ehr mit Freuden singt, S, A, T, B; N vi)
253 Lobet, ihr Himmel, Gott den Herrn, S, A, T, B, bc (org); G xvi, N vi
(253*a* Lobet, ihr Himmel, Gott den Herrn, S, A, T, B)
254 Die heilige Gemeine, S, A, T, B, bc (org); G xvi, N vi
255 Lobt Gott in seinem Heiligtum, S, A, T, B, bc (org); G xvi

SWV		
(255*a*	Lobt Gott in seinem Heiligtum, 2 S, A, B; N vi)	
256	Alles was Odem hat, lobe den Herrn (Responsorium), S, A, T, B, bc (org); G xvi	
(256*a*	Alles was Odem hat, lobe den Herrn (Responsorium), 2 S, A, B; N vi)	
	Symphoniae sacrae, [op.6] (Venice, 1629)†	25, 79, 80, 90–92, 96, 100
257	Paratum cor meum, S/T, 2 vn, bc (org); G v, N xiii	
258	Exultavit cor meum in Domino, S, 2 vn, bc (org); G v, N xiii	91
259	In te, Domine, speravi, A, vn, bn/trbn, bc (org); G v, N xiii	
260	Cantabo Domino in vita mea, T, 2 vn, bc (org); G v, N xiii	91
261	Venite ad me, T, 2 vn, bc (org); G v, N xiii	
262	Jubilate Deo omnis terra, B, 2 flautinos/vn, bc (org); G v, N xiii	
263	Anima mea liquefacta est, 2 T, 2 fl/cornettinos, bc (org) (on 263/4*a*); G v, N xiii	17
264	2p.: Adjuro vos, filiae Hierusalem, 2 T, 2 fl/cornettinos, bc (org) (as no.263); G v, N xiii	17
(263/4*a*	Anima mea liquefacta est, 2 T, 2 vn, [vle], bc, *Dlb* Pi 8 [1625–30, Pirna], Pi 57 [1630–*c*1635, Pirna])	17
(263/4*b*	Anima mea liquefacta est, [4 S], bc, *Dlb* 54*a* [*c*1635, Pirna], probably unauthentic arr. of 263/4*a*)	17
265	O quam tu pulchra es, T, Bar, 2 vn, bc (org); G v, N xiii	90, 92, 94–5
266	2p.: Veni de Libano, T, Bar, 2 vn, bc (org); G v, N xiii	91
267	Benedicam Dominum in omni tempore, S, T, B, cornett/vn, bc (org); G v, N xiv	
268	2p.: Exquisivi Dominum, S, T, B, cornett/vn, bc (org); G v, N xiv	
269	Fili mi, Absalon, B, 2 trbn/vn, 2 trbn, bc (org); G v, N xiv	90
270	Attendite, popule meus, B, 2 trbn/vn, 2 trbn, bc (org); G v, N xiv	
271	Domine, labia mea aperies, S, T, cornett/vn, trbn, bn, bc (org); G v, N xiv	
272	In lectulo per noctes, S, A, 3 bn/viole, bc (org); G v, N xiv	
273	2p.: Invenerunt me custodes, S, A, 3 bn/viole, bc (org); G v, N xiv	
274	Veni, dilecte mi, I: S, trbn/T, 2 trbn; II: S, T; bc (org, theorbo) (text for I: T added by Schütz to his copy); G v, N xiv	91
275	Buccinate in neomenia tuba, 2 T, B, cornett, tpt/cornett, bn, bc (org); G v, N xiv	
276	2p.: Jubilate Deo in chordis et organo, 2 T, B, cornett, tpt/cornett, bn, bc (org); G v, N xiv	

SWV		
	Verba D. Pauli ... beatis manibus Dn. Johannis-Hermanni Scheinii ... consecrata (Dresden, 1631)††	26, 76
277	Das ist je gewisslich wahr, 2 S, A, 2 T, B, bc (org) ad lib, on death of Johann Hermann Schein, Leipzig, 19 Nov 1630 (earlier version of no.388); G xii, N xxxi	
	An hoch printzlicher Durchläuchtigkeit zu Dennenmarck ... Beylager: Gesang der Venus-Kinder in der Invention genennet Thronus Veneris (Copenhagen, 1634)	29, 76
278	O der grossen Wundertaten (Canconetta), 4 S [III lost], [2 vn], bc, for procession at wedding festivities of Prince Christian of Denmark and Princess Magdalena Sibylla of Saxony, Copenhagen, 13 Oct 1634; G xviii, N xxxvii	
	Musicalische Exequien ... dess ... Herrn Heinrichen dess Jüngern und Eltisten Reussen, [op.7] (Dresden, 1636)††; G xii, N iv, S viii	31, 78, 92–5
279	Nacket bin ich vom Mutterleibe kommen (Concert ... in Form einer teutschen Missa), 2 S, A (alternating with B II), 2 T, B (+ cap. ad lib), bc (org, vle), for burial of Prince Heinrich Posthumus of Reuss, Gera, 4 Feb 1636	93
280	Herr, wenn ich nur dich habe (Motet), I: S, A, T, B; II: S, A, T, B; bc (org, vle) ad lib (as no.279)	93
281	Herr, nun lässest du deinen Diener/Selig sind die Toten (Canticum B. Simeonis), I: 2 A, 2 T, B; II: 2 S, B; bc (org, vle) (as no.279)	93
	Erster Theil kleiner geistlichen Concerten, [op.8] (Leipzig, 1636)†	31–2, 76, 90, 96–8
282	Eile mich, Gott, zu erretten, S, bc (org); G vi, N x, S	96
283	Bringt her dem Herren, Mez, bc (org); G vi, N x	
284	Ich danke dem Herrn, A, bc (org); G vi, N x	
285	O süsser, o freundlicher, T/S, bc (org); G vi, N x	
286	Der Herr ist gross, 2 S, bc (org); G vi, N x	
287	O lieber Herre Gott, 2 S, bc (org); G vi, N x, S	
(287*a*	O lieber Herre Gott, *D-Kl 59i* [1635, Dresden])	31
288	Ihr heiligen, lobsinget dem Herren, 2 S, bc (org); G vi, N x	
289	Erhöre mich, wenn ich rufe, 2 S, bc (org); G vi, N x	
(289*a*	Erhöre mich, wenn ich rufe, 2 S [II lost], [B], bc, *Dlb* Pi 8 [1625–30, Pirna], Pi 57 [1630–*c*1635, Pirna])	17
290	Wohl dem, der nicht wandelt, S, A, bc (org); G vi, N x	
291	Schaffe in mir, Gott, ein reines Herz, S, T, bc (org); G vi, N xi	33, 76
292	Der Herr schauet vom Himmel, S, B, bc (org); G vi, N xi	

293 Lobet den Herren, der zu Zion wohnet, 2 A, bc (org) (?earlier version, lost, in inventory of Kassel Hofkapelle, 1638); G vi, N x — 31
294 Eins bitte ich vom Herren, 2 T, bc (org); G vi, N x
295 O hilf, Christe, Gottes Sohn [Christe Deus adjuva], 2 T, bc (org) (Lat. text added by Schütz to his copy); G vi, N x, S
296 Fürchte dich nicht, ich bin mit dir, 2 B, bc (org); G vi, N x
(296*a* Fürchte dich nicht, ich bin mit dir, *Kl* 49*y* [1635, Dresden]) — 31
297 O Herr hilf, 2 S, T, bc (org) (earlier version of no.402); G vi, N xi, S
298 Das Blut Jesu Christi, 2 S, B, bc (org) (?earlier version, lost, in inventory of Kassel Hofkapelle, 1638); G vi, N xi, S — 31
299 Die Gottseligkeit ist zu allen Dingen nütz, 2 S, B, bc (org); G vi, N xi, S
300 Himmel und Erde vergehen, 3 B, bc (org) (?earlier version, lost, in inventory of Kassel Hofkapelle, 1638); G vi, N x — 31
301 Nun komm, der Heiden Heiland [Veni, Redemptor gentium], 2 S, 2 B, bc (org) (Lat. text added by Schütz to his copy); G vi, N xi, S — 98, *99*
(301*a* Nun komm, der Heiden Heiland, *Kl* 49*g* [1635, Dresden]) — 31
302 Ein Kind ist uns geboren, S, A, T, B, bc (org); G vi, N xi, S
(302*a* Ein Kind ist uns geboren, *Kl* 50*c* [1635, Dresden]) — 31
303 Wir gläuben all an einen Gott, 2 S, T, B, bc (org); G vi, N xi — 97
304 Siehe, mein Fürsprecher ist im Himmel, S, A, T, B (+ cap.), bc (org) (later version of no.84); G vi, N xi, S
(304*a* Siehe, mein Fürsprecher ist im Himmel, S, A, T, B, bc (org), *Kl* 50*a* [1635, Dresden]) — 31
305 Ich hab mein Sach Gott heimgestellt [Meas dicavi res Deo] (Aria), 2 S, A, T, B (+ cap.), bc (org) (later version of no.94; Lat. text added by Schütz to his copy); G vi, N xii — 97–8

Anderer Theil kleiner geistlichen Concerten, [op.9] (Dresden, 1639)† — 34–5, 90, 96–8
306 Ich will den Herren loben allezeit, S, bc (org); G vi, N x, S
307 Was hast du verwirket, A, bc (org); G vi, N x, S — 96
308 O Jesu, nomen dulce, T, bc (org); G vi, N x
309 O misericordissime Jesu, T, bc (org); G vi, N x
310 Ich liege und schlafe, B, bc (org); G vi, N x
311 Habe deine Lust an dem Herren, 2 S, bc (org); G vi, N x — 97
312 Herr, ich hoffe darauf, 2 S, bc (org); G vi, N x, S — 97
313 Bone Jesu, verbum Patris, 2 S/T, bc (org); G vi, N x — 97
314 Verbum caro factum est, 2 S, bc (org); G vi, N x
315 Hodie Christus natus est, S, T, bc (org); G vi, N xi
316 Wann unsre Augen schlafen ein [Quando se claudunt lumina], S, B, bc (org); G vi, N xi — 80
(316*a* Wann unsre Augen schlafen ein, *Kl* 59*k* [1635, Dresden]) — 31
317 Meister, wir haben die ganze Nacht gearbeitet, 2 T, bc (org) (earlier version, lost, in inventory of Kassel Hofkapelle, 1638); G vi, N x — 31
318 Die Furcht des Herren, 2 T, bc (org); G vi, N x, S
319 Ich beuge meine Knie, 2 B, bc (org); G vi, N x
320 Ich bin jung gewesen, 2 B, bc (org); G vi, N x
321 Herr, wann ich nur dich habe, 2 S, T, bc (org); G vi, N xi, S
322 Rorate coeli desuper, 2 S, B, bc (org); G vi, N xi — 98
323 Joseph, du Sohn David, 2 S, B, bc (org); G vi, N xi, S
324 Ich bin die Auferstehung, 2 T/S, B, bc (org); G vi, N x, S
325 Die Seele Christi heilige mich, A, T, B, bc (org) (earlier version, lost, in inventory of Kassel Hofkapelle, 1638); G vi, N xi — 98
326 Ich ruf zu dir, Herr Jesu Christ [Te Christe supplex invôco] (Hymnus), 3 S, Bar, bc (org); G vi, N xi, S — 31, 97
(326*a* Ich ruf zu dir, Herr Jesu Christ, 3 S, bc, *Dlb* Pi 8 [1625–30, Pirna], Pi 57 [1630–*c*1635, Pirna]) — 17
327 Allein Gott in der Höh sei Ehr (Hymnus), 2 S, 2 T, bc (org); G vi, N xi — 98
328 Veni, Sancte Spiritus (in concerto), 2 S, 2 T, bc (org); G vi, N xi — 98
329 Ist Gott für uns, S, A, T, B, bc (org); G vi, N xi, S
330 Wer will uns scheiden von der Liebe Gottes, S, A, T, B, bc (org); G vi, N xi, S
331 Die Stimm des Herren, S, A, T, B, bc (org); G vi, N xi — 98
(331*a* Die Stimm des Herren, *Kl* 52*h* [1635, Dresden]) — 31
332 Jubilate Deo omnis terra, S, A, T, B, bc (org); G vi, N xi
333 Sei gegrüsset, Maria (Dialogus), 2 S, A, T, B, 5 insts, bc (org) (see also no.334); G vi, N xii, S — 98
334 Ave Maria, gratia plena (Dialogus), 2 S, A, T, B, 5 insts, bc (org) (Lat. version of no.333); G vi, N xii, S — 98

SWV		
335	Was betrübst du dich, 2 S, A, T, B, bc (org); G vi, N xii, S	
336	Quemadmodum desiderat cervus, S, A, 2 T, B, bc (org); G vi, N xii	
337	Aufer immensam, Deus, aufer iram, S, A, 2 T, B, bc (org); G vi, N xii	
in *RISM* 1641³:		43, 76
338	Teutoniam dudum belli [Adveniunt pascha pleno], 2 S, A, T, B, 2 vn, bc (org); G xv, N xxxviii	17, 43
339	Ich beschwere euch, ihr Töchter zu Jerusalem (Dialogus), 2 S, 2 S/vn, A, T, B, bc (org); G xiv	43
in *RISM* 1646⁴:		43, 76
340	O du allersüssester und liebster Herr Jesu, 2 S, A, T, B, 2 vn, bc (org); G xiv	43
Symphoniarum sacrarum secunda pars, op.10 (Dresden, 1647)†		37, 42–3, 44, 75, 79, 98, 100, 102
341	Mein Herz ist bereit, S/T, 2 vn, bc (org, vle); G vii, N xv	
(341*a*	Mein Herz ist bereit, [S], 2 vn, bc (org); *Kl* 49*k* [1635, Dresden; title on wrapper is autograph]; N xv)	31
342	Singet dem Herren ein neues Lied, S/T, 2 vn, bc (org, vle); G vii, N xv	
343	Herr, unser Herrscher, S/T, 2 vn, bc (org, vle); G vii, N xv	
344	Meine Seele erhebt den Herren, S, 2 vn (alternating ad lib with 2 viole/trbn, 2 cornetts/tpt, 2 flautinos, 2 cornettinos/vn), bc (org, vle); G vii, N xv	100
345	Der Herr ist meine Stärke, S/T, 2 vn, bc (org, vle); G vii, N xv	
346	Ich werde nicht sterben, S/T, 2 vn, bc (org, vle); G vii, N xv	37
(346*a*	Ich werde nicht sterben, S/T, 2 vn, bc, *Kl* 49*t* [1640–47]; N xv)	37
347	2p. of no.346: Ich danke dir, Herr, S/T, 2 vn, bc (org, vle); G vii, N xv	37
348	Herzlich lieb hab ich dich, o Herr, A, 2 vn, bc (org, vle); G vii, N xv	
(348*a*	Herzlich lieb hab ich dich, o Herr, A, 2 vn, bc (org), *Kl* 49*d* [1635, Dresden; title on wrapper is autograph]; N xv)	31
349	Frohlocket mit Händen, T, 2 vn, bc (org, vle) (earlier version, lost, in inventory of Kassel Hofkapelle, 1638); G vii, N xv	31
350	Lobet den Herrn in seinem Heiligtum, T, 2 vn, bc (org, vle); G vii, N xv	100
351	Hütet euch, dass eure Herzen, B, 2 vn, bc (org, vle); G vii, N xv	

SWV		
352	Herr, nun lässest du deinen Diener, B, 2 vn, bc (org, vle); G vii, N xv	
(352*a*	Herr, nun lässest du deinen Diener (Canticum Simeonis), B, 2 vn, bc (org), *Kl* 50*e* [1635, Dresden; title and dedication to Christoph Cornet on wrapper are autograph]; N xv)	31
353	Was betrübst du dich, 2 S/T, 2 vn, bc (org, vle); G vii, N xvi	100, 101
354	Verleih uns Frieden genädiglich, 2 S/T, 2 vn, bc (org, vle); G vii, N xvi	
355	2p.: Gib unsern Fürsten, 2 S/T, 2 vn, bc (org, vle); G vii, N xvi	
356	Es steh Gott auf, 2 S/T, 2 vn, bc (org, vle) (as stated in preface, based on Monteverdi: Armato il cor and Zefiro torna e di soavi accenti); G vii, N xvi	24, 100
357	Wie ein Rubin in feinem Golde leuchtet, S, A, 2 vn, bc (org, vle); G vii, N xvi	
358	Iss dein Brot mit Freuden, S, B, 2 vn, bc (org, vle); G vii, N xvi	
359	Der Herr ist mein Licht, 2 T, 2 vn, bc (org, vle); G vii, N xvi	100
360	Zweierlei bitte ich, Herr, 2 T, 2 vn, bc (org, vle); G vii, N xvi	
361	Herr, neige deine Himmel, 2 B, 2 vn, bc (org, vle); G vii, N xvi	
(361*a*	Herr, neige deine Himmel, 2 B, 2 vn, bc (org), *Kl* 49*i* [1635, Dresden; title on wrapper is autograph]; N xvi)	31
362	Von Aufgang der Sonnen, 2 B, 2 vn, bc (org, vle); G vii, N xvi	
363	Lobet den Herrn, alle Heiden, A, T, B, 2 vn, bc (org, vle); G vii, N xvii	
364	Die so ihr den Herren fürchtet, A, T, B, 2 vn, bc (org, vle); G vii, N xvii	
365	Drei schöne Dinge seind, 2 T, B, 2 vn, bc (org, vle); G vii, N xvii	
366	Von Gott will ich nicht lassen, 2 S, B, 2 vn, bc (org, vle); G vii, N xvii	100
367	Freuet euch des Herren, A, T, B, 2 vn, bc (org, vle); G vii, N xvii	
Danck-Lied: für die hocherwiesene fürstl. Gnade in Weymar (Gotha, 1647)		
368	Fürstliche Gnade zu Wasser und Lande (C. T. Dufft), T, 2 insts, bc, for birthday celebration for Duchess Eleonora Dorothea of Saxe-Weimar, perf. Weimar, 12 Feb 1647, according to diary of Duke Wilhelm; G xv, N xxxvii	42
Musicalia ad chorum sacrum, das ist: Geistliche Chor-Music . . . erster Theil, op.11 (Dresden, 1648)†; G viii, N v, S		45, 75, 100–02
369	Es wird das Szepter von Juda, S, A, 2 T, B, bc	
370	2p.: Er wird sein Kleid in Wein waschen, S, A, 2 T, B, bc	

371 Es ist erschienen die heilsame Gnade Gottes, 2 S, A, T, B, bc
372 Verleih uns Frieden genädiglich, 2 S, A, T, B, bc
373 2p.: Gib unsern Fürsten, 2 S, A, T, B, bc
374 Unser keiner lebet ihm selber, 2 S, A, T, B, bc
375 Viel werden kommen, S, A, 2 T, B, bc
376 Sammlet zuvor das Unkraut, S, A, 2 T, B, bc
377 Herr, auf dich traue ich, 2 S, A, T, B, bc
378 Die mit Tränen säen, 2 S, A, T, B, bc
379 So fahr ich hin zu Jesu Christ, 2 S, A, T, B, bc
380 Also hat Gott die Welt geliebt (Aria), S, A, 2 T, B, bc
381 O lieber Herre Gott, 2 S, A, 2 T, B, bc
382 Tröstet, tröstet mein Volk, 2 S, A, 2 T, B, bc
383 Ich bin eine rufende Stimme, 2 S, A, 2 T, B, bc
384 Ein Kind ist uns geboren, 2 S, A, 2 T, B, bc
385 Das Wort ward Fleisch, 2 S, A, 2 T, B, bc
386 Die Himmel erzählen die Ehre Gottes, 2 S, A, 2 T, B (+ tutti [ad lib]), bc (later version of no.455) — 102
387 Herzlich lieb hab ich dich, o Herr (Aria), 2 S, A, 2 T, B, bc
388 Das ist je gewisslich wahr, 2 S, A, 2 T, B (+ tutti [ad lib]), bc (later version of no.277)
389 Ich bin ein rechter Weinstock, 2 S, A, 2 T, B, bc
390 Unser Wandel ist im Himmel, 2 S, A, 2 T, B, bc
391 Selig sind die Toten, 2 S, A, 2 T, B, bc — 102
392 Was mein Gott will, das gscheh allzeit, A, T, 4 insts, bc (earlier version, lost, in inventory of Kassel Hofkapelle, 1638; not identical with anon. setting, 2 S, A, 2 T, B, *Kl* 53*u* [*c*1610])
393 Ich weiss, dass mein Erlöser lebt, 3 S, A, T, 2 B, bc
394 Sehet an den Feigenbaum, S, T, 5 insts, bc
395 Der Engel sprach zu den Hirten (super Angelus ad pastores, Andreae Gabrielis), S, T, inst/B, *4 insts, bc
396 Auf dem Gebirge, 2 A, 5 insts, bc
397 Du Schalksknecht, T, 6 insts, bc (earlier version, lost, in inventory of Kassel Hofkapelle, 1638)

Symphoniarum sacrarum tertia pars, op.12 (Dresden, 1650)† — 43, 48, 75, 79, 90, 102–3

398 Der Herr ist mein Hirt, S, A, T, 2 vn, compl. 4vv and insts ad lib, bc (org, vle); G x, S
(398*a* Der Herr ist mein Hirt, S, A, T, 2 vn, 3 trbn ad lib, bc (org), *Kl* 49*s* [1640–50], trbn parts ?unauthentic)
399 Ich hebe meine Augen auf, A, T, B, 2 vn, compl. 4vv and insts ad lib, bc (org, vle); G x, S
400 Wo der Herr nicht das Haus bauet, 2 S, B, vn, cornettino/vn, compl. 4vv and insts ad lib, bc (org, vle); G x, S — 102
401 Mein Sohn, warum hast du uns das getan (in dialogo), S, Mez, B, 2 vn, compl. 4vv and insts ad lib, bc (org, vle); G x, S — 103
(401*a* Mein Sohn, warum hast du uns das getan (Dominica 1. post Epiphan.[ias], in dialogo), S, Mez, B, 2 vn, bc (org), *Kl* 49*w* [1640–50]; G x) — 43
402 O Herr hilf, 2 S, T, 2 vn, bc (org, vle) (later version of no.297); G x, S
403 Siehe, es erschien der Engel des Herren, S, 2 T, B, 2 vn, compl. 4vv and insts ad lib, bc (org, vle); G x, S
404 Feget den alten Sauerteig aus, S, A, T, B, 2 vn, bc (org, vle); G x
405 O süsser Jesu Christ, 2 S, A, T, 2 vn, compl. 4vv and insts ad lib, bc (org, vle); G x, S
406 O Jesu süss, wer dein gedenkt (super Lilia convallium, Alexandri Grandis), 2 S, 2 T, 2 vn, bc (org, vle); G x, S
(406*a* O Jesu süss, wer dein gedenkt (super Lillium convallium Alexandri Grandi), [2 S, 2 T], 2 vn, bc (+ vle), *Kl* 59*q* [1640–50]) — 43
407 Lasset uns doch den Herren, unsern Gott, loben, 2 S, T, B, 2 vn, compl. 4vv and insts ad lib, bc (org, vle) (?earlier version, without compl., lost, in inventory of Weimar Hofkapelle, 1662); G x, S
408 Es ging ein Sämann aus zu säen, S, A, T, B, 2 vn, bn, compl. 4vv and insts ad lib, bc (org, vle); G xi, S — 103
409 Seid barmherzig, S, A, T, B, 2 vn, bn (+ B), compl. 4vv and insts ad lib, bc (org, vle); G xi, S — 103
410 Siehe, dieser wird gesetzt zu einem Fall, 2 S, A, T, B, 2 vn, bc (org, vle); G xi, S
411 Vater unser, der du bist im Himmel, S, Mez, 2 T, B, 2 vn, compl. 4vv and insts ad lib, bc (org, vle); G xi, S — 103
412 Siehe, wie fein und lieblich ist, 2 S, A, T, B, 2 vn, bn, compl. 2 insts ad lib, bc (org, vle) (later version of no.48); G xi — 102
413 Hütet euch, dass eure Herzen, 2 S, A, 2 T, B, 2 vn, bc (org, vle); G xi, S
414 Meister, wir wissen, dass du wahrhaftig bist, 2 S, A, T, B, 2 vn,

SWV

bn, compl. 4vv and insts ad lib, bc (org, vle); G xi, S

415 Saul, Saul, was verfolgst du mich, 2 S, A, T, 2 B, 2 vn, 2 cap. S, A, T, B ad lib, bc (org, vle); G xi, S — 102, 103

416 Herr, wie lang willt du mein so gar vergessen, 2 S, A, 2 T, B, 2 vn, compl. 4 viole ad lib [IV not pr.], bc (org, vle); G xi, S — 102, 103

(416*a* Herr, wie lang willt du mein so gar vergessen, 2 S, A, 2 T, B, 2 vn, 4 viole ad lib, bc, *Kl* 49*l* [1640–50]) — 43

417 Komm, heiliger Geist, S, Mez, 2 T, Bar, B, 2 vn, 2 cap. S, A, T, B ad lib, bc (org, vle); G xi, S — 102, 103

418 Nun danket alle Gott, 2 S, A, 2 T, B, 2 vn, compl. 4vv and insts ad lib, bc (org, vle); G xi, S

(418*a* Nun danket alle Gott, 2 S, A, 2 T, B, 2 vn, bc, *Kl* 52*c* [1640–50]) — 43

in *RISM* 1652[6a]: — 76

419 O meine Seel, warum bist du betrübet (Trauer-Lied) (C. Brehme), S, A, T, B, on death of Anna Margaretha Brehme, Dresden, 21 Sept 1652; G xviii, N xxxvii — 52

Zwölff geistliche Gesänge, op.13 (Dresden, 1657)†; G xii, S xv — 60, 75, 79, 104

420 Kyrie, Gott Vater in Ewigkeit (super Missam Fons bonitatis), S, A, T, B, bc ad lib — 104

421 All Ehr und Lob soll Gottes sein (Das teutsche Gloria in excelsis), S, A, T, B, bc ad lib — 104

422 Ich gläube an einen einigen Gott (Der nicönische Glaube), S, A, T, B, bc ad lib — 104

423 Unser Herr Jesus Christus, in der Nacht (Die Wort der Einsetzung des heiligen Abendmahls), 2 S, A, B, bc ad lib — 104

424 Ich danke dem Herrn von ganzem Herzen (Der 111. Psalm), 2 S, A, B, bc ad lib

425 Danksagen wir alle Gott, 2 S, A, T, bc ad lib

426 Meine Seele erhebt den Herren (Magnificat), S, A, T, B, bc ad lib

427 O süsser Jesu Christ (Des H. Bernhardi Freudengesang), 2 S, A, B/I: 2 S, A, B; II: 2 S, A, B; bc ad lib — 104

428 Kyrie eleison, Christe eleison (Die teutsche gemeine Litaney), S, A, T, B, bc ad lib

429 Aller Augen warten auf dich (Das Benedicite vor dem Essen), S, A, T, B, bc ad lib

(429*a* Aller Augen warten auf dich, 2 S, B, *Dlb* Pi 8 [1625–30, Pirna]) — 17

SWV

430 Danket dem Herren, denn er ist freundlich (Das Gratias nach dem Essen), S, A, T, B, bc ad lib

(430*a* Danket dem Herren, denn er ist freundlich, 2 S, B, *Dlb* Pi 8 [1625–30, Pirna]) — 17

431 Christe fac ut sapiam (Hymnus pro vera sapientia), S, A, T, B/I: S, A, T, B; II: S, A, T, B; bc ad lib

Canticum B. Simeonis . . . nach dem hochseligsten Hintritt . . . Johann Georgen (Dresden, 1657) — 58, 76

432–3 Herr, nun lässest du deinen Diener, [S], S, A, 2 T, B, bc ad lib, on death of Elector Johann Georg I of Saxony, Dresden, 8 Oct 1657; G xii, N xxxi, S

in David Schirmers . . . poetische Rauten-Gepüsche (Dresden, 1663): — 52, 53, 76

434 Wie wenn der Adler sich aus seiner Klippe schwingt, S, bc, on engagement of Princess Magdalena Sibylla of Saxony and Duke Friedrich Wilhelm of Saxe-Altenburg, Dresden, 1651; G xv, N xxxvii

Historia, der freuden- und gnadenreichen Geburth Gottes und Marien Sohnes, Jesu Christi (Dresden, 1664) — 61, 78, 79, 104

435 Die Geburt unsers Herren Jesu Christi, Chor des Evangelisten: T, bc (org, vle); [Chor der Concerten in die Orgel: S, 3 A, 3 T, 4 B, chorus 4vv, chorus 6vv (+ viole ad lib), 2 vn, 2 violettas, 2 va (? = violettas), vle, 2 rec, bn, 2 tpt/cornetts, 2 trbn, bc (org)] (only Chor des Evangelisten printed; Chor der Concerten in MS, possibly = no.435*a* (ii)); G i, N i — 65, 75–6, 104–5

(435*a* Die Geburt unsers Herren Jesu Christi (2 versions): (i) [S, 3 A], T, [3 T, 4 B, chorus 4vv, chorus 6vv (+ viole ad lib), 2 vn, va, 3 viole, 2 rec, bn, 2 tpt/cornetts, 2 trbn], bc (org, hpd, viola), *S-Uu* Caps. 71 [1660–64, Dresden] (probably = die Geburth Christi, in stilo recitativo, perf. Dresden, 25 Dec 1660, according to court diaries); G xvii (recitatives only); (ii) [Chor des Evangelisten: T, bc]; (Chor der Concerten in die Orgel:) 2 S, 2 A, A/T, 3 T, 4 B, 2 vn, va, 2 violettas/viole, 6 viole [II–VI lost], 2 rec, bn, 2 tpt, 2 trbn [II lost], bc (org), *S-Uu* Caps.71 [?1664, Dresden] (possibly = Chor der Concerten of no.435; 2 trbn in final chorus spurious); G xvii, N i)

(435*b* Die Geburt unsers Herren Jesu Christi, 2 S, 3 A, 3 T, 4 B, 2 vn, [va], 2 violettas, vle, 2 rec, bn, 2 tpt, 2 trbn, bc, formerly Singakademie, Berlin [origins unknown] (inc.; recit = 435, Chor des Evangelisten, incorporating MS corrections to unique copy in *D-B*; intermedia possibly unauthentic); N i)

436 Ego autem sum, [B], bc, formerly Marienkirche, Helmstedt [1638, Helmstedt; MS now *W* 323 Mus.Handschr., but partbook containing no.436 no longer extant]; G xviii

437 Veni, Domine, [S], bc, formerly Marienkirche, Helmstedt [1638, Helmstedt; MS now *W* 323 Mus.Handschr., but partbook containing no.437 no longer extant]; G xviii

438 Die Erde trinkt für sich (Madrigal) (Opitz), A, T, bc, *Kl 59l* [1640–50]; G xv, N xxxvii — 43

439 Heute ist Christus der Herr geboren, 3 S, bc, *Kl 52g* [1632–8]; G xiv, S

440 Güldne Haare, gleich Aurore (Canzonetta), 2 S, 2 vn, [bc], *Kl 58i* [before 1650, ?] (contrafactum of Monteverdi: Chiome d'oro); G xv, N xxxvii — 100

441 Liebster, sagt in süssem Schmerzen (Opitz), 2 S, 2 vn, bc, *Kl 49h* [1627–32, Dresden; insts are autograph]; G xv, N xxxvii — 21, 77

442 Tugend ist der beste Freund (Opitz), 2 S, 2 vn, bc, *Kl 49f* [1640–50]; G xv, N xxxvii — 43

443 Weib, was weinest du (Dialogo per la pascua), 2 S, A, T [+ rip.], bc, *Kl 49x/2* [?*c*1645, Dresden; text is autograph]; G xiv, facs. edn. (Kassel, 1965)

(443*a* Weib, was weinest du, [2 S, A, T, B + rip.], bc, *Dlb* 1479/E/502 [?. ?Grimma]; *Kl 49x/1* [origins unknown; title-page only])

444 Es gingen zweene Menschen hinauf (in dialogo), 2 S, A, Bar, bc, *Kl 49u* [1640–50]; G xiv — 43

445 Ach bleib mit deiner Gnade, see 'Doubtful and spurious works'

446 In dich hab ich gehoffet, Herr, see 'Doubtful and spurious works'

447 Erbarm dich mein, o Herre Gott, S, 2 viole/vn, 2 viole, vle, bc, *S-Uu* 34:1 [before 1665, ?]; G xviii, N xxxii, S

448 Gelobet seist du, Herr (Gesang der dreyer Menner im feurig Ofen), 2 S, A, T, B, 2 cornettinos ad lib, 3 trbn ad lib, [cap. 2 S, A, T, B ad lib], cap. 2 vn, 2 viole, vle ad lib, bc, formerly Staats- und Universitätsbibliothek, Königsberg [1652, ?Naumburg]; G xiii, S

449 Herr, unser Herrscher (Psalmus 8), 2 S, A, T, B (+ cap. ad lib), cornettino/vn ad lib, vn/cornett ad lib, 4 trbn ad lib, bc (org, vle), *D-Kl 50d* [1635, Dresden; title is autograph] (contrary to N, insts authentic); G xiii, N xxvii — 31

450 Ach Herr, du Schöpfer aller Ding (Madrigale spirituale), S, A, 2 T, B, bc, *Kl 52k* [1615–27, Dresden] (on Marenzio: Deh poi ch'era ne' fati); G xiv, N xxxii, S — 14, 17

(450*a* Ach Herr, du Schöpfer aller Ding (sopra Deh poi ch'era ne' fati del Marentio), 2 S [I lost], B, bc, *Dlb* Pi 8 [1625–30, Pirna], Pi 57 [1630–*c*1635, Pirna]) — 14

451 Nachdem ich lag in meinem öden Bette (Opitz), S, B, 2 vn, 2 insts, bc, formerly Stadtbibliothek, Breslau [origins unknown]; G xv, N xxxvii

452 Lässt Salomon sein Bette nicht umgeben (Opitz), S, B, 2 vn, 2 insts, bc, formerly Stadtbibliothek, Breslau [origins unknown]; G xv, N xxxvii

453 Freue dich des Weibes deiner Jugend, S, A, T, B (+ tutti [ad lib]), tpt, cornett, 3 trbn ad lib, bc, *D-Dlb* Löb 56 [*c*1640, Löbau] (= Freue dich des Weibes, à 6 et 9, listed as pr. work in catalogue of A. Unger's collection, Naumburg, 1657); G xiv — 76

454 Nun lasst uns Gott dem Herren, see 'Doubtful and spurious works'

455 Die Himmel erzählen die Ehre Gottes (Psalmus 19), 2 S, A, 2 T, B (+ cap. ad lib), bc (+ vle), *Kl 50f* [1635, Dresden; title on wrapper and revision are autograph] (earlier version of no.386); G xiv, N xxvii, S — 31, 102

456 Hodie Christus natus est, 2 S, A, 2 T, B, bc, *Kl 49c* [1640–50]; G xiv, S — 43

457 Ich weiss, dass mein Erlöser lebet, 2 S, A, 2 T, B, *FBo* XI 8 47 [origins unknown]; formerly Dreikönigskirche, Dresden [before 1628, ?]; S — 17

458 Kyrie eleison, Christe eleison (Litania), 2 S, A, 2 T, B, bc, formerly Staats- und Universitätsbibliothek, Königsberg [before 1657, ?Naumburg]; G xii

459 Saget den Gästen (Dominica XX. post Trinitatis), S, A, T, B, 2 vn, bn, bc, *D-Kl 52u* [1623–7]; G xiv, S — 17

460 Itzt blicken durch des Himmels Saal (Madrigal) (Opitz), 2 S, A, T, B, 2 vn, bc (+ vle), *Kl 49e* [1635, Dresden]; G xv, N xxxvii — 31

461 Herr, der du bist vormals genädig gewest, 2 S, 2 T, B, 2 vn, 3 trbn ad lib, cap. S, A, T, B ad lib, [bc], *Kl 49n* [*c*1650] (contrary to N, insts authentic); G xiii, N xxviii — 47

462 Auf dich, Herr, traue ich (Psalmus 7), I: S, A, T, B; II: S, A, T, B; — 19

SWV

coro aggiunto (2 vn, viola, cornett, 3 trbn) ad lib; bc (org), *Kl* 49*q* [1627–32] (contrary to N, insts authentic); G xiii, N xxvii

463 Cantate Domino canticum novum, I: S, A, T, B; II: S, A, T, [B], bc (org), *Kl* 51*q* [1650–51] (rev. version of G. Gabrieli: Cantate Domino, 1615); S

464 Ich bin die Auferstehung, I: S, A, T, B; II: S, A, T, B, *WRh* AW B 1334 [origins unknown] (probably = Ich bin die Auferstehung . . . mit 8 Stimmen, listed in Leipzig fair catalogue, aut. 1620); N xxxi, S — 17, 76

465 Da pacem, Domine, I: 5 viole + 1/2vv; II: S, A, T, B; bc, formerly Staats- und Universitätsbibliothek, Königsberg [before 1657, ?Naumburg], for electoral assembly, Mühlhausen, 4 Oct–5 Nov 1627; G xv, N xxxviii, S — 22

466 Herr, wer wird wohnen in deiner Hütten (Psalmus 15), I: A, B, 2 vn, vle; II: S, T, 3 trbn; bc, *D-Kl* 49*p* [1627–32]; G xiii, N xxvii — 19

467 Wo Gott der Herr nicht bei uns hält, I: S, lutes; II: S, 3 viols; III: S, 3 trbn, *Kl* 49*m* [1615–18] (bc in source added *c*1625, unauthentic); G xiii, N xxxii — 7, 79

(467*a* Wo Gott der Herr nicht bei uns hält, see 'Doubtful and spurious works')

468 Magnificat anima mea, S, A, T, B, 2 vn, 3 trbn, 2 cap. S, A, T, B ad lib, bc (+ vle), *S-Uu* 34:4 [before 1665, Dresden]; G xviii, S — 43

469 Surrexit pastor bonus, 2 S, A, 2 T, B, 2 vn, 3 trbn, 2 cap. S, A, T, B ad lib, bc (org), *D-Kl* 49*a* [1640–50]; G xiv — 43

470 Christ ist erstanden von der Marter alle, S, A, T, 4 viole [III lost], 4 trbn, [2 cap. a 4], bc (+ lutes, org piccolo, org grande), *Kl 52b* [1614–15; bc and lute are autograph] (variant sinfonia in late bc part [*c*1650], pr. in critical reports to G and N, from A. Hammerschmidt, Nehmet hin und esset, 1645 (*DTÖ*, xvi)); G xiv, N xxxii — 7, 77

(470*a* Christ ist erstanden von der Marter alle, S, [?parts], *Kl* 52*b* [*c*1613]; facs. ed. in Breig, 1984)

471 O bone Jesu, fili Mariae, 2 S, 2 A, T, B [+ rip. ad lib], 2 vn/viole, 4 viole, [vle], bc (org), *S-Uu* 34:5 [?1666]; G xviii, S

472 Herr Gott, dich loben wir, see 'Doubtful and spurious works'

SWV

473 Wo der Herr nicht das Haus bauet (Psalmus 127), I: 2 S, A, T, B, 2 vn, 3 trbn; II (cap.): S, A, T, B; bc (org), *D-Kl* 50*b* [1627–32, Dresden] (contrary to N, insts authentic); G xiii, N xxviii — 21

474 Ach wie soll ich doch in Freuden leben, I: [S], lutes; II: [S], 3 viole; III: [S], 3 trbn; cap. A, T, B, vn, cornett; bc, *Kl* 56*d* [1614–15; vn is autograph]; G xviii, N xxxviii — 7, 77

475 Veni, Sancte Spiritus, I: 2 S, bn; II: B, 2 cornetts; III: 2 T, 3 trbn; IV: A, T, vn, fl, vle; bc (org), *Kl* 49*b* [*c*1620, Dresden; summary of scoring in bc and text for A are autograph] (swv, G, N and S give various alternative scorings based on unauthentic additions to MS); G xiv, N xxxii, S — 17

476 Domini est terra, I: S, A, T, B (+ tutti [ad lib]); II: S, A, T, B (+ tutti [ad lib]); cornett, *cornett, bn, *4 bn, vn, *vn, 4 trbn; bc, formerly Staats- und Universitätsbibliothek, Königsberg [before 1657, ?Naumburg] (contrary to N, insts authentic; tutti indications, texting of insts in source probably unauthentic); G xiii, N xxvii, S — 21

477 Vater Abraham, erbarme dich mein (Dialogus divites Epulonis cum Abrahamo), 2 S, A, T, B, 2 vn (alternating with 2 fl), vle, bc, *D-Kl* 53*y* [1640–50]; G xviii, S — 43

Die sieben Wortte unsers lieben Erlösers und Seeligmachers Jesu Christi — 78, 105–6

478 Da Jesus an dem Kreuze stund, S, A, 2 T, B [+ cap. ad lib], 5 insts, bc, *Kl* 48 [origins unknown] (final chorus listed in catalogue of A. Unger's collection, Naumburg, 1657); G i, N ii, S

Historia des Leidens und Sterbens unsers Herrn und Heylandes Jesu Christi nach dem Evangelisten S. Matheum, 1666 — 66, 78, 79, 104, 105, 106–7

479 Das Leiden unsers Herren Jesu Christi, wie es beschreibet der heilige Evangeliste Matthaeus, 2 S, A, 3 T, 2 B, chorus 4vv, *LEm* II 2, 15 [*c*1700, Dresden] (= die Passion unsers Herrn Jesu Christi aus dem Evangelisten S. Matthaeo des Capelmeisters Schützes neue Composition, perf. Dresden, 1 April 1666, according to court diaries; final chorus ?M. G. Peranda); G i, N ii, S, facs. edn. (Leipzig, 1981)

Historia des Leidens und Sterbens . . . Jesu Christi nach dem Evangelisten St. Lucam — 66, 78, 79, 104, 105, 106–9

480 Das Leiden unsers Herren Jesu Christi, wie uns das beschreibet der heilige Evangeliste Lucas, S, A, 3 T, 2 B, chorus 4vv, *LEm* II 2, 15 [*c*1700, Dresden] (anon., but transmitted with nos.479 and 481; presumably = die Passion unsers Herrn Jesu Christi aus dem Evangelisten S. Luca, des Cap. Schützens Neue Composit., perf. Dresden, 8 April 1666, according to court diaries); G i, N ii, S, facs. edn. (Leipzig, 1981)

Historia des Leidens und Sterbens . . . Jesu Christi nach dem Evangelisten St. Johannem — 78, 79, 104, 105, 106–9

481 Das Leiden unsers Herren Jesu Christi, wie uns das beschreibet der heilige Evangeliste Johannes, S, 3 T, 2 B, chorus 4vv, *LEm* II 2, 15 [*c*1700, Dresden] (presumably = version perf. Dresden, 13 April 1666, according to court diaries); G i, N ii, S, facs. edn. (Leipzig, 1981) — 66

(Historia dess Leidens und Sterbens . . . Jesu Christi aus dem Evangelisten S Johanno)

(481*a* Das Leiden unsers Herren Jesu Christi, wie uns das beschreibet der heilige Evangeliste Johannes, *W* 1.11.1 Aug 2° [1665, Weissenfels] (= die Passion aus dem Evangelisten Johanne nach der neuen Composit Cappelm. Heinrich Schützens, perf. Dresden, 24 March 1665, according to court diaries); G i, S) — 65, 66

Königs und Propheten Davids hundert und neunzehender Psalm . . . nebenst dem Anhange des 100. Psalms . . . und eines deutschen Magnificats, *Dlb* 1479/E/504, Stefan Zweig's private collection, London [1671, Dresden] (title-page and index pr. Dresden, 1671; dedication to Johann Georg II, other notes and corrections are autograph); ed. W. Steude (Leipzig and Kassel, 1984) — 68–9, 76

482 Wohl denen, die ohne Wandel leben (Psalm 119: Aleph et Beth), I: S, A, T, B; II: [S], A, [T], B; bc (org)

483 Tue wohl deinem Knechte (Psalm 119: Gimel et Daleth), I: S, A, T, B; II: [S], A, [T], B; bc (org)

484 Zeige mir, Herr, den Weg deiner Rechte (Psalm 119: He et Vau), I: S, A, T, B; II: [S], A, [T], B; bc (org)

485 Gedenke deinem Knechte an dein Wort (Psalm 119: Dsain et Chet), I: S, A, T, B; II: [S], A, [T], B; bc (org)

486 Du tust Guts deinem Knechte (Psalm 119: Thet et Jod), I: S, A, T, B; II: [S], A, [T], B; bc (org)

487 Meine Seele verlanget nach deinem Heil (Psalm 119: Caph et Lamed), I: S, A, T, B; II: [S], A, [T], B; bc (org)

488 Wie habe ich dein Gesetze so lieb (Psalm 119: Mem et Nun), I: S, A, T, B; II: [S], A, [T], B; bc (org)

489 Ich hasse die Flattergeister (Psalm 119: Samech et Aiin), I: S, A, T, B; II: [S], A, [T], B; bc (org)

490 Deine Zeugnisse sind wunderbarlich (Psalm 119: Pe et Zade), I: S, A, T, B; II: [S], A, [T], B; bc (org)

491 Ich rufe von ganzem Herzen (Psalm 119: Koph et Resch), I: S, A, T, B; II: [S], A, [T], B; bc (org)

492 Die Fürsten verfolgen mich ohne Ursach (Psalm 119: Schin et Thau), I: S, A, T, B; II: [S], A, [T], B; bc (org)

493 Jauchzet dem Herrn alle Welt (Psalm 100), I: S, A, T, B; II: [S], A, [T], B; bc (org) (probably = composition perf. Dresden, 15 Oct 1665, according to court diaries: Zum Introitu, intonierte der mittlere Hofprediger . . . Jauchzet dem Herren, worauff der Chor musicaliter antwortete und den 100. Psalm vollents absolvierte, dessen Composition hatt der Capellmeister Schüze izo hierzu von neuem gemacht) — 66, 69

494 Meine Seele erhebt den Herren (Teutsch Magnificat), I: S, A, T, B; II: [S], A, [T], B; bc (org); N xxviii, S — 69

(494*a* Meine Seele erhebt den Herren (Teutzsch Magnificat), I: S, A, T, B; II: S, A, T, B; bc, *D-Dlb* 1479/E/501 [1671–8, Grimma]; N xxviii)

495 Unser Herr Jesus Christus, in der Nacht, I: S, A, T, B; II: S, A, T, B, *Dlb*1/C/2 [*c*1630, ?Sangerhausen]; ed. W. Braun (Kassel, 1961)

496 Esaja, dem Propheten, das geschah, B, [8vv], 2 cornetts (alternating with 2 rec), [7vv], [bc], *WF* partbooks without call no. [origins unknown]; N xxxii

497 Ein Kind ist uns geboren, 2 T [II lost], bc, *Dlb* Pi 57 [1630–*c*1635, Pirna] — 17

498 Stehe auf, meine Freundin, I: S, A, T, B; II: S, A, T, B, *BIB* VI/2:2 [*c*1650]; ed. W. Steude (Leipzig, 1972)

499 Tulerunt Dominum, 4 tpt [I, II lost], [20 parts], *Dlb* Pi 50 [*c*1635, Pirna]

500 An den Wassern zu Babel (Psalmus 137), I: T, 4 trbn [IV losII: 2 S, B; bc (org, lutes, hpd), *Kl* 49*o* [1627–32]; N xxviii — 21

SWV

in Christliche LeichPredigt, beim Begräbnis . . . der . . . Frawen Magdalenen, Herrn Heinrich Schützens . . . ehelicher Haussfrawen (Leipzig, 1625)

501 Mit dem Amphion zwar mein Orgel und mein Harfe (Klag-Lied) (?Schütz), T, bc (lutes/hpd), on death of Magdalena Schütz, Dresden, 6 Sept 1625; ed. (with facs.) E. Möller (Leipzig and Kassel, 1984) — 19

A1 Vier Hirtinnen, gleich jung, gleich schon, 2 S, A, T, bc, *Kl* 58*f* [1615–*c*1620, Dresden; music and part of text are autograph] (anon., but identifiable by description in inventory of Friedrich Emanuel Praetorius's private collection, Lüneburg, 1684, and by character of source); G xviii, N xxxvii — 14, 77, *84*

A2 Ach Herr, du Sohn David, see 'Doubtful and spurious works'

A3 Der Gott Abraham, see 'Doubtful and spurious works'

A4 Stehe auf, meine Freundin, see 'Doubtful and spurious works'

A5 Benedicam Dominum in omni tempore, see 'Doubtful and spurious works'

A6 Freuet euch mit mir, see 'Doubtful and spurious works'

A7 Herr, höre mein Wort, see 'Doubtful and spurious works' — 21

A8 Machet die Tore weit, see 'Doubtful and spurious works'

A9 Sumite psalmum, see 'Doubtful and spurious works'

A10 Dominus illuminatio mea, see 'Doubtful and spurious works'

A11 Es erhub sich ein Streit, see 'Doubtful and spurious works'

A*a* Das Leiden unsers Herrn Jesu Christi, wie es uns St. Marcus beschreibet, see 'Doubtful and spurious works'

A*b* Zeuchst du nun von hinnen, see 'Doubtful and spurious works'

A*c* Wo seid ihr so lang gelieben, see 'Doubtful and spurious works'

A*d* Deus in nomine tuo, see 'Doubtful and spurious works'

A*e* Tancredi, der Clorindam vor ein Manns Person, see 'Doubtful and spurious works'

A*f* Kyrie eleison, Christe eleison, see 'Doubtful and spurious works'

A*g* Domine Deus virtutem, see 'Doubtful and spurious works'

A*h* Damit, dass diese Gsellschaft wert, see 'Doubtful and spurious works'

A*i* O höchster Gott, see 'Doubtful and spurious works'

A*k* Jesu dulcissime, see 'Doubtful and spurious works' — 17

SWV

DOUBTFUL AND SPURIOUS WORKS

445 Ach bleib mit deiner Gnade, S, A, T, B, formerly Preussische Staatsbibliothek, Berlin [*c*1650, ?] (contrafactum of C. Cramer: Sag, was hilft alle Welt, 1641); G xvi, N xxxii

446 In dich hab ich gehoffet, Herr, [S, A], T, B, *B* 40 200 [origins unknown]; G xviii, N xxxii

454 Nun lasst uns Gott dem Herren, 2 S, A, 2 T, B, formerly Staats- und Universitätsbibliothek, Königsberg [origins unknown] (contrafactum of section of no.279); G xvi, N xxxii

467*a* Wo Gott der Herr nicht bei uns hält, 3 S, bc, *Kl* 59*n* [1640–50] (spurious arr. of no.467); N xxxii

472 Herr Gott, dich loben wir, I: [S] (+ cornett/vn/trbn ad lib), [A, T, B]; II: S, A, T, B (+ 3 trbn [III lost]); 2 clarinos, 2 tpt, timp, 2 vn/cornetts, bc (org), *D-B* 20 374 [1677, Erfurt]; G xviii, N xxxii

A2 Ach Herr, du Sohn David, 3 S, T, B, bc, formerly Stadtbibliothek, Breslau [origins unknown] (anon., attrib. by Moser, 'Unbekannte Werke' (1935), on grounds of presumed identity with lost work of this title; not by Schütz); S

A3 Der Gott Abraham, A, T, B, *2 vn, *3 trbn, bc, *D-Kl* 52*s* [1640–50] (anon., attrib. by Engelbrecht, 1958, on stylistic grounds; not by Schütz); S

A4 Stehe auf, meine Freundin, I: 2 S, A, T; II: A, 2 T, B, formerly Preussische Staatsbibliothek, Berlin [1643, ?] (anon., attrib. by Moser, 'Unbekannte Werke' (1935), on grounds of presumed identity with lost work of this title, since discovered, see no. 498; A4 not by Schütz); S

A5 Benedicam Dominum in omni tempore, I: S, A, T, B; II: S, A, T, B, bc, *Kl* 55*c* [*c*1660, ?Vienna] (anon., attrib. by Engelbrecht, 1958, on stylistic grounds; not by Schütz); ed. C. Engelbrecht (Kassel, 1959)

A6 Freuet euch mit mir, S/T, T/S, bc (org), or I: S, A, T, B; II: S, A, T, B; bc (org), *Kl* 52*t* [1640–50] (anon., attrib. by Engel on stylistic grounds; possibly = Dialogus vom verlohrnen Schaaff, und Groschen, H. Heinrich Schützens, listed in inventory of Katherinenkirche, Zwickau, 1634–61); ed. H. Engel (Berlin, 1950, 2/1960)

A7 Herr, höre mein Wort (Psalmus 5), I: S, A, T, B (+ rip. *4 insts); II: S, A, T, B (+ rip. *4 insts); bc (org), *Kl* 52*o* [1627–32] (anon., attrib. by Engel on stylistic grounds; — 21 probably by Schütz); ed. H. Engel (Berlin, 1950); N xxvii

A8 Machet die Tore weit, I: S, A, T, B; II: S, A, T, B, *Dlb* Löb 8, Löb 70 [1624–5, Löbau], *KMs* 2920–27 [after 1637] (both attrib. Schütz), formerly Stadtbibliothek, Breslau [origins unknown] (attrib. S. Rüling); ed. H. J. Moser (Leipzig, 1935)

A9 Sumite psalmum, 2 S, A, T, B, 2 vn, 3 trbn, bc, *D-Kl* 53*p* [*c*1660, ?Vienna] (anon., attrib. by Engelbrecht, 1958, on stylistic grounds; not by Schütz); ed. C. Engelbrecht (Kassel, 1959)

A10 Dominus illuminatio mea, S, A, 2 T, B, 2 vn, 2 va, bn, bc, *Kl* 53*q* [*c*1660, ?Vienna] (anon., attrib. by Engelbrecht, 1958, on stylistic grounds; not by Schütz)

A11 Es erhub sich ein Streit (in Festo S. Michaelis angeli), I: S, A, T, B; II: S, A, T, B; III: T, 3 cornetts; IV: T, tpt, 3 bn; bc (org), *Kl* 53*g* [*c*1630–1638] (anon., attrib. by H. Spitta chiefly on stylistic grounds, see G xviii; probably not by Schütz); G xviii, S

Historia des Leidens und Sterbens . . . Jesu Christi nach dem Evangelisten St. Marcum

A*a* Das Leiden unsers Herrn Jesu Christi, wie es uns St. Marcus beschreibet, S, A, 3 T, 2 B, chorus 4vv, *D-LEm* II 2, 15 [*c*1700, Dresden] (anon., transmitted with nos.479–81 and sometimes considered possibly authentic; identifiable through Dresden court diaries as work of M. G. Peranda); G i, facs. edn. (Leipzig, 1981)

A*b* Zeuchst du nun von hinnen, S, bc, *D-DS* 1196 [*c*1630] (anon., attrib. suggested by F. Noack; probably not by Schütz); ed. in Noack, 1924

Frewden-Lied bey und über dem . . . fürstl. Willkommen zu gebrauchen, Geschehen auff dem Friedenstein den 10. Christmonats im Jahr 1646 (Gotha, 1646)

A*c* Wo seid ihr so lang geblieben (C. T. Dufft), S, bc (anon., attrib. by Thiele; not by Schütz); ed. in Thiele, 'Thüringer Meister' (1954)

A*d* Deus in nomine tuo, [4vv], B, [3 insts, bc], *Kl* 62*f* [*c*1660, ?Vienna] (anon., attrib. by Engelbrecht, 1958, on basis of common origin with A5, 9 and 10; not by Schütz)

Tancredus et Clorinda

A*e* Tancredi, der Clorindam vor ein Manns Persons, [S], T, [T, str, bc], *J-Tm* without call no. [?Reinsdorf, 1638–57] (contrafactum of Monteverdi: Il combattimento di Tancredi e Clorinda; anon., attrib. by W. Osthoff; by D. von dem Werder); facs. in Osthoff, 1961

A*f* Kyrie eleison, Christe eleison (Litania), 3 S, A, T, B, bc (org), *S-Uu* Caps.69:7 [?1660–64, Dresden] (anon., attrib. by Grusnick, 1969, on grounds of style and evident Saxon origin; written on Schütz's own paper and hence very probably by him); S

A*g* Domine Deus virtutem, I: A, T, B, 2 vn; II: 2 S, 3 trbn/viole ad lib [III lost; ? = fagotto grosso/vle]; III: cap. S, A, T, B ad lib; bc (+ fagotto grosso/vle), *S-Uu* 40: 13 [before 1665, Dresden] (anon., attrib. by Grusnick, 1966, on provenance of source and stylistic grounds; possibly by Schütz); S

A*h* Damit, dass diese Gsellschaft wert (Intrada Apollinis) (Schütz), 2 S, [?parts], *DS* Mus. 1194 [origins unknown] (anon., attrib. by E. Noack, 1967, chiefly on basis of textual identity with section of Schütz's Wunderlich Translocation, 1617; probably not by Schütz)

A*i* O höchster Gott, 2 S/T, bc, *D-Dlb* Pi 8 [1625–30, Pirna], Pi 57 [1630–*c*1635, Pirna] (anon., attrib. by Steude, 1967, chiefly on stylistic grounds; not by Schütz)

A*k* Jesu dulcissime, S, A, 3 T, B, *Kl* 52*k* [1615–27, Dresden] (on G. Gabrieli: O Jesu Christe; anon., attrib. by Breig, 1974, chiefly on stylistic grounds; probably by Schütz); ed. W. Breig (Kassel, 1974) — 17

— [without text], [?parts], bc, *Dlb* Gri 7 [*c*1640, Grimma] (attrib. 'Ex Sagittario'; ?by J. Sagittarius)

— Die nur vertraulich stellen (C. Becker), S, [A, T, B], S pr. as top part of setting for 5vv in Geistlicher Lieder . . . Ander Theil (Gotha, 1648) with attrib. 'Melod. Schützii'; no relation to no.230

— Ein feste Burg ist unser Gott, [I: S, A], T, [B; II: S], A, [T], B, Z 80.3 [1617–22]

LOST WORKS

References to lost works appear principally in inventories; other information is found in surviving librettos and in the Dresden court diaries; see also nos. 293, 298, 300, 317, 325, 349, 392, 397, 407, 464, 493, A6 — 18, 29, 46, 47, 53, 54, 77, 78

Inventories:

Gotha – inventory of works performed at the dedication of the palace church, 1646 (see Schneider, 1905–6)
Grimma – inventory of Christian Andreas Schulze's private collection, 1699 (see Steude, 1967)
Kassel – inventory of the Hofkapelle, 1638 (see Zulauf, 1902)
Leipzig – inventory of Gottfried Kühnel's private collection, 1684 (see Schering, 1918–19)
Lüneburg – inventory of Friedrich Emanuel Praetorius's private collection, 1695 (see Seiffert, 1907–8) — 77
Naumburg – inventory of Andreas Unger's private collection, 1657 (see Werner, 1926)
Pirna – inventory of Cantorey und Musicorum Gesellschaft, 1654 (see Nagel, 1896)
Weimar – inventory of the Hofkapelle, 1662 (see Aber, 1921) — 46, 77

(all probably incl. bc)

Ach Herr, du Sohn Davids, a 6, Naumburg (not A2)
Ach Herr, strafe mich nicht, S, T, in index of *D-W* 323 Mus.Handschr. [1638, Helmstedt]
Alleluia, lobet den Herrn (Psalm 150), with tpts, timp, first perf. Dresden, 1 Jan 1667; ?another setting perf. Dresden, 22 July 1668 — 67
Alleluia, lobet ihn in seinem Heiligtum, a 16 or 18, Weimar (catalogue includes incipit)
Anima mea liquefacta est, a 3, Kassel
Aquae tuae Domine, perf. Dresden, 15 June, 23 Nov 1662 — 61, 63
Audite coeli, Kassel
Auf, auf, meine Harfe, a 10, Weimar
Auf dich, Herr, traue ich, a 16 or 24, Naumburg
Benedicite omnia opera Domini, a 20, Leipzig
Canite, psallite, plaudite, a 12, Kassel
Christ lag in Todesbanden, Kassel — 31
Confitebor tibi, 5vv, 5 insts, Weimar
Dafne (Opitz), opera, perf. Torgau, 13 April 1627, for marriage of Landgrave Georg II of Hessen-Darmstadt and Princess Sophia Eleonora of Saxony, text extant — 21, 22, 24
Das ander Maria, a 6, Weimar (listed with Maria, sei gegrüsset; ? inaccurate reference to no.334, or one of 'Zwey deutsche geystl. Madrial H. Sag.' listed elsewhere in Weimar inventory)
Der Herr ist mein Hirt, 5vv, 5 insts, Weimar (? inaccurate reference to no.398*a*)
Der Herr ist mein Hirt, a 8 (Strasbourg, 1657) (? = no.33 or no.398*a*) — 60
Der Herr sprach zu meinem Herren, a 11, Weimar, a 17, Naumburg
Der Wind beeist das Land, d Dorian, 2 T, Lüneburg
Dies ist der Tag des Herrn, a 6, Naumburg
Die, so ihr den Herren fürchtet, Weimar
Dies Ort, mit Bäumen ganz umgeben, a, S, bc, Lüneburg
Domine, exaudi orationem, a 7 or 10, Naumburg, a 7, 10 or 14, Weimar
Dorinda, Weimar
Du bist aller Dinge schöne, 'finalis G', 2 S, A, T, B, 2 vn, 3 trbn, Grimma
Du hast mir mein Herz, a 8, Naumburg
Ego dormio, a 5, Kassel
Ein Kindelein so lobelich, Kassel
Ein Kind ist uns geboren, a 8, Naumburg
Einsmals der Hirte Coridon, g Dorian, 2 S, 2 vn, Lüneburg (? = anon. work with same incipit, Kassel)
Einsmals in einem schönen Tal, d Dorian, a 2 and a 6, Lüneburg
Ein wunder Löwe, Kassel
Erhör mich, wenn ich rufe, a 8, Naumburg
Esaja, dem Propheten, a 8, Naumburg (? variant version of no.496; see also Mattheson, 1740)
Es ist erschienen, 3vv, Weimar
Es ist Zeit, die Stund ist da, 4vv, Weimar
Es sei denn eure Gerechtigkeit, a 8, Naumburg
Es stehe Gott auf, a 13, Weimar
Factum est praelium magnum (in festo Mich.[aelis]), C, a 9, Lüneburg (probably not A11)
Fröhlich auf, ihr Himmels Volk (J. G. Schottelius), see Theatralische neue Vorstellung
Gelobet sei der Herr, a 5, 10, 11 or 20, Weimar (? inaccurate reference to no.448)
Glückwündschung des Apollinis und der neun Musen, 12vv, 12 cornetts, tpts, timp, for birthday of Johann Georg I, Dresden, 5 March 1621, text extant; facs. edn. of text (Kassel, 1929) — 17
Gott, man lobet dich in der Stille, a 8, Naumburg
Herr, komm hinab (Dominica XXI. post Trin.[itatem]), a 9, Kassel, Naumburg

Herr, warum trittest du so ferne (Psalm 10), e Phrygian, a 8, 12 or 18, Lüneburg

Heut ist Christus der Herr geboren, a 6, Kassel (? variant version of no.439 or 456)

Jauchzet dem Herrn (Psalm 100), with tpts, first perf. Dresden, 28 Sept 1662 — 61

Jauchzet dem Herrn, a 6, Gotha

Jauchzet, jauchzet, a 4, Weimar

Jesus trat in ein Schiff, a 8, Naumburg

Kyrie, Weimar (probably = one of following works or no.458)

Kyrie eleison (Littaney), a 5, Weimar (? = Af, see 'Doubtful and spurious works')

[Kyrie eleison] (Deutsche Littaney), a 12 or 18, Weimar (catalogue includes incipit)

Kyrie, Gott Vater in Ewigkeit, e, 6vv, 6vv in rip., 6 insts, Lüneburg

Lobsinget Gott, ihr Männer von Galilea (in fest.[o] Ascens.[ionis] Christi), d Dorian, a 5 or 10, Naumburg, a 10, Lüneburg

Machet die Tore weit, a 20, Naumburg

Magnificat, 3vv, 2 vn, Weimar

Magnificat, d Dorian, S, T, B, 2 vn, 3 trbn, bn, cap. a *4, Lüneburg, (? = preceding work with ad lib parts)

Maria, sei gegrüsset, a 5, Weimar (listed with Das ander Maria; ? = no.333, or one of 'Zwey deutsche geystl. Madrial H. Sag.' listed elsewhere in Weimar inventory)

Mein Freund, ich tu dir nicht Unrecht, a 6, Weimar

Mein Freund, komme, a 6, Pirna

Meister, wir haben die ganze Nacht, a 8, Naumburg

Misericordias Domini, a 6, Naumburg

Nun hat recht die Sünderin (Schottelius), see Theatralische neue Vorstellung

Orpheus und Euridice (A. Buchner), opera-ballet, for marriage of Prince Johann Georg of Saxony and Princess Magdalena Sybilla of Brandenburg, Dresden, 20 Nov 1638, text extant — 34

Preise, Jerusalem, den Herrn (Psalm 147), a 6 in concerto, Gotha

Renunciate Johannis quae audistis, perf. Dresden, 16 Dec 1665 — 66

Saget den Gästen, a 4, Naumburg, a 9, Pirna (? both inaccurate references to no.459)

Sag, o Sonne meiner Seelen, G Mixolydian, a 4, Lüneburg

Siehe, wie fein und lieblich ist, a 6, Kassel, Naumburg (? inaccurate reference to no.48)

Singet dem Herrn ein neues Lied (Psalm 149), 2 choirs, Gotha

Theatralische neue Vorstellung von der Maria Magdalena, Wolfenbüttel, Dec 1644, text of 2 numbers, Fröhlich auf, ihr Himmels Volk and Nun hat recht die Sünderin, in J. G. Schottelius: Fruchtbringender Lustgarten (Lüneburg, 1647/*R*1967) (Schütz's authorship of the rest uncertain) — 39

Tröste uns, Gott, Weimar (anon., but following 2 works by Schütz and headed 'so er gesezet'; catalogue includes incipit)

Unser Leben währet siebzig Jahr, a 5, Naumburg

Venus, du und dein Kind, Weimar

Wenn der Herr die gefangenen Zion, 6vv, 6 insts, Weimar

Wer ist der, so von Edom kömmt, d Dorian, a 10 or 18, Lüneburg

Wer sich dünken lasset, a 4, Weimar

Wer unter dem Schirm des Höchsten, 5vv, 2 vn, Weimar (? = anon. work with same incipit, d Dorian, 2 S, A, T, B (+ cap.), 2 vn, Lüneburg)

Wie ein Rubin, a 3, Weimar (? inaccurate reference to no.357)

Wunderlich Translocation des weitberümbten und fürtrefflichen Berges Parnassi (Schütz), for visit of Emperor Matthias, Dresden, 25 July 1617, music lost, text in Panegyrici Caesario-Regio-Archiducales (Dresden, 1617); source does not name composer but music presumably by Schütz (see also Damit, dass diese Gsellschaft wert, Ah, listed under 'Doubtful and spurious works') — 14

Zwei wunderschöne Täublein zart (Madrigal), for wedding of Reinhart von Taube and Barbara Sybilla von Carlowitz, Dresden, 10 Feb 1624, text extant — 18

'9 Madrigalien oder weltliche Stükke, H. S. darunter das Jägerliedt. A[dam] D[rese]', Weimar, incl. Das Zielbachische Jägerliedt, 1 work attrib. Schütz; also 7 anon. titles possibly by Schütz: Ach liebste, lass uns eilen (Opitz), a 4; Der Kuckuck hat sich zum Tode, a 4; Distel und Dorn stechen sehr, a 4; Gehet meine Seufzer hin, a 5; So bist du nun, mein Lieb, a 6; Täglich geht die Sonne unter (Opitz), a 6; Wenn dich, o Sylvia, a 6

'28 zusammengebundene Kirchen Stükke, H. Schützens', Weimar, incl. 20 works attrib. Schütz, 1 work attrib. ? Michael Cracowit; also 6 anon. titles: Bleib bei uns, 5vv, 5 insts; Es gingen zweene Menschen hinauf, 5vv, 2 va; Ich freue mich des, 5vv, 5 insts; Jesus Christus, unser Heiland, a 6; Kyrie, Gott Vater in Ewigkeit, 'ex E', 6vv, 11 parts (? = work of this title attrib. Schütz in Lüneburg, see above); O du allersüssester Herr Jesu, a 7 (? = no.340)

Schütz may also be presumed to have composed most, if not all, of the theatrical music for the Danish royal wedding of 1634. — 27–9

ALPHABETICAL KEY
(*German*)

(*Latin*)

Anima mea liquefacta est, lost; Aquae tuae Domine, lost; Aspice, Pater, 73; Attendite, popule meus, 270; Audite coeli, lost; Aufer immensam, Deus, 337; Ave Maria, 334; Benedicam Dominum, 267, A5; Benedicite omnia opera Domini, lost; Bone Jesu, 313; Buccinate in neomenia tuba, 275; Calicem salutaris accipiam, 60; Canite, psallite, lost; Cantabo Domino in vita mea, 260; Cantate Domino canticum novum, 81, 463; Christe Deus adjuva, 295; Christe fac ut sapiam, 431; Confitebor tibi, lost; Confitemini Domino, 91; Da pacem, Domine, 465; Deus in nomine tuo, A*d*; Deus misereatur nostri, 55; Discedite a me, 87; Domine Deus, pater, 90; Domine Deus virtutem, A*g*; Domine, exaudi, lost; Domine, labia mea, 271; Domine, ne in furore, 85; Domine, non est exaltatum, 78; Domini est terra, 476; Dominus illuminatio mea, A10; Dulcissime et benignissime Christe, 67

Ecce advocatus meus, 84; Ego autem sum, 436; Ego dormio, 63; Ego dormio, lost; Ego enim inique egi, 58; Ego sum tui plaga doloris, 57; En novus Elysiis, 49; Et ne despicias, 54; Exquisivi Dominum, 268; Exultavit cor meum, 258; Factum est praelium, lost; Fili mi, Absalon, 269; Gratias agimus tibi, 93; Heu mihi, Domine, 65; Hodie Christus natus est, 315, 456; In lectulo per noctes, 272; In te, Domine, speravi, 66, 259; Inter brachia, 82; Invenerunt me, 273; Jesu dulcissime, A*k*; Jubilate Deo in chordis, 276; Jubilate Deo omnis terra, 262, 332; Magnificat (2 works), lost; Magnificat anima mea, 468; Meas dicavi res Deo, 305; Misericordias, lost

Nonne hic est, 74; O bone Jesu, 471; O bone, o dulcis, 53; Oculi omnium in te sperant, 88; O Jesu, nomen dulce, 308; O misericordissime Jesu, 309; O quam tu pulchra es, 265; Paratum cor meum, 257; Pater noster, 89, 92; Pro hoc magno mysterio, 77; Quando se claudunt lumina, 316; Quemadmodum desiderat cervus, 336; Quid commisisti, 56; Quid detur tibi, 72; Quo, nate Dei, 59; Quoniam ad te clamabo, 62; Quoniam non est in morte, 86; Reduc, Domine Deus meus, 75; Renunciate Johanni, lost; Rorate coeli desuper, 322; Sicut Moses serpentem, 68; Si non humiliter sentiebam, 79; Speret Israel in Domino, 80; Spes mea, Christe Deus, 69; Sumite psalmum, A9; Supereminet omnem scientiam, 76; Surrexit pastor bonus, 469

Te Christe supplex invoco, 326; Teutoniam dudum belli, 338; Tulerunt Dominum, 499; Turbabor, sed non perturbabor, 70; Veni de Libano, 266; Veni, dilecte mi, 274; Veni, Domine, 437; Veni, Redemptor gentium, 301; Veni, rogo, in cor meum, 83; Veni, Sancte Spiritus, 328, 475; Venite ad me, 261; Verba mea auribus percipe, 61; Verbum caro factum est, 314; Vulnerasti cor meum, 64

(*Italian*)

Alma afflitta, che fai, 4; Così morir debb'io, 5; Di marmo siete voi, 17; D'orrida selce alpina, 6; Dunque à Dio, 15; Feritevi, ferite, 9; Fiamma ch'allacia, 10; Fuggi, fuggi, o mio core, 8; Giunto è pur, 18; Io moro, 13; Mi saluta costei, 12; O dolcezze amarissime, 2; O primavera, 1; Quella damma son io, 11; Ride la primavera, 7; Selve beate, 3; Sospir che del bel petto, 14; Tornate, o cari baci, 16; Vasto mar, nel cui seno, 19

BIBLIOGRAPHY

LIFE, DOCUMENTS, SOURCES

GerberNL; *WaltherML*

M. Hoe von Hoenegg: *Chur sächsische evangelische Jubel Frewde* (Leipzig, 1618)

Heinrich Schütz: Autobiographie (*Memorial 1651*), facs. edn. (Leipzig, 1972)

Heinrich Schütz, Memorial 1651, ed. K. Vötterle (Kassel, 1936)

M. Geier: *Kurtze Beschreibung des . . . Herrn Heinrich Schützens . . . Lebens-Lauff* (Dresden, 1672, repr. 1935/*R*1972)

J. Mattheson: *Critica musica*, ii (Hamburg, 1725/*R*1964)

——: *Grundlage einer Ehren-Pforte* (Hamburg, 1740); ed. M. Schneider (Berlin, 1910/*R*1965)

C. von Winterfeld: *Johannes Gabrieli und sein Zeitalter* (Berlin, 1834/*R*1965)

C. von Rommel: *Geschichte von Hessen*, vi (Kassel, 1837)

K.-A. Müller: *Forschungen auf dem Gebiete der neueren Geschichte*, i (Dresden and Leipzig, 1838)

M. Fürstenau: *Beiträge zur Geschichte der königlich sächsischen musikalischen Kapelle* (Dresden, 1849)

W. Schäfer: 'Einige Beiträge zur Geschichte der kurfürstlichen musikalischen Capelle oder Cantorei unter den Kurfürsten August, Christian I. u. II. u. Johann Georg I.', *Sachsen-Chronik*, i (1854), 404

——: 'Heinrich Schütz: kurfürstlich sächsischer Kapellmeister, 1617 bis 1672', *Sachsen-Chronik*, i (1854), 500

M. Fürstenau: *Zur Geschichte der Musik und des Theaters am Hofe zu Dresden*, i (Dresden, 1861/*R*1971)

F. Chrysander: 'Geschichte der Braunschweig-Wolfenbüttelschen Capelle und Oper vom 16. bis zum 18. Jahrhundert', *Jb für musikalische Wissenschaft*, i (1863), 147

M. Fürstenau: 'Fürstlicher Gottesdienst im 17. Jahrhunderte', *MMg*, iii (1871), 58

L. H. Fischer: 'Heinrich Schütz und Christoph Kaldenbach', *MMg*, xv (1883), 91

R. Eitner: 'Heinrich Schütz (Sagittarius): Verzeichnis seiner bis heute aufbewahrten Werke', *MMg*, xviii (1886), 47, 57, 65

J. Sittard: *Zur Geschichte der Musik und des Theaters am württembergischen Hofe*, i (Stuttgart, 1890/*R*1970)

P. Spitta: 'Schütz, Heinrich', *ADB*; rev. as 'Heinrich Schütz' Leben und Werke', *Musikgeschichtliche Aufsätze* (Berlin, 1894/*R*1976), 3–60

A. Hammerich: *Musiken ved Christian den fjerdes hof* (Copenhagen, 1892)

C. Elling: 'Die Musik am Hofe Christians IV. von Dänemark: nach

Angul Hammerich', *VMw*, ix (1893), 62; rev. S. A. E. Hagen, *Historisk tidsskrift*, vi (1892–4), 420
W. Nagel: 'Die Kantoreigesellschaft zu Pirna', *MMg*, xxviii (1896), 148
W. Quanter: ('Die Anfänge des Berliner Theaters'), *Mittheilungen des Vereins für die Geschichte Berlins*, xiii (1896), 26
M. Seiffert: 'Anecdota Schütziana', *SIMG*, i (1899–1900), 213
A. Werner: 'Samuel und Gottfried Scheidt: neue Beiträge zu ihrer Biographie', *SIMG*, i (1899–1900), 401
F. Spitta: 'Neu entdeckte Schützsche Werke', *Monatschrift für Gottesdienst und kirchliche Kunst*, v (1900), 122
O. Doering: *Des Augsburger Patriciers Philipp Hainhofer Reisen nach Innsbruck und Dresden* (Vienna, 1901)
A. Göhler: *Verzeichnis der in den Frankfurter und Leipziger Messkatalogen der Jahre 1564 bis 1759 angezeigten Musikalien* (Leipzig, 1902/*R*1965)
E. Zulauf: *Beiträge zur Geschichte der landgräflich-hessischen Hofkapelle zu Cassel bis auf die Zeit Moritz des Gelehrten* (Kassel, 1902); repr. in *Zeitschrift des Vereins für hessische Geschichte und Landeskunde*, xxvi (1903), 1
E. Schuster: *Kunst und Künstler in den Fürstenthümern Calenberg und Lüneburg in der Zeit von 1636 bis 1727* (Hanover and Leipzig, 1905)
M. Schneider: 'Die Einweihung der Schlosskirche auf dem "Friedenstein" zu Gotha im Jahre 1646', *SIMG*, vii (1905–6), 308
C. Valentin: *Geschichte der Musik in Frankfurt am Main vom Anfange des XIV. bis zum Anfang des XVIII. Jahrhunderts* (Frankfurt am Main, 1906/*R*1972)
M. Seiffert: 'Die Chorbibliothek der St. Michaelisschule in Lüneburg zu Seb. Bach's Zeit', *SIMG*, ix (1907–8), 593
A. Schering: 'Ein wiederaufgefundenes Werk von Heinrich Schütz', *ZIMG*, x (1908–9), 68
R. Wustmann: *Musikgeschichte Leipzigs*, i: *Bis zur Mitte des 17. Jahrhunderts* (Leipzig and Berlin, 1909/*R*1975)
A. Werner: *Städtische und fürstliche Musikpflege in Weissenfels bis zum Ende des 18. Jahrhunderts* (Leipzig, 1911)
B. Engelke: 'Geschichte der Musik im Dom von den ältesten Zeiten bis 1631', *Geschichts-Blätter für Stadt und Land Magdeburg*, xlviii (1913), 264
A. Pirro: *Schütz* (Paris, 1913, 2/1924/*R*1975)
A. Hantzsch: 'Hervorragende Persönlichkeiten in Dresden und ihre Wohnungen', *Mitteilungen des Vereins für Geschichte Dresdens*, xxv (1918)
K. Lütge: 'Heinrich Schütz, *Kläglicher Abschied*', *Festschrift Hermann Kretzschmar* (Leipzig, 1918/*R*1973), 85

A. Schering: 'Die alte Chorbibliothek der Thomasschule in Leipzig', *AMw*, i (1918–19), 275
A. Werner: 'Ein Albumblatt von Heinrich Schütz', *NZM*, lxxxvi (1919), 181
A. Aber: *Die Pflege der Musik unter den Wettinern und wettinischen Ernestinern* (Bückeburg and Leipzig, 1921)
E. H. Müller von Asow: *Heinrich Schütz: Leben und Werke* (Berlin and Dresden, 1922)
A. Werner: *Städtische und fürstliche Musikpflege in Zeitz bis zum Anfang des 19. Jahrhunderts* (Bückeburg and Leipzig, 1922)
A. Einstein: 'Schütz-Miszellen, i', *ZMw*, v (1922–3), 432
H. Biehle: *Musikgeschichte von Bautzen* (Leipzig, 1924)
F. Noack: 'Die Tabulaturen der hessischen Landesbibliothek zu Darmstadt', *Kongressbericht: Basel 1924*, 276
A. Einstein: 'Heinrich Schütz', *Ganymed*, v (1925), 104; pubd separately (Kassel, 1928); repr. in *Von Schütz bis Hindemith* (Zurich and Stuttgart, 1957), 9
E. H. Müller von Asow: *Heinrich Schütz* (Leipzig, 1925)
A. Werner: 'Die alte Musikbibliothek und die Instrumentensammlung an St. Wenzel in Naumburg a. d. S.', *AMw*, viii (1926), 390
F. A. Drechsel: 'Ein unbekanntes Schreiben von Heinrich Schütz', *ZMw*, ix (1926–7), 627
G. Kinsky: 'Ein Schütz-Fund', *ZMw*, xii (1929–30), 597
P. Uhle: 'Zur Lebensgeschichte des Tonschöpfers Heinrich Schütz', *Mitteilungen des Vereins für Chemnitzer Geschichte*, xxvii (1929–30), 13
C. Cassel: *Geschichte der Stadt Celle*, i (Celle, 1930)
C. Mahrenholz: 'Heinrich Schütz und das erste Reformations-Jubiläum 1617', *Musik und Kirche*, iii (1931), 149; repr. in *Musicologica et liturgica* (Kassel, 1960), 196
E. H. Müller von Asow, ed.: *Heinrich Schütz: Gesammelte Briefe und Schriften* (Regensburg, 1931/*R*1976)
H. Rauschning: *Geschichte der Musik und Musikpflege in Danzig* (Danzig, 1931)
M. Gondolatsch: 'Richtiges und Falsches um einen Heinrich-Schütz-Brief', *ZMw*, xv (1932–3), 428
L. Krüger: 'Johann Kortkamps Organistenchronik, eine Quelle zur Hamburgischen Musikgeschichte des 17. Jahrhunderts', *Zeitschrift des Vereins für Hamburgische Geschichte*, xxxiii (1933), 188
M. Schneider: 'Zum Weihnachtsoratorium von Heinrich Schütz', *Theodor Kroyer: Festschrift zum sechzigsten Geburtstage* (Regensburg, 1933), 140
H. Birtner: 'Heinrich Schütz und Landgraf Moritz von Hessen', *Hessenland*, xlvi (1935), 100

F. Blume: 'Heinrich Schütz 1585–1672', *Die grossen Deutschen*, i (Berlin, 1935), 627
W. Dane: 'Briefwechsel zwischen dem landgräflich hessischen und dem kurfürstlich sächsischen Hof um Heinrich Schütz (1614–1619)', *ZMw*, xvii (1935), 343
F. Dietrich: 'Aus der Predigt beim Leichenbegängnis Heinrich Schützens am 17. November 1672', *Musik und Kirche*, vii (1935), 199
W. Gurlitt: 'Heinrich Schütz: zum 350. Geburtstag am 8. Oktober 1935', *JbMP 1935*, 64; repr. in *Musikgeschichte und Gegenwart*, i (Wiesbaden, 1966), 140
F. Keil: 'Die Familie Schütz in Weissenfels', *25 Jahre Städtisches Museum Weissenfels* (Weissenfels, 1935), 63
O. Michaelis: *Heinrich Schütz: eine Lichtgestalt des deutschen Volkes* (Leipzig and Hamburg, 1935)
H. J. Moser: 'Neues über Heinrich Schütz', *AcM*, vii (1935), 146
——: 'Unbekannte Werke von Heinrich Schütz', *ZMw*, xvii (1935), 332
A. Schering: 'Aus der Selbstbiographie eines deutschen Kantors (Elias Nathusius, †1676)', *Festschrift Max Schneider zum 60. Geburtstag* (Halle, 1935), 84
K. Hartmann: 'Musikpflege in Alt-Bayreuth', *Archiv für Geschichte und Altertumskunde von Oberfranken*, xxxiii/1 (1936), 1
H. J. Moser: *Heinrich Schütz: sein Leben und Werk* (Kassel, 1936, rev. 2/1954; Eng. trans., 1959)
E. Reinhardt: *Benjamin Schütz* (Erfurt, 1936)
A. Schröder: 'Das Heinrich-Schütz-Bildnis und sein Maler Christoph Spetner', *Die Heimat* (Weissenfels), cxxiv (1936)
R. Mohr: 'Ein Brief von Heinrich Schütz an die Stadt Frankfurt am Main', *Deutsche Musikkultur*, i (1936–7), 103
G. Schünemann: 'Ein neues Bildnis von Heinrich Schütz', *Deutsche Musikkultur*, i (1936–7), 47
B. Maerker: 'Rembrandts Bildnis eines Musikers – ein Schützportrait?', *Deutsche Musikkultur*, ii (1937–8), 329
R. Casimiri: 'Enrico Sagittario (Heinrich Schütz) alla scuola di Giovanni Gabrieli', *NA*, xv (1938), 88
B. Martin: 'Hat Rembrandt ein Bild von Schütz geschaffen?', *Musik und Kirche*, x (1938), 78
G. Ilgner: *Matthias Weckmann: sein Leben und seine Werke* (Wolfenbüttel and Berlin, 1939)
W. Serauky: *Musikgeschichte der Stadt Halle*, ii/1 (Halle and Berlin, 1939/*R*1970)
H. J. Moser: *Kleines Heinrich-Schütz-Buch* (Kassel, 1940, 3/1952; Eng. trans., 1967)

——: 'Heinrich Schütz und Dresden', *Dresdner Kapellbuch*, ed. G. Hausswald (Dresden, 1948), 77
H. Schnoor: *Dresden: 400 Jahre Musikkultur* (Dresden, 1948)
I. Becker-Glauch: *Die Bedeutung der Musik für die Dresdener Hoffeste bis in die Zeit Augusts des Starken* (Kassel, 1951)
W. Tappolet: 'Das Leben von Heinrich Schütz', *Musik und Gottesdienst*, v (1951), 33, 65
W. Schramm: 'Bückeburg', *MGG*
H. Sievers: 'Braunschweig', *MGG*
G. Weizsäcker: *Heinrich Schütz: Lobgesang eines Lebens* (Stuttgart, 1952, 2/1956)
O. Wessely: 'Zur Frage nach dem Geburtstag von Heinrich Schütz', *Anzeiger der phil.-hist. Klasse der Österreichischen Akademie der Wissenschaften*, xc (1953), 231
——: 'Zwei unveröffentlichte Heinrich-Schütz-Dokumente', *Musikerziehung*, vii (1953–4), 7
I. Becker-Glauch: 'Dresden', §I, *MGG*
G. Fock: 'Till frågen om "Schütz och Norden"', *STMf*, xxxvi (1954), 102
E. P. Kretzschmer: 'Schütz und Gera', *Festschrift zur Ehrung von Heinrich Schütz (1585–1672)* (Weimar, 1954), 57
R. Petzoldt: 'Zur sozialen Lage des Musikers der Schütz-Zeit', *Festschrift zur Ehrung von Heinrich Schütz (1585–1672)* (Weimar, 1954), 20
W. Senn: *Musik und Theater am Hof zu Innsbruck* (Innsbruck, 1954)
A. Thiele: 'Heinrich Schütz und Weimar', *Festschrift zur Ehrung von Heinrich Schütz (1585–1672)* (Weimar, 1954), 62
——: 'Thüringer Meister im Umkreis von Heinrich Schütz und Heinrich Albert', *Festschrift zur Ehrung von Heinrich Albert (1604–1651)* (Weimar, 1954), 65
A. Werner: 'Begegnungen mit Heinrich Schütz', *Festschrift zur Ehrung von Heinrich Schütz (1585–1672)* (Weimar, 1954), 86
H. C. Worbs: *Heinrich Schütz: Lebensbild eines Musikers* (Leipzig, 1956)
H. Sievers: 'Hannover', *MGG*
F. Blume: 'Geistliche Musik am Hofe des Landgrafen Moritz von Hessen', *Zeitschrift des Vereins für hessische Geschichte und Landeskunde*, lxviii (1957), 131
W. Gurlitt: 'Heinrich Schütz 1585–1672', *Die grossen Deutschen*, v (Berlin, 2/1957), 109
W. Braun: 'Theodor Schuchardt und die Eisenacher Musikkultur im 17. Jahrhundert', *AMw*, xv (1958), 291
W. Brennecke and C. Engelbrecht: 'Kassel', *MGG*
C. Engelbrecht: *Die Kasseler Hofkapelle im 17. Jahrhundert und ihre anonymen Musikhandschriften aus der Kasseler Landesbibliothek*

(Kassel, 1958); rev. H. H. Eggebrecht, *Mf*, xiii (1960), 81
A. Schmiedecke: 'Die Familie Schütz in Weissenfels', *Mf*, xii (1959), 180
P. Várnai: *Heinrich Schütz* (Budapest, 1959)
W. Bittinger, ed.: *Schütz-Werke-Verzeichnis* (*SWV*): *Kleine Ausgabe* (Kassel, 1960)
W. Haacke: *Heinrich Schütz: eine Schilderung seines Lebens und Wirkens* (Königstein, 1960)
H. R. Jung: 'Ein neuaufgefundenes Gutachten von Heinrich Schütz aus dem Jahre 1617', *AMw*, xviii (1961), 241
W. Osthoff: 'Monteverdis *Combattimento* in deutscher Sprache und Heinrich Schütz', *Festschrift Helmuth Osthoff* (Tutzing, 1961), 195
E. Schmidt: *Der Gottesdienst am kurfürstlichen Hofe zu Dresden* (Berlin and Göttingen, 1961)
H. Sievers: *Die Musik in Hannover* (Hanover, 1961)
O. Wessely: 'Ein unbekanntes Huldigungsgedicht auf Heinrich Schütz', *Anzeiger der phil.-hist. Klasse der Österreichischen Akademie der Wissenschaften*, xcviii (1961), 132
H. R. Jung: 'Ein unbekanntes Gutachten von Heinrich Schütz über die Neuordnung der Hof-, Schul- und Stadtmusik in Gera', *BMw*, iv (1962), 17
O. Benesch: 'Schütz und Rembrandt', *Festschrift Otto Erich Deutsch* (Kassel, 1963), 12; repr. in *Sagittarius*, iii (1970), 49; Eng. trans. in O. Benesch: *Collected Writings*, i (London, !970), 228
H. Burose: 'Was ein altes Stammbuch erzählt', *Allgemeiner Harz-Berg-Kalender 1963*, 34
K. Gudewill and W. Bittinger: 'Schütz, Heinrich', *MGG*
F. Krummacher: *Die Überlieferung der Choralbearbeitungen in der frühen evangelischen Kantate* (Berlin, 1965)
B. Grusnick: 'Die Düben-Sammlung: ein Versuch ihrer chronologischen Ordnung', *STMf*, xlvi (1964), 27–82; xlviii (1966), 63–186
H. Haack: 'Heinrich Schütz und Lodovico Viadana', *Mf*, xix (1966), 28
J. H. Schmidt: 'Heinrich Schützens beziehungen zu Celle: ein Beitrag zur Schütz-Biographie', *AMw*, xxiv (1966), 274; repr. in *Sagittarius*, ii (1969), 36
E. Stimmel: 'Die Familie Schütz: ein Beitrag zur Familiengeschichte des Georgius Agricola', *Abhandlungen des Staatlichen Museums für Minerologie und Geologie zu Dresden*, xi (1966), 377
E. Noack: *Musikgeschichte Darmstadts vom Mittelalter bis zur Goethezeit* (Mainz, 1967)
W. Steude: 'Neue Schütz-Ermittlungen', *DJbM*, xii (1967), 40
M. Ruhnke: 'Wolfenbüttel', *MGG*
R. Tellart: *Heinrich Schütz: l'homme et son oeuvre* (Paris, 1968)
W. Braun: 'Mitteldeutsche Quellen der Musiksammlung Gotthold in Königsberg', *Musik des Ostens*, v (1969), 84

B. Grusnick: 'Litania Upsaliensis: eine unbekannte Litanei von Heinrich Schütz', *Sagittarius*, ii (1969), 39
J. P. Larsen: 'Schütz und Dänemark', *Sagittarius*, ii (1969), 9; repr. in *BMw*, xiv (1972), 215
W. Steude: 'Die Markuspassion in der Leipziger Passionen-Handschrift des Johann Zacharias Grundig', *DJbM*, xiv (1969), 96
W. Breig: 'Neue Schütz-Funde', *AMw*, xxvii (1970), 59
K. Hofmann: 'Zwei Abhandlungen zur Weihnachtshistorie von Heinrich Schütz', *Musik und Kirche*, xl (1970), 325; xli (1971), 15
E. K. Sass: *Comments on Rembrandt's Passion Paintings and Constantijn Huygens's Iconography* (Copenhagen, 1971)
D. Arnold: 'Schütz in Venice', *Music and Musicians*, xxi/1 (1972), 30
I. Bach: 'Bildnisse von Heinrich Schütz', *Heinrich Schütz 1585–1672: Festtage 1972*, ed. W. Siegmund-Schultze (Gera, 1972), 29
O. Brodde: *Heinrich Schütz: Weg und Werk* (Kassel, 1972)
H. Eppstein: *Heinrich Schütz* (Stockholm, 1972; Ger. trans., 1975)
H. Haase, ed.: *Heinrich Schütz (1585–1672) in seinen Beziehungen zum Wolfenbütteler Hof* (Wolfenbüttel, 1972)
H. R. Jung: 'Zwei unbekannte Briefe von Heinrich Schütz aus den Jahren 1653/54', *BMw*, xiv (1972), 231
R. Petzoldt and D. Berke: *Heinrich Schütz und seine Zeit in Bildern* (Kassel, 1972)
U. Prinz: 'Anmerkungen zur Neuausgabe des "Beckerschen Psalters" von Heinrich Schütz', *Mf*, xxv (1972), 175
A. Schmiedecke: 'Heinrich Schütz' Beziehungen zu Köstritz, Gera, Weissenfels und Zeitz', *Heinrich Schütz 1585–1672: Festtage 1972*, ed. W. Siegmund-Schultze (Gera, 1972), 24
O. Biba: 'Heinrich Schütz und Österreich', *Singende Kirche*, xx (1972–3), 73
H. Haase: 'Nachtrag zu einer Schütz-Ausstellung', *Sagittarius*, iv (1973), 85
R. A. Leaver: 'The Funeral Sermon for Heinrich Schütz', *Bach*, iv/4 (1973), 3; v/1 (1974), 9; v/2 (1974), 22; v/3 (1974), 13
K. Petzoldt: 'Das Schicksal des Grabes von Heinrich Schütz', *Sagittarius*, iv (1973), 34
H. Walter: 'Ein unbekanntes Schütz-Autograph in Wolfenbüttel', *Musicae scientiae collectanea: Festschrift Karl Gustav Fellerer* (Cologne, 1973), 621
W. Breig: 'Heinrich Schütz' Parodiemotette "Jesu dulcissime" ', *Convivium musicorum: Festschrift Wolfgang Boetticher* (Berlin, 1974), 13
H. Glahn and S. Sørensen, eds.: *Musikhåndskrifterne fra Clausholm* (Copenhagen, 1974)
R. Petzoldt, 'Studien zum Sozialstatus des Musikers im 17. und 18. Jahrhundert', *Heinrich Schütz und seine Zeit: Bericht über die wis-*

senschaftliche Konferenz des Komitees für die Heinrich-Schütz-Festtage der Deutschen Demokratischen Republik 1972 (Berlin, 1974), 74
D. Paisey: 'Some Occasional Aspects of Johann Hermann Schein', *British Library Journal*, i (1975), 171
T. Seebass, ed.: *Musikhandschriften in Basel aus verschiedenen Sammlungen* (Basel, 1975)
M. Ruhnke: 'Zur Hochzeit: die Psalmen Davids: ein Brief von Heinrich Schütz an die Stadt Braunschweig', *Beiträge zur Musikgeschichte Nordeuropas: Kurt Gudewill zum 65. Geburtstag* (Wolfenbüttel, 1977)
K. Gudewill: *Der 'Gesang der Venuskinder' von Heinrich Schütz (1634)* (Kiel, 1978); repr. *SJb*, vi (1984), 72
Schütz-Jahrbuch (SJb) (1979–)
W. Breig: 'Schützfunde und -zuschreibungen seit 1960: auf dem Wege zur Grossen Ausgabe des Schütz-Werke-Verzeichnisses', *SJb*, i (1979), 63
R. Brunner: 'Bibliographie des Schütz-Schrifttums 1926–1950', *SJb*, i (1979), 93
J. J. Berns: ' "Theatralische neue Vorstellung von der Maria Magdalena": ein Zeugnis für die Zusammenarbeit von Justus Georg Schottelius und Heinrich Schütz', *SJb*, ii (1980), 120
R. Brunner: 'Bibliographie des Schütz-Schrifttums 1926–1950, *SJb*, iii (1981), 64
E. Linfield: 'A New Look at the Sources of Schütz's Christmas History', *SJb*, iv–v (1982–3), 19
W. Steude: 'Das wiedergefundene Opus ultimum von Heinrich Schütz: Bemerkungen zur Quelle und zum Werk', *SJb*, iv–v (1982–3), 9
E. Möller: 'Ein unbekanntes Bild von Heinrich Schütz in der Ratsschulbibliothek Zwickau', *BMw*, xxv (1983), 297
W. Breig: 'Schütz' Osterkonzert "Christ ist erstanden" SWV 470, seine Kasseler Ersatz-Symphonia und Hammerschmidts "Dialogi" von 1645', *SJb*, vi (1984), 52
R. Brunner: 'Bibliographie des Schütz-Schrifttums 1672–1925', *SJb*, vi (1984), 102
J. U. Fechner: 'Ein unbekanntes weltliches Madrigal von Heinrich Schütz', *SJb*, vi (1984), 23
——: ' "Wie die Sonne unter den Planeten in der Mitte leuchtet, so die Musik unter den freien Künsten": zu Heinrich Schütz' Eintrag in das Stammbuch des Andreas Möring', *SJb*, vi (1984), 93
M. Gregor-Dellin: *Heinrich Schütz: sein Leben, sein Werk, seine Zeit* (Munich, 1984)
E. Möller: 'Neue Schütz-Funde in der Ratsschulbibliothek und im Stadtarchiv Zwickau', *SJb*, vi (1984), 5

OTHER STUDIES

F. Spitta: *Die Passionen nach den vier Evangelisten von Heinrich Schütz* (Leipzig, 1886)

P. Spitta: 'Händel, Bach und Schütz', *Zur Musik* (Berlin, 1892/*R*1976), 59–92

——: *Die Passionsmusiken von Sebastian Bach und Heinrich Schütz* (Hamburg, 1893)

A. Pirro: 'Les formes de l'expression dans la musique de Heinrich Schütz', *Tribune de Saint-Gervais*, vi (1900), 314

E. W. Naylor: 'Some Characteristics of Heinrich Schütz (1585–1672)', *PMA*, xxxii (1905–6), 23

J. Smend: 'Zur Wortbetonung des Lutherischen Bibeltextes bei Heinrich Schütz', *ZMw*, v (1922–3), 75

F. Blume: *Das monodische Prinzip in der protestantischen Kirchenmusik* (Leipzig, 1925/*R*1975)

J. M. Müller-Blattau, ed.: *Die Kompositionslehre Heinrich Schützens in der Fassung seines Schülers Christoph Bernhard* (Leipzig, 1926, 2/1963)

A. Schering: 'Zur Metrik der "Psalmen Davids" von Heinrich Schütz', *Festschrift Peter Wagner* (Leipzig, 1926/*R*1969), 176

H. Spitta: *Heinrich Schütz' Orchester- und unveröffentlichte Werke* (diss., U. of Göttingen, 1927)

R. Gerber: 'Wort und Ton in den "Cantiones sacrae" von Heinrich Schütz', *Gedenkschrift für Hermann Abert* (Halle, 1928/*R*1974), 57

W. Schuh: *Formprobleme bei Heinrich Schütz* (Leipzig, 1928/*R*1976); rev. H. Engel, *ZMw*, xv (1932–3), 281

R. Gerber: *Das Passionsrezitativ bei Heinrich Schütz und seine stilgeschichtlichen Grundlagen* (Gütersloh, 1929/*R*1973)

H. J. Moser: 'Schütz und das evangelische Kirchenlied', *Jb der Staatlichen Akademie für Kirchen- und Schulmusik Berlin*, iii (1929–30), 7; repr. in *Musik in Zeit und Raum* (Berlin, 1960), 75

F. Blume: 'Heinrich Schütz in den geistigen Strömungen seiner Zeit', *Musik und Kirche*, ii (1930), 245

H. J. Moser: *Die mehrstimmige Vertonung des Evangeliums*, i (Leipzig, 1931/*R*1968)

R. Gerber: 'Die "Musikalischen Exequien" von Heinrich Schütz', *Musik und Kirche*, vi (1934), 296

A. Heller: *Der Deutsche Heinrich Schütz in seinen italienischen Madrigalen* (diss., U. of Prague, 1934)

W. Kreidler: *Heinrich Schütz und der Stile concitato von Claudio Monteverdi* (Kassel, 1934)

A. A. Abert: *Die stilistischen Voraussetzungen der "Cantiones sacrae" von Heinrich Schütz* (Wolfenbüttel and Berlin, 1935)

W. Blankenburg: 'Heinrich Schütz und der protestantische Choral', *Musik und Kirche*, vii (1935), 219

——: 'Zu Heinrich Schütz' Passionen', *Musik und Kirche*, vii (1935), 2
H. Hoffmann: 'Die Gestaltung der Evangelisten-Worte bei Heinrich Schütz und J. S. Bach', *Festschrift Max Schneider zum 60. Geburtstag* (Halle, 1935), 49
E. Wachten: 'Die Symbolgestaltung in der "Auferstehungshistorie" von Heinrich Schütz', *ZMw*, xvii (1935), 393
K. Gudewill: *Das sprachliche Urbild bei Heinrich Schütz* (Kassel, 1936)
F. Blume: 'Heinrich Schütz: Gesetz und Freiheit', *Deutsche Musikkultur*, i (1936–7), 36
L. Reitter: *Doppelchortechnik bei Heinrich Schütz* (Derendingen, 1937)
H. Birtner: 'Fragen der Aufführungspraxis, insbesondere Continuo-Besetzung bei Heinrich Schütz', *Deutsche Musikkultur*, iii (1938–9), 269
H. Hoffmann: 'Bearbeitungsfragen bei Schütz', *AMz*, lxvi (1939)
——: *Heinrich Schütz und Johann Sebastian Bach: zwei Tonsprachen und ihre Bedeutung für die Aufführungspraxis* (Kassel, 1940)
R. Engländer: 'Zur Frage der "Dafne" (1671) von G. A. Bontempi und M. G. Peranda', *AcM*, xiii (1941), 59
C. Roskowski: *Die 'Kleinen geistlichen Konzerte' von Heinrich Schütz* (diss., U. of Münster, 1947)
J. Piersig: *Das Weltbild des Heinrich Schütz* (Kassel, 1949)
G. Toussaint: *Die Anwendung der musikalisch-rhetorischen Figuren in den Werken von Heinrich Schütz* (diss., U. of Mainz, 1949)
F. Schöneich: 'Zum Aufbau des Gloria-Teils in Schützens Musikalischen Exequien', *Musik und Kirche*, xx (1950), 182
W. Ehmann: 'Besetzungsfragen bei Heinrich Schütz', *Musik und Gottesdienst*, vi (1952), 97
H. Federhofer: 'Die Figurenlehre nach Christoph Bernhard und die Dissonanzbehandlung in Werken von Heinrich Schütz', *GfMKB, Bamberg 1953*, 132
G. Abraham: 'Passion Music from Schütz to Bach', *MMR*, lxxxiv (1954), 115, 152, 175
A. Adrio: 'Heinrich Schütz und Italien', *Bekenntnis zu Heinrich Schütz* (Kassel, 1954), 55
W. Ehmann: 'Heinrich Schütz in unserer musikalischen Praxis', *Bekenntnis zu Heinrich Schütz* (Kassel, 1954), 9
T. G. Georgiades: *Musik und Sprache* (Berne and Heidelberg, 1954, 2/1974)
H. J. Moser: 'Heinrich Schütz in unserem Gottesdienst', *Bekenntnis zu Heinrich Schütz* (Kassel, 1954), 40; repr. as 'Heinrich Schütz als Lehrer, Prediger und Prophet', *Musik in Zeit und Raum* (Berlin, 1960), 93
L. Schrade: 'Heinrich Schütz and Johann Sebastian Bach in the Protestant Liturgy', *The Musical Heritage of the Church* (Valparaiso,

Ind., 1954), 31; repr. in *De scientia musicae studia atque orationes* (Berne and Stuttgart, 1967), 396
C. B. Agey: *A Study of the 'Kleine geistliche Konzerte' and 'Geistliche Chormusik' of Heinrich Schütz* (diss., Florida State U., 1955)
W. Ehmann: 'Heinrich Schütz: Die Psalmen Davids, 1619, in der Aufführungspraxis', *Musik und Kirche*, xxvi (1956), 145
J. Heinrich: *Stilkritische Untersuchungen zur 'Geistlichen Chormusik' von Heinrich Schütz* (diss., U. of Göttingen, 1956)
W. Ehmann: 'Die geistliche Chormusik von Heinrich Schütz in ihrer musikalischen Darstellung', *Kirchenmusik*, ed. W. Ehmann (Eberstadt, nr. Darmstadt, 1958), 75
P. Gümmer: 'Der Gesangstil in den "Kleinen geistlichen Konzerten" von Heinrich Schütz', *Kirchenmusik*, ed. W. Ehmann (Eberstadt, nr. Darmstadt, 1958), 84
A. Roeseler: *Studien zum Instrumentarium in den Vokalwerken von Heinrich Schütz: die obligaten Instrumente in den Psalmen Davids und in den Symphoniae sacrae I* (diss., Free U. of Berlin, 1958)
H. H. Eggebrecht: *Heinrich Schütz: Musicus poeticus* (Göttingen, 1959)
——: 'Zum Figur-Begriff der Musica poetica', *AMw*, xvi (1959), 57
S. Hermelink: 'Rhythmische Struktur in der Musik von Heinrich Schütz', *AMw*, xvi (1959), 378
G. Kirchner: *Der Generalbass bei Heinrich Schütz* (Kassel, 1960)
A. Mendel: 'A Brief Note on Triple Proportion in Schuetz', *MQ*, xlvi (1960), 67
A. E. Rowley: *A Bibliography of Heinrich Schütz* (diss., U. of London, 1960)
H. H. Eggebrecht: 'Ordnung und Ausdruck im Werk von Heinrich Schütz', *Musik und Kirche*, xxxi (1961), 1; pubd separately (Kassel, 1961)
W. S. Huber: *Motivsymbolik bei Heinrich Schütz* (Basle, 1961)
L. Schrade: *Das musikalische Werk von Heinrich Schütz in der protestantischen Liturgie* (Basle, 1961)
M. Geck: 'Ein textbedingter Archaismus im Werke von Heinrich Schütz', *AcM*, xxxiv (1962), 161
G. A. Trumpff: 'Die "Musikalischen Exequien" von Heinrich Schütz', *NZM*, cxxiii (1962), 120
W. Ehmann: 'Die "Kleinen geistlichen Konzerte" von Heinrich Schütz und unsere musikalische Praxis', *Musik und Kirche*, xxxiii (1963), 9
G. Mittring: 'Totendienst und Christuspredigt: zum Text der Musikalischen Exequien von Heinrich Schütz', *Musik als Lobgesang: Festschrift für Wilhelm Ehmann* (Darmstadt, 1964), 43
W. Blankenburg: 'Vom Verhältnis von Musik und Sprache bei Heinrich Schütz', *ÖMz*, xxi (1966), 163
K. Gudewill: 'Die textlichen Grundlagen der geistlichen Vokalmusik bei Heinrich Schütz', *Sagittarius*, i (1966), 52

J. Mertin: 'Zur Aufführungspraxis in der Musik von Heinrich Schütz', *ÖMz*, xxi (1966), 166
H. Reich: 'Händels Trauer-Hymne und die Musikalischen Exequien von Schütz', *Musik und Kirche*, xxxvi (1966), 74
H. Sittner: 'Heinrich Schütz, Brücke zwischen Süd und Nord', *ÖMz*, xxi (1966), 156
O. Söhngen: 'Heinrich Schütz und die zeitgenössische Musik', *Sagittarius*, i (1966), 15
C. Ganz: 'Der Beckerpsalter von Heinrich Schütz', *Musica sacra*, lxxxvii/2 (1967), 46; lxxxvii/3 (1967), 71
T. G. Georgiades: *Schubert: Musik und Lyrik* (Göttingen, 1967) [incl. 'Schubert und Schütz']
D. McCulloch: 'Heinrich Schütz (1585–1672) and Venice', *Church Music*, ii (1967), no.20, p.8; no.21, p.4
H. Wichmann-Zemke: *Untersuchungen zur Harmonik in den Werken von Heinrich Schütz* (diss., U. of Kiel, 1967)
H. Drude: *Heinrich Schütz als Musiker der evangelischen Kirche* (diss., U. of Göttingen, 1969)
H. H. Eggebrecht: *Schütz und Gottesdienst: Versuch über das Selbstverständliche* (Stuttgart, 1969)
D. Manicke: 'Heinrich Schütz als Lehrer', *Sagittarius*, ii (1969), 17
W. Braun: 'Musikalische Inspiration – zwischen systematischer und historischer Forschung', *Mf*, xxiii (1970), 4
W. Gerstenberg: 'Heinrich Schütz im Konzert der europäischen Nationen', *Sagittarius*, iii (1970), 9
G. Newton: 'Heinrich Schuetz', *National Association of Teachers of Singing: Bulletin*, xxvi/3 (1970), 2
R. Bray: 'The "Cantiones sacrae" of Heinrich Schütz Re-examined', *ML*, lii (1971), 299
S. Hermelink: 'Bemerkungen zur Schütz-Edition', *Musikalische Edition im Wandel des historischen Bewusstseins*, ed. T. G. Georgiades (Kassel, 1971), 203
O. Wessely: 'Zur Ars inveniendi im Zeitalter des Barock', *Orbis musicae*, i (1971–2), 113
D. Arnold: 'Schütz's "Venetian" Psalms', *MT*, cxiii (1972), 1071
W. Blankenburg: 'Die Dialogkompositionen von Heinrich Schütz', *Musik und Kirche*, xlii (1972), 121
——: 'Heinrich Schütz 1672–1972', *Musik und Kirche*, xlii (1972), 3
——: 'Schütz und Bach', *Musik und Kirche*, xlii (1972), 219
K. Blum and M. Elste: *Internationale Heinrich-Schütz-Diskographie 1928–1972* (Bremen, 1972)
F. Blume: 'Heinrich Schütz nach 300 Jahren', *Neue Zürcher Zeitung* (5 Nov 1972); repr. in *Syntagma musicologicum*, ii (Kassel, 1976), 139; Eng. trans. in *SMA*, vii (1973), 1

W. Breig: 'Zum Parodieverfahren bei Heinrich Schütz', *Musica*, xxvi (1972), 17
H. H. Eggebrecht: 'Musikalische Analyse (Heinrich Schütz)', *MZ*, viii (1972), 17
I. Hempel: 'Literatur zu Ehren von Heinrich Schütz', *Börsenblatt für den deutschen Buchhandel*, cxxxix/35 (1972), 669
J. Rifkin: 'Schütz and Musical Logic', *MT*, cxiii (1972), 1067
J. Roche: 'What Schütz Learnt from Grandi in 1629', *MT*, cxiii (1972), 1074
S. Schmalzriedt: *Heinrich Schütz und andere zeitgenössische Musiker in der Lehre Giovanni Gabrielis* (Neuhausen, 1972)
M. Seelkopf: 'Italienische Elementen in den Kleinen geistlichen Konzerten von Heinrich Schütz', *Mf*, xxv (1972), 452
I. Allihn: 'Bericht: Schütz-Konferenz', *BMw*, xv (1973), 277
H. H. Eggebrecht: 'Heinrich Schütz: Rede 1972', *Musik und Kirche*, xliii (1973), 58
T. G. Georgiades: 'Heinrich Schütz zum 300. Todestag', *Sagittarius*, iv (1973), 57; repr. in *Kleine Schriften* (Tutzing, 1977), 177
R. Henning: 'Zur Textfrage der "Musicalischen Exequien" von Heinrich Schütz', *Sagittarius*, iv (1973), 44
R. A. Leaver: 'Heinrich Schütz as a Biblical Interpreter', *Bach*, iv/3 (1973), 3
R. Gerlach: 'Lateinische und deutsche Komposition bei Heinrich Schütz: eine vergleichende Interpretation zweier Psalmvertonungen', *Convivium musicorum: Festschrift Wolfgang Boetticher* (Berlin, 1974), 83
R. Grunow, ed.: *Begegnungen mit Heinrich Schütz: Erzählungen über Leben und Werk* (Berlin, 1974)
S. Köhler: 'Heinrich Schütz in Venedig: die Bedeutung der musikalischen Renaissance für sein Werk', *Heinrich Schütz und seine Zeit: Bericht über die wissenschaftliche Konferenz des Komitees für die Heinrich-Schütz-Festtage der Deutschen Demokratischen Republik 1972* (Berlin, 1974), 26
H. Krause-Graumnitz: 'Heinrich Schütz' schöpferische Gestaltung der zyklischen Grossform dargestellt an seinen "Musikalischen Exequien" des Jahres 1636', *Heinrich Schütz und seine Zeit* (Berlin, 1974), 38
E. H. Meyer: 'Gedanken zu Heinrich Schütz' musikgeschichtlicher Rolle', *Heinrich Schütz und seine Zeit* (Berlin, 1974), 18
W. Osthoff: *Heinrich Schütz: l'incontro storico fra lingua tedesca e poetica musicale italiana nel seicento* (Venice, 1974); Ger. trans. in *SJb*, ii (1980), 78
W. Rackwitz: 'Festansprache', *Heinrich Schütz und seine Zeit* (Berlin, 1974), 13

W. Siegmund-Schultze: 'Das Spätschaffen von Heinrich Schütz', *Heinrich Schütz und seine Zeit* (Berlin, 1974), 50
F. Streller: 'Anmerkungen zur werkgerechten Interpretation Schützscher Musik', *Heinrich Schütz und seine Zeit* (Berlin, 1974), 103
O. Wessely: 'Der Fürst und der Tod', *Beiträge 1974/75*, 60
Q. Faulkner: *The "Symphoniae Sacrae" of Heinrich Schütz: a Manual for Performance* (New York, 1975)
A. Heller: 'Heinrich Schütz in seinen italienischen Madrigalen', *Gustav Becking zum Gedächtnis* (Tutzing, 1975), 373
P. Steinitz: 'German Church Music', *NOHM*, v (1975), 557–776
F. P. Constantini: 'Über die Ausführung des Falsobordone in der Auferstehungs-Historia von Heinrich Schütz', *Musik und Kirche*, xlvii (1977), 55
H. E. Smither: *A History of the Oratorio*, ii: *The Oratorio in the Baroque Era: Protestant Germany and England* (Chapel Hill, 1977)
F. Krummacher: *Die Choralbearbeitung in der protestantischen Figuralmusik zwischen Praetorius und Bach* (Kassel, 1978)
M. Lang: 'Begegnung von Sprache und Musik: dargestellt an Beispiel einer Schütz-Mottete', *Bachstunden Festschrift für Helmut Walcha* (Frankfurt, 1978), 129
S. Kunze: 'Instrumentalität und Sprachvertonung in der Musik von Heinrich Schütz', *SJb*, i (1979), 9
A. Botti Caselli: 'La "Storia della resurrezione": un debito di Heinrich Schütz nei confronti di Antonio Scandello?', *Heinrich Schütz e il suo tempo: Atti del 1° convegno internazionale di studi*, ed., G. Rostirolla (Rome, 1981), 59
W. Braun: *Neues Handbuch der Musikwissenschaft*, iv: *Die Musik des 17. Jahrhunderts* (Laaber, 1981)
W. Breig: 'Höfische Festmusik im Werk von Heinrich Schütz', *Daphnis: Zeitschrift für mittlere deutsche Literatur*, x (1981), 711
——: 'Mehrchörigkeit und individuelle Werkkonzeption bei Heinrich Schütz', *SJb*, iii (1981), 24
P. E. Carapezza: 'Scelte poetiche e vincoli stilistici del Sagittario madrigalista', *Heinrich Schütz e il suo tempo* (Rome, 1981), 105; Ger. trans. in *SJb*, i (1979), 44
K. Gudewill: 'Heinrich Schütz und Italien', *Heinrich Schütz e il suo tempo* (Rome, 1981), 19
A. Kirwan-Mott: *The Small-scale Sacred Concertato in the Early seventeenth Century* (Ann Arbor, 1981), 314 [discussion of Schütz's *Kleine geistliche Concerte*]
S. Kunze: 'Rhythmus, Sprache, musikalische Raumvorstellung: zur Mehrchörigkeit Giovanni Gabrielis', *SJb*, iii (1981), 12
F. Piperno: 'La sinfonia strumentale nel 1° volume delle "Symphoniae

sacrae" di Heinrich Schütz (1629)', *Heinrich Schütz e il suo tempo* (Rome, 1981), 187

A. B. Skei: *Heinrich Schütz: a Guide to Research* (New York and London, 1981)

W. Steinbeck: 'Sprachvertonung bei Heinrich Schütz als analytisches Problem', *SJb*, iii (1981), 51

W. Witzenmann: 'Problemi di modalità nelle Passioni di Schütz', *Heinrich Schütz e il suo tempo* (Rome, 1981), 129; Ger. trans. in *SJb*, ii (1980), 103

A. Forchert: 'Heinrich Schütz als Komponist evangelischer Kirchenliedtexte', *SJb*, iv–v (1982–3), 57

S. Kunze: 'Sprachauslegung und Instrumentalität in der Musik von Schütz', *SJb*, iv–v (1982–3), 39

U. Siegele: 'Musik als Zeugnis der Auslegungsgeschichte: Heinrich Schützens Motette "Die mit Tränen säen" aus der "Geistlichen Chormusik" ', *SJb*, iv–v (1982–3), 50

H. Birtner: 'Grundsätzliche Bemerkungen zur Schütz-Pflege in unser Zeit (1935)', *Musica*, xxxviii (1984), 532

W. Blankenburg: 'Zur Bedeutung der Andachtstexte im Werk von Heinrich Schütz', *SJb*, vi (1984), 62

W. Breig: 'Schütz-Aspekte im Vorfeld des Jubiläumsjahres 1985', *Musica*, xxxviii (1984), 527

B. Smallman: *The Music of Heinrich Schütz* (Leeds, 1985)

JOHANN JACOB FROBERGER

George J. Buelow

CHAPTER ONE

Life

Johann Jacob (or Jakob) Froberger, the pre-eminent German composer of keyboard music in the mid-17th century, was born in Stuttgart; he was baptized on 19 May 1616. As a pupil of Frescobaldi, court organist at Vienna and frequent traveller and performer in the Low Countries, England, France and Germany, he forged a distinctive personal idiom out of stylistic features of Italian, French and German keyboard music. He left a legacy of keyboard works which exerted a significant influence on German music until well into the 18th century.

Despite Froberger's prominent role in the history of keyboard music and the high regard in which he was held by composers and theorists of the later 17th and early 18th centuries, his life cannot be documented in detail. 18th-century biographical accounts, including those of Walther and Mattheson (*Grundlage einer Ehren-Pforte*), are largely inaccurate. Walther, for example, stated that he had been assured by 'a certain relative' of Froberger's that he had died in Mainz. Mattheson's biographical entry is, with its several errors, specially disappointing. It was he who wrongly gave Froberger's birthplace as Halle and his birthdate 19 years too late; these 'facts' remained uncorrected until the 1930s (see especially Seidler). Halle, though not Froberger's birthplace, was where the family line apparently originated: his grandfather, Simon, lived there, and his father, Basilius, was born there in 1575. Basilius moved to Stuttgart, where eventually, in 1621, he became court Kapellmeister; at least four of his six surviving sons, Johann Jacob's elder brothers, were later employed in the Stuttgart Hofkapelle.

Conjecture at best permits one to surmise that Froberger received his early musical training from his father or elder brothers. The Stuttgart court enjoyed a rich musical life, provided by a wide variety of musicians. These included a number of Englishmen, and Froberger may have received instruction from one of the English lutenists, such as Andrew Borell, who is known to have been paid for teaching one of Basilius Froberger's sons in 1621–2, and John and David Morell. It has been suggested that he studied with the prominent Stuttgart organists Johann Eckhardt and J. U. Steigleder, but there is no proof that he did so. He certainly went to Vienna, probably in about 1634 and perhaps to join the Hofkapelle. According to Mattheson he was taken there by the Swedish ambassador, who was impressed with his voice. No later documents connect him with Stuttgart, though he probably maintained some connection with its royal family, since in later life his patron was Princess Sibylla of Württemberg-Montbéliard, at whose estate he died. Court records in Vienna do not include his name until 1637, when from 1 January until 30 October he was employed as an organist. The Emperor Ferdinand III's great enthusiasm for Italian culture may have made possible Froberger's eventually successful request, after several earlier attempts, for leave to study in Rome with Frescobaldi; he was granted 200 gulden to enable him to do so. He remained in Rome until the end of 1640 or early in 1641. Records show that he again worked at the Viennese court from April 1641 until October 1645. There is no further evidence of his activity there until 1653.

Few documents throw light on Froberger's life between 1645 and 1653. A few clues suggest the following chronology. In a letter to Kircher, Froberger alluded to a second visit to Italy before 1649 (see Scharlau). He mentioned a psalm setting, now lost, which he hoped would be performed at S Apollinare, Rome, and he waited anxiously to learn what the 'master' thought of it. The 'master' to whom he referred here is Carissimi, *maestro di cappella* of S Apollinare (the German collegiate church), with whom he may have studied before 1649. He also reported to Kircher about performances he had given at the courts of Florence and Mantua. That he had returned to Vienna by the late summer of 1649 is evident from his description to Kircher of an audience that he had with the emperor shortly before the death of the Empress Maria Leopoldine on 19 August 1649. It was at her funeral that he became acquainted with William Swann, envoy of the Prince of Orange. Swann wrote to Constantijn Huygens, the prince's secretary, about him, and this in turn led to a friendship between Huygens and Froberger and probably to efforts on Huygens's part to make Froberger's music well known in the Netherlands. A further indication of his presence in Vienna at this time is his dedication, dated 29 September 1649, of a manuscript book of keyboard works to the emperor. It was possibly at the end of 1649 that he embarked on fairly extensive travels, perhaps while in the employ of the emperor's brother, Archduke

Leopold Wilhelm, who had been governor of the Spanish Netherlands since 1647 and maintained a residence in Brussels. Froberger was apparently in Brussels in 1650; at least one of his manuscripts bears the inscription that the music in it was composed there in that year. Another journey was in 1652 to Paris (see Krebs), where he gave a highly acclaimed concert and almost certainly met a number of prominent French musicians, including Chambonnières, Louis Couperin, Jacques Gallot and Denis Gaultier. At some point, probably before this, he also travelled to England: in another letter to Kircher, dated 9 February 1654, he mentioned his visits to Germany, the Netherlands, England and France. Mattheson's fanciful account, unconfirmed in any English source, of the visit to England, states that he was beset by highwaymen and pirates, arrived penniless and unknown, accepted work as an organ blower and only later received due honour as court organist to Charles II – who did not become king until 1660! The only evidence supporting any element in this tale is his harpsichord piece *Plainte faite à Londres*, which refers to a pirate attack between Calais and Dover. He may also at this period have gone to Dresden; again according to Mattheson he participated at court in a keyboard contest with Matthias Weckmann, who later became a friend of his.

On 1 April 1653 Froberger returned to Vienna as court organist, a position he retained until 1658, when, Walther and Mattheson suggested, he was released because of 'royal displeasure'. The nature of Froberger's difficulties with the new emperor, Leopold I, is unknown; that they occurred at all seems surprising, for he dedicated a book of capriccios and ricercares to

Leopold in August that year. He may well have gone to Mainz then, as Walther and Mattheson stated, but they were wrong in stating that he died there. He was given final asylum on the estate of Princess Sibylla of Württemberg-Montbéliard at Héricourt, where he spent his last years apparently in isolation and solitude as the princess's friend and tutor. She corresponded with Huygens about him, expressing great admiration for his impressive personality and unmatched keyboard artistry. It is in a letter to Huygens that she described the circumstances of his sudden death, on 6 or 7 May 1667, while attending Vespers. He was buried a Catholic; he apparently became a convert while studying in Rome with Frescobaldi.

CHAPTER TWO

Works

With only two exceptions, all of Froberger's extant compositions are for the keyboard. They are relatively few in number and are typical of Italian and French keyboard genres of the 17th century. (The numbering used in this discussion is that of the DTÖ edition.) Among the pieces in Italianate forms are the toccatas, ricercares, canzonas, fantasias and capriccios, which are either for organ or harpsichord; his best works, the suites for harpsichord (or clavichord), are French-influenced. Froberger wrote, as far as is known, only one set of variations on a popular melody – his famous *Partita auff die Mayerin* (Suite no.6) – of the type frequently composed by Frescobaldi (see fig.6). Moreover, his reputation as the composer of several character-pieces bearing descriptive titles was well established as early as the 18th century. They are in fact in the style of allemandes and comprise *Tombeau fait à Paris sur la mort de Monsieur Blancheroche*; *Lamentation faite sur la mort très douloureuse de Sa Majesté Imperiale, Ferdinand le troisième, An. 1657*; *Lamento sopra la dolorosa perdità della Real Mstà de Ferdinando IV* (allemande in Suite no.12); *Plainte faite à Londres, pour passer le Mélancholie* (allemande in Suite no.30); and *Lamentation sur ce que j'ay été volé, et se joue à discretion et encore mieux que les Soldats m'ont traité. Allemande. NB. Cum D. Froberger Bruxellis Lovanium iter faciens à militibus Lotharingis tunc grassantibus verberibus male tractatus fuisset imo (quamvis ceteroquin Patentes Caesareas inspexissent) spoliatus saucius tandem dimissus: hanc Lamentationen pro animi afflicti solatione composuit* (allemande in Suite no.14). Mattheson and others referred too to a now lost work: *Allemande, faite en passant le Rhin dans une barque en grand péril.*

The many works of Froberger clearly influenced by the music of his teacher Frescobaldi are in style closely related to one another. They include 25 toccatas (in the DTÖ edition of 26, no.14 = no.22), 15 ricercares, eight fantasias (no.8 is of questionable authenticity), six canzonas and 17 capriccios (no.11 is contained within Canzona no.1). All except the toccatas are laid out with a separate staff for each polyphonic line. Unlike Frescobaldi, however, Froberger made no real distinction between canzona and capriccio, both of which he treated as fugal genres, frequently multi-sectional and constructed from successive variations of the opening subject. The variation technique is simple in being restricted almost completely to recasting the subject in fresh rhythmic guises, often in a new metre or in reduced note values. The prevailing fugal texture includes subject–answer relationships at the 4th or 5th but very little use of episodic material; thus even the best examples tend to be thematically monotonous. Most of the canzonas are in three sections; the capriccios range from unified wholes to works in two, three, four and six parts. Both genres employ essentially similar types of lively fugal subjects, often with leaps and interesting rhythmic patterns, which lend vitality and even at times brilliance to the contrapuntal display.

6. Autograph MS (dated 1649) of the beginning of Froberger's 'Partita auff die Mayerin'

One senses in this kind of thematic invention the initial stage of a development that was to lead directly to the great fugue subjects of Bach; if Adlung was correct in stating that Bach held Froberger in 'high esteem', it must have been at least partly because of his thematic invention in these works. Froberger's contrapuntal technique here, as in all his fugal works, is conservative and at times conventional.

Froberger's fantasias and ricercares are even more scholastic and strict in contrapuntal design. They are works in *alla breve* style, based on slow-moving subjects in semibreves and minims and worked out according to principles of the *prima prattica* derived from 16th-century Italian sacred polyphony. Two of the fantasias, nos.1 and 4, employ a favourite type of 16th-century fugal subject, the solmization theme. The former, the so-called 'hexachord fantasia', the only music by Froberger published during his lifetime, was printed by Kircher as an example worthy of emulation.

Froberger's toccatas, notated on a treble staff of six lines and a bass staff of seven, show many characteristics of Frescobaldi's works in the genre. They are, however, different in being organized into large contrasting sections. An opening, freely improvisatory display is juxtaposed to a section of imitative counterpoint, which may or may not be interrupted at least once by a return, usually abbreviated, of the improvisatory section. As in the other fugal forms, the succeeding contrapuntal sections are usually variations on the first subject. Apel (*Geschichte der Orgel- und Klaviermusik bis 1700*, Eng. trans., p.552) traced this type of toccata to those of Michelangelo Rossi, another of Frescobaldi's pupils. In Froberger's toccatas the opening free-fantasia sections seem at first glance to be similar to passages in the same style found in Frescobaldi's toccatas. Long-sustained chords on pedal notes lead straight into brilliant runs, arpeggios and other keyboard figurations passed between the hands, often with exposed dissonant clashes. Frescobaldi's toccatas, however, are not only more dramatic but much more varied in their textures and motivic writing; many too are in a great number of individual sections between which there is little thematic unity. His toccatas are both more disruptive and explosive and essentially more improvisatory in the strictest sense of the word, denoting a freely developing series of musical ideas. Only two of Froberger's – nos.5 and 6 – belong to the type of toccata *alla levatione* for liturgical use frequently found in Frescobaldi's published collections. They are probably early works and are the ones that most strongly resemble his teacher's compositional style, completely eschewing sections of fugal imitation and freely exploring daring dissonant clashes and highly affective melodic writing.

Froberger's historical importance has always rested on his 30 harpsichord suites. His reputation in this field stemmed largely from the mistaken belief that he was the first composer to organize the keyboard suite as a unified form in one tonality and in the standard 18th-century grouping of allemande–courante–sarabande–gigue. Unfortunately Adler furthered this myth by regrouping the dance movements in his edition of the suites (in DTÖ) to conform to this preconceived pattern, even though such an arrangement is not found in the original sources. Suites nos.1–5 and 7–11 survive in two autograph volumes presented to the Emperor Ferdinand III in Vienna, one dated 1649 and the other written before 1656. The earlier volume contains suites with only three movements, in the order allemande–courante–sarabande, except for Suite no.2, which also includes a gigue as the final dance. The second autograph contains only four-movement suites but arranged in the order allemande–gigue–courante–sarabande. Suites nos.13–20 survive only in a volume published in Amsterdam some 30 years after Froberger's death. Although these suites are arranged in the classic order allemande–courante–sarabande–gigue, the publisher carefully stated on the title-page that he had placed the dances 'in a better order'. Since no manuscript from the last decade of Froberger's life is known, it cannot be determined whether he eventually approved of the sequence in which the gigue occurs at the end of a suite.

Froberger's dance movements are largely homophonic, while developing to a considerable extent the *style brisé* of the French lutenists to maintain an interesting keyboard texture suggesting contrapuntal independence of parts. The dances are all in symmetrical binary forms, with each half repeated. Froberger relied to some extent on the concept of the variation suite found in earlier instrumental dance collections by German and Austrian composers, but thematic connections are obvious only between the allemandes and courantes. While the individual dances often resemble the music of Louis Couperin and Chambonnières, Froberger expanded significantly the expressive dimensions of the allemande, and in most of the suites it is the most elaborate piece. His allemandes are filled with interesting keyboard figures moving between the two hands and obscuring the foursquare metre found in French examples. The melodic writing is particularly expressive, and the pseudo-polyphonic texture is effective in maintaining rhythmic variety. The courantes, although some also emphasize interior part-writing, are generally more orientated towards the highest part. They are written in 3/2, 6/4 or 3/4 time. The degree of thematic relationship with the preceding allemande is normally limited to the opening bars. It is Froberger's sarabandes that resemble most closely the French models on which he based his style. They are regular in design, usually in 3/2 time, with greater use of trochaic rhythm at the expense of the iambic rhythm more characteristic of sarabandes. The melodies are conservative in style and usually confined within rather a narrow range, but they are not inexpressive, especially when, as often happens, they are supported by affective dissonances. The gigues contain conventional contrapuntal writing, with frequent fugal entries, which, however, soon lapse into a more homophonic texture, often in dotted rhythms. They are written in a variety of metres, with compound duple, simple triple or quadruple signatures. As a whole the suites afford striking evidence of the French influence on the style of mid-17th-century German keyboard music. Nevertheless, the German penchant for intensity of expression, lively textures and colourful harmonic tensions transforms the French models into suites that for almost a century provided much of the basis of further development of the form in Germany.

CHAPTER THREE

Achievement and influence

There is much in Froberger's music that is conventional. The contrapuntal forms especially tend to lack the free-flowing contrapuntal variety of Frescobaldi and the rich harmonic colouring, rhythmic variety and contrapuntal ingenuity of north German organ music. The harpsichord suites are graceful and often elegant, but not as weighty as the impressive French and German suites of about 1700. Why, then, does Froberger hold such a prominent place in music history? At least in part his importance results from a personal gift for stylistic synthesis exercised at a crucial moment in the development of German keyboard music. His music was not published during his lifetime, with the one exception already noted, and the three known autographs contain less than half of his output. In the 1690s, however, a series of publications of his music began to appear, and his music also circulated in numerous manuscript copies throughout Europe. It was popular because he had created keyboard works that were distinctly German at a time in the 17th century when a need and desire for national traits became more evident; and the major ingredients of their German style are the amalgamation and considerable transformation of Italian and French characteristics. The toccatas, for example, usually combine the Italian improvisatory practice with distinct contrapuntal sections in a rational arrangement that leads the way to the later north German form of the toccata and fugue. His fugal works are not far removed from the ultimate concept of fugues in the late Baroque period in Germany. The dance suites, though not in fact the models for 18th-century suites, as was once thought, are nevertheless among the earliest examples of a rational organization of French dances within a common tonality for keyboard performance. Whether or not he actually 'invented' this concept is irrelevant; he was widely admired for these suites – for their distinctive keyboard style and for their heightened variety and expressiveness of texture growing out of the *style brisé* – and they were studied and imitated by many composers until well into the 18th century. Many keyboard sources (see Riedel, p.98) show that his style was so closely imitated that when his dances are found combined in pasticcio fashion with other anonymous pieces it is difficult to distinguish their composers from him. Sometimes too his music provided models for parodies; his *Partita auff die Mayerin*, for example, seems to be the model for J. A. Reincken's work with the same title. His fame can be gauged by the frequency with which his music was held up for emulation, as in the already cited instance from Kircher. In the preface to his *Musicalische Vorstellung einiger biblischer Historien* (1700), Kuhnau singled out the 'famous Froberger and other excellent composers' for having created the programmatic keyboard piece, such as the *tombeau*. Mattheson (*Der vollkommene Capellmeister*) chose examples from two of his keyboard works to illustrate the improvisational keyboard style. Even Mozart copied his 'hexachord fantasia' for study (see Einstein's edition of Köchel's Mozart catalogue, Anh.109 VII–Anh.292).

In 1758, though, Adlung could call his music 'somewhat old-fashioned'. Yet this judgment, inevitable with the passing of time and the emergence of new styles, cannot gainsay the fact that he was recognized by many of his successors as a crucial forerunner of the rise of virtuoso keyboard music in the 18th century as well as a fine composer in his own right.

WORKS

Editions: *J. J. Froberger: Orgel- und Klavierwerke*, ed. G. Adler, DTÖ, viii, Jg.iv/1; xiii, Jg.vi/2; xxi, Jg.x/2 (1897–1903) [A]	158, 159, 162
J. J. Froberger: Oeuvres complètes pour clavecin, i: Livres de 1649, 1656 et 1658, ed. H. Schott, Le pupitre, lvii (Paris, 1979)	

Numbers in the right-hand margin denote references in the text.

KEYBOARD	156, 158–67
Diverse . . . curiose partite, di toccate, canzone, ricercate, alemande, correnti, sarabande e gique, hpd/org/insts (Mainz, 1693); A	
Divese curiose e rare partite musicali . . . prima continuatione (Mainz, 1696) [incl. 2 capriccios repr. from 1693]; A	163
10 suittes de clavessin . . . mis en meilleur ordre et corrigée d'un grand nombre de fautes (Amsterdam, *c*1697); A	155, 158, *160*, 162, 165, 166
Libro secondo di toccate, fantasie, canzone, allemande, courante, sarabande, gigue et altre partite, 1649, *A-Wn* [incl. Partita auff die Mayerin]; A	162–3, 165
Libro quarto di toccate, ricercari, capricci, allemande, gigue, courante, sarabande, 1656, *Wn*; A	
Libro di capricci e ricercate, *c*1658, *Wn*; A	

VOCAL	158, 165
Alleluia, absorpta est mors, 3vv, 2 vn, bc, *S-Uu*	
Apparuerunt apostolis, 3vv, 2 vn, bc, *Uu*	
See also DTÖ, viii, xiii, xxi (1897–1903) and Riedel	

BIBLIOGRAPHY

WaltherML

A. Kircher: *Musurgia universalis* (Rome, 1650/*R*1970)

N. Binninger: *Observationum et curationum medicinalium centurias quinque* (Montbéliard, 1673)

J. Mattheson: *Der vollkommene Capellmeister* (Hamburg, 1739/*R*1954)

——: *Grundlage einer Ehren-Pforte* (Hamburg, 1740); ed. M. Schneider (Berlin, 1910/*R*1969)

J. Adlung: *Anleitung zu der musikalischen Gelahrtheit* (Erfurt, 1758/*R*1953, 2/1783)

L. von Köchel: *Die kaiserliche Hof-Musikkapelle in Wien von 1543–1867* (Vienna, 1869)

E. Schebek: *Zwei Briefe über J. J. Froberger, kaiserliche Kammer-Organist in Wien* (Prague, 1874)

W. J. A. Jonckbloet and J. P. N. Land: *Musique et musiciens du XVII*[e] *siècle: correspondance et oeuvres musicales de Constantin Huygens* (Leiden, 1882)

F. Beier: *Über J. J. Frobergers Leben und Bedeutung für die Geschichte der Klaviersuite* (Leipzig, 1884)

O. Fleischer: 'Denis Gaultier', *VMw*, ii (1886), 110–80

G. Adler: 'Die Kaiser Ferdinand III., Leopold I., Joseph I., und Karl VI. als Tonsetzer und Förderer der Musik', *VMw*, viii (1892), 252

K. Krebs: 'J. J. Froberger in Paris', *VMw*, x (1894), 232

M. Seiffert: *Geschichte der Klaviermusik* (Leipzig, 1899)

A. Pirro: *Les clavecinistes* (Paris, 1925)

G. Ziegler: 'Ist Froberger in Halle geboren?', *ZMw*, viii (1925–6), 109

K. Seidler: *Untersuchungen über Biographie und Klavierstil Johann Jakob Froberger* (diss., U. of Königsberg, 1930)

G. Frotscher: *Geschichte des Orgel-Spiels und der Orgel-Komposition* (Berlin, 1935–6, enlarged 3/1966)

M. Reimann: *Untersuchungen zur Formgeschichte der französischen Klavier-Suite* (diss., U. of Regensburg, 1940)

F. W. Riedel: *Quellenkundliche Beiträge zur Geschichte der Musik für Tasteninstrumente in der zweiten Hälfte des 17. Jahrhunderts (vornehmlich in Deutschland)* (Kassel, 1960)

W. Apel: 'Neu aufgefundene Clavierwerke von Scheidemann, Tunder, Froberger, Reincken und Buxtehude', *AcM*, xxxiv (1962), 65

A. Somer: *The Keyboard Music of Johann Jakob Froberger* (diss., U. of Michigan, 1963)

W. Apel: *Geschichte der Orgel- und Klavermusik bis 1700* (Kassel, 1967; Eng. trans., rev. 1972)

U. Scharlau: 'Neue Quellenfunde zur Biographie Johann Jakob Froberger', *Mf*, xxii (1969), 47

G. Beechey: 'Johann Jakob Froberger (1616–1667)', *The Consort*, xxvii (1972), 32
D. Starke: *Frobergers Suitentänze* (Darmstadt, 1972)
H. Siedentopf: *Johann Jakob Froberger: Leben und Werk* (Stuttgart, 1977)
H. M. Schott: *A Critical Edition of the Works of J. J. Froberger with Commentary* (diss., U. of Oxford, 1978)

DIETRICH BUXTEHUDE

Kerala J. Snyder

CHAPTER ONE

Life

Dietrich Buxtehude was born in about 1637, possibly at Helsingborg, Denmark (now Hälsingborg, Sweden) or at Oldesloe (now Bad Oldesloe, Germany). He is best known as a composer of organ music, of which he was one of the most important composers before J. S. Bach. He also left an equally impressive body of sacred vocal music, composed during a formative stage in the development of the Protestant church cantata.

No documents exist to verify the date and place of Buxtehude's birth, and even his nationality has been disputed. The only contemporary information comes from a notice (in *Nova literaria Maris Balthici*) shortly after his death: 'he recognized Denmark as his native country, whence he came to our region; he lived about 70 years'. Although his family must originally have come from the town of Buxtehude, south-west of Hamburg, his ancestors had settled at Oldesloe in the Duchy of Holstein early in the 16th century. His father, Johannes (1601/2–74), migrated from Oldesloe to the Danish province of Scania; his presence there as organist of the Mariekirke in Helsingborg is documented for the year 1641. The hypothesis advanced by Pedersen and by Stahl (1951) that he could be identified with a German schoolmaster named Johannes, present in Oldesloe in 1638, and that Dietrich was therefore born in Oldesloe, now appears questionable in the light of a review of the archives there. The death notice does not exclude Oldesloe as a birthplace, however, since Holstein was under Danish control at the time. In 1641 or 1642 Johannes moved across the sound to Helsingør, Denmark, to become organist of the St Olai Kirke a position he held until his retirement in 1671. A son Peiter was born there to him and his wife, Helle Jespers Daater, in 1645; it is unknown whether Helle was also the mother of Dietrich. There were two daughters in the family, Anna and Cathrine, both presumably older than Dietrich.

Dietrich Buxtehude most likely attended the Latin school at Helsingør and received his music education from his father. In 1657 or 1658 he became organist at his father's former church at Helsingborg and in 1660 moved back to Helsingør as organist of the Marienkirche, a German-speaking congregation. With the death of Franz Tunder on 5 November 1667 the position of organist of the Marienkirche at Lübeck, one of the most important in north Germany, became vacant. After several other organists had applied for the post and been rejected, Buxtehude was chosen on 11 April 1668. At the same time he was appointed Werkmeister, a post encompassing the duties of secretary, treasurer and business manager of the church; it carried a separate salary but at this period was given to the organist. Buxtehude became a citizen of Lübeck on 23 July 1668, and a few days later, on 3 August 1668, he married Anna Margarethe Tunder, the younger daughter of his predecessor. It is not known whether this was a condition of his employment, as it was to be with his successor, but the practice was not unusual at the time. Seven daughters were born of this union, four of whom survived to adulthood: Magdalena (or Helena) Elisabeth (*b* 1670), Anna Margreta (*b* 1675), Anna Sophia (*b* 1678) and Dorothea Catrin (*b* 1683). Buxtehude's father joined him at Lübeck in 1673; his brother Peter (Peiter), a barber, followed in 1677.

Buxtehude's official duties required him to play for the main morning service and the afternoon service on Sundays and feast days and for Vespers on the preceding afternoon. In addition to the customary preludes to the congregational chorales and the musical offerings of the choir, Buxtehude supplied music during Communion, often with the participation of instrumentalists or vocalists, or both, who were paid by the church. Part of his fame, however, rested on an activity which was totally outside his official church duties: soon after arriving in Lübeck he reinstated the practice of giving concerts in the church, called 'Abendmusik', which Tunder had held on a weekday but which Buxtehude moved to 4 p.m., immediately after the afternoon service, on five Sundays a year – the last two in Trinity and the second, third and fourth in Advent. No music by him known to have been performed at these concerts has survived, though some of his extant vocal music may have been used.

There is little record of travel, but a recently discovered painting from 1674 (see fig.7) documents his close friendship with the Hamburg organist Jan Adam Reincken and suggests frequent visits to Hamburg, where he would also have known Christoph Bernhard and Matthias Weckmann. His friendship with Johann Theile is attested by a poem that he contributed to Theile's *St Matthew Passion* (Lübeck, 1673) and his help in financing the publication of Theile's masses

7. A group of musicians including Dietrich Buxtehude (centre right) and Jan Adam Reincken (centre left, at the harpsichord): painting (1674), 'Häusliche Musikszene', by Johannes Voorhout

(Wismar, 1673). The claim that Theile was Buxtehude's teacher (J. Mattheson, *Critica musica*, ii, 1725/*R*1964) must be discounted, in view of Buxtehude's greater skill in composition at that time. Poems by Buxtehude also appear in the *Harmonologia musica* (1702) of Andreas Werckmeister; it was Werckmeister who conveyed many of Buxtehude's organ compositions to J. G. Walther, whose copies are still extant. Buxtehude was also friendly with the Düben family in Stockholm. It is to the large manuscript collection of sacred music of the elder Gustaf Düben that is owed the preservation of most of his vocal music. It is also the main source for the music of Kaspar Förster, which Buxtehude seems to have known, either by a personal connection or through Düben.

Among the younger generation of organists, Nicolaus Bruhns was Buxtehude's pupil, and Pachelbel dedicated his *Hexachordum Apollonis* (1699) to him. Mattheson and Handel visited him in Lübeck on 17 August 1703; Mattheson was being considered as a successor to him, but at the mention of the condition relating to marriage described above he quickly lost interest. The documentary evidence for Bach's famous trip to Lübeck rests on the proceedings of the Arnstadt consistory of 21 February 1706, where it is noted that he 'has been to Lübeck in order to learn one thing and another about his art' and that he had requested a leave of four weeks but had stayed 'about four times as long'. Thus he was probably present at the 'extraordinary' Abendmusik performances of 2 and 3 December 1705, commemorating the death of the Emperor Leopold I and the accession of Joseph I; it is clear from Bach's obituary, however, that the purpose of his visit was to hear Buxtehude's organ playing.

Buxtehude died at Lübeck on 9 May 1707 and was buried on 16 May in the Marienkirche, beside his father and four daughters who had predeceased him. A successor agreeable to the 'marriage condition', J. C. Schieferdecker, had been serving as his assistant; he was appointed organist and Werkmeister on 23 June and married Anna Margreta Buxtehude on 5 September 1707.

CHAPTER TWO

Vocal works

Of Buxtehude's 128 vocal compositions that have survived complete, all except eight wedding pieces have sacred texts. They are often called cantatas, although strictly speaking most of them are not. During the 17th century this term was generally used only for secular vocal music; modern terminology has extended its use to cover sacred works made up of independent movements, but even in this sense it applies to only a minority of Buxtehude's works.

Buxtehude used three different types of text, either alone or in combination: (*a*) prose, mostly biblical but also including Latin devotional prose, (*b*) German chorales and (*c*) other strophic poetry. The systematic divisions of the vocal works by Blume, Sørensen and Geck are all ultimately based on this textual criterion, so it is not surprising that their results are rather similar, although they each give their categories different names. Since Buxtehude generally adhered to well-established German tradition in the style of the music to which he set each of these types of text, the textual categories translate easily into their corresponding musical categories: concertos, chorale settings and arias. In the following discussion the terminology is that developed by Krummacher and used by Geck, since it is the most musical; in the list of works, however, the more objective textual categories are retained.

I Concertos

Buxtehude drew all his German prose texts from the Luther Bible, with a strong preference for excerpts from psalms. In setting these biblical texts as vocal concertos, he followed the procedure, inherited from the motet, of dividing the text into short phrases and giving each a new musical motif closely tied to the words. The musical sections thus generated are often strongly contrasting, but they remain dependent parts of a larger whole and cannot usually be considered separate movements. The prevailing style is concertato, where voice or voices and instruments, or the voices alone with continuo, toss these word-bound musical motifs back and forth in a manner ultimately derived from the Venetian polychoral style. Sections of arioso are often introduced as well, and changes of metre provide additional contrast. Within this group of works, which are mostly for solo voice, there is a considerable range in the degree of sectionalization. At one extreme is BUXWV38, which is very homogeneous in its total organization over an ostinato bass, while a work such as BUXWV98 shows strong tendencies towards the cantata in its inclusion of a section of recitative and one which is quite aria-like in its preference for a purely musical organization over the usual word-bound motifs. Further tendencies towards the cantata can be seen in the inclusion of a separate introductory sonata at the beginning or a closed 'Alleluia' or 'Amen' section at the end of many, or both.

The concerto is the only category where Latin texts predominate over German. The majority are from the Vulgate, with an even stronger preference for psalms, including settings of two complete Sunday vesper psalms with doxology (BUXWV17 and 69). In addition there are five non-biblical texts (BUXWV11, 82–3, 92, 94) in a highly subjective, sometimes mystical, devotional prose, which was a popular element of both Catholic and Protestant piety in the 17th century. Many of these Latin concertos are similar to their German counterparts; however, there is a somewhat greater freedom from word-bound motifs, particularly in the non-biblical works.

II Chorale settings

Although in purely poetic terms a chorale text is identical to a strophic poem, an important musical difference is that the chorale text is identified with a specific melody, and with only a few exceptions Buxtehude used this as a cantus firmus. Three different compositional styles can be seen in his chorale settings. The simplest and the most frequent is a purely homophonic setting with the melody in the soprano, known as cantional style after the hymnals set in this way. Buxtehude always added instruments when he used this style, both to accompany the voices and to provide interludes between phrases. The most traditional method of chorale setting in figural music, yet used relatively seldom by Buxtehude, is the concertato method derived from the 16th-century chorale motet. This is similar to a concerto on a biblical text, except that the motif for each section is derived from the chorale melody and not just the text. Finally, the influence of his organ chorale settings can be seen in a few vocal works (e.g. BUXWV32 and 42), where the cantus firmus appears in only one voice, while the instrumental parts weave counterpoint around it.

With the exception of BUXWV3, which may be a fragment, Buxtehude always set several verses of the chorale, and, as in the concertos, there is a considerable variation in the degree of homogeneity. The majority are highly unified, with all stanzas set in cantional style, either purely strophically or with slight variations in harmony or instrumental participation. A few, however (e.g. BUXWV21, 41, 78, 100), show strong tendencies towards the chorale cantata in differentiating the separate verses by scoring and cantus firmus treatment. *Jesu meine Freude* (BUXWV60) could in fact be considered a pure chorale cantata, with three of its stanzas set as arias which are only distantly related to the cantus firmus.

III Arias

The most important category of text in Buxtehude's vocal music is the strophic poem. With the exception of a few medieval Latin hymns, which are used without cantus firmus (BUXWV68, 89, 91, 99), some Latin poems of more recent origin (BUXWV6, 28, 93), three excerpts from the popular *Jesu dulcis memoria* (BUXWV56–7 and 88) and one Swedish poem (BUXWV8), they are all 17th-century German poems, most of them products of the early years of the Pietistic movement. Many of these texts are in fact drawn from contemporary sacred songbooks published between 1659 and 1686. Although Buxtehude did not use the accompanying music, these simple strophic songs for voice and continuo exerted a powerful influence on his aria style.

The formal range in Buxtehude's treatment of strophic texts is even greater than that seen in the concertos and chorale settings. At one end of the scale is simple strophic form, which is found in all of the wedding arias

8. Autograph MS of the beginning of Buxtehude's chorale setting 'Nimm von uns, Herr' (BUXWV78) *in German organ tablature*

and in some of the sacred arias as well. Most of his arias, however, are in varied strophic form, providing changes in the music from strophe to strophe while still maintaining an overall unity and highlighting the strophic nature of the text with a ritornello or sometimes a vocal refrain. Although the scoring may change, the metre is constant throughout, and one or more strophes of music recur as the piece progresses. Finally, there are settings of strophic texts where the melody is completely through-composed. Some are highly unified by means of an ostinato bass (BUXWV57, 62, 70) or a ritornello or simply by a homogeneous style. A few show concerto-like sectional contrast which does not always correspond to the strophic structure (e.g. BUXWV87 and 101), and a small group could be considered aria cantatas (e.g. BUXWV6, 63, 93, 108). Here the first strophe is a closed section for all the voices in concertato style, with the succeeding strophes set as an aria for solo voices, unified by a ritornello or a refrain. A closing concertato movement uses the final strophe or an appended 'Amen' or 'Alleluia', or both.

The melodies of Buxtehude's arias are often very similar in style to the arias that are found in the song-books: they are strongly periodic in their phrase structure and flow gently through a declamation of the text that moves in a combination of quavers and paired semiquavers, often in common time. A more melismatic and Italianate aria style, often in triple metre, is also found, but Buxtehude's arias are more akin to those of Monteverdi and Carissimi than to those of his own Italian contemporaries, whose music was moving inexorably towards total domination of the text in the da capo aria. There is not the slightest tendency towards the da capo aria in Buxtehude's surviving works. His strong preference for strophic form has been emphasized, and his use of the strophic-bass aria and other ostinato forms is another example of his kinship with Monteverdi. While it is quite possible that Kaspar Förster was the intermediary in this stylistic relationship, as Sørensen (1963) has suggested, one cannot be certain of this in view of the large amount of Italian music available in Germany (including Lübeck) at that time.

IV Cantatas

While there are tendencies towards independent movements that can be seen in each of the three preceding categories, Buxtehude's cantata form emerged most clearly when he combined two or more of these textual–musical types in the same piece of music. His commonest practice, accounting for 20 cantatas, was to combine in the music, concerto and aria and, in the text, biblical words and modern, subjective poetic commentary on them. Each cantata begins with a concerto movement, usually preceded by an instrumental sonata; beyond this there is considerable formal variety. The aria, however, can always be perceived as the core of the cantata and is quite highly unified in either pure or varied strophic form. Concertato style almost always returns as a framing element at the end, usually by means of a simple repetition of the opening movement, sometimes with a movement on a different biblical text or an 'Amen' or 'Alleluia'. Sometimes the concertato writing appears, in the manner of a rondo, between the strophes of the aria. All Buxtehude's cantatas have German texts except for *Membra Jesu* (BUXWV75), a cycle of seven concerto-aria cantatas dedicated to Gustaf Düben in 1680. Only in isolated instances did Buxtehude combine chorale and aria (BUXWV86) or concerto and chorale (BUXWV29) to form a cantata, but there are five examples of the 'older mixed cantata', which combines all three elements (BUXWV4, 34, 43, 51, 112). His method of building cantatas by drawing together these previously diverse elements was shared by many of his contemporaries, providing the foundation for the 'newer mixed cantata' of the 18th century, with its addition of recitative set to madrigalesque poetry.

V Miscellaneous

A few pieces by Buxtehude do not fit well into any of the above categories. The music for his father's funeral on 29 January 1674 (BUXWV76) might be called a chorale-aria cantata, but it is more likely that its two parts were performed separately and that the first part, a chorale setting, was performed on the organ. The chorale setting was in fact composed earlier (in 1671) for another funeral, and in its extremely learned contrapuntal style it is unlike any other of Buxtehude's chorale settings; it is modelled on a similar work by Christoph Bernhard (see Snyder, 1980). The aria, to a text undoubtedly written by Buxtehude himself, is in simple strophic form with string accompaniment but is also more contrapuntal than any of his other arias. The *Missa alla brevis* (BUXWV114), if authentic, is Buxtehude's only surviving work in the *stile antico*; its manuscript can be dated *c*1675. These works in learned counterpoint, including the two canons (BUXWV123–4) entered in autograph books, reflect Buxtehude's friendship with Theile, Reincken and Bernhard in the early

1670s. Finally, *Benedicam Dominum* (BUXWV113), scored for six choirs – two vocal and four instrumental – is his one extant contribution to the genre of the 'colossal' Baroque style.

VI Abendmusik

From the oratorios that Buxtehude presented at his numerous Abendmusiken, three librettos are the only sure survivals: *Die Hochzeit des Lamms*, a two-part oratorio from 1678, and the two 'extraordinary' presentations of 1705, *Castrum doloris* and *Templum honoris*. The libretto for 1700, consisting of four programmes of shorter selections and a repeat of the music from the preceding New Year, is summarized in the literature but was lost in World War II. The surviving librettos indicate that the oratorio-like Abendmusiken consisted of a mixture of choruses, recitatives, strophic arias and chorale settings, with considerable instrumental participation. The mixed programmes of 1700 were made up of arias, chorale settings and concertos; unfortunately, none of Buxtehude's extant music is set to those specific texts.

The titles or themes from some other years are also known: in 1688 the theme was the prodigal son, and the catalogues of the Frankfurt and Leipzig fairs for the spring of 1684 listed the future publication of two works by Buxtehude described as Abendmusiken, *Himmlische Seelenlust auf Erden über die Menschwerdung und Geburt unsers Heylandes Jesu Christi* and *Das Allerschröcklichste und Allererfreulichste, nemlich, Ende der Zeit und Anfang der Ewigkeit*. Whether these were to be librettos or music and whether they were in fact published is unclear. Maxton (1927–8) claimed to have found *Das Allerschröcklichste* in Uppsala in a set of anonymous parts for an oratorio with an eschatological theme and published an abridged arrangement of it under the title *Das jüngste Gericht* (BUXWV suppl. 3). Its authenticity as a work of Buxtehude has, however, been a subject of continued discussion (see Geck, 1961 and 1963; Maxton, 1962; Krummacher, 1978; Karstädt, 1979; Snyder, 1981 and Ruhle) which remains unresolved, although there is now general agreement that this work is not the lost *Allerschröcklichste*. Ruhle's dissertation (1982) includes a complete libretto and an edition of those movements omitted by Maxton.

CHAPTER THREE

Instrumental works

I Organ music

The music Buxtehude composed for the organ must be approached through the characteristic sound of the north German organ, with its contrasting tonal colours and its independent and strongly voiced pedal. The 52-stop organ which Buxtehude played in Lübeck had 15 stops in the pedal, more than in any of its three manual divisions; this included two 32′ stops and a full complement of principals, mixtures and reeds. Thus it is not surprising to find that in his organ music the pedal goes far beyond what had been its traditional role – as slow harmonic support or bearer of the cantus firmus – to participate fully in the fabric of the music, including its share of virtuoso display. Another characteristic of the north German organ was its *Brustwerk* and *Rückpositiv* divisions, featuring solo reeds and many upper partials which could be combined to produce a sharply differentiated melodic line. This type of sound is particularly well suited to a solo voice, such as a highly ornamented cantus firmus, with the other voices played on another manual, using a contrasting registration. If there are three accompanying voices the pedal becomes almost mandatory for the bass line, because, while four voices fall easily between the two hands on one manual, it is much more difficult to play three voices with the left hand alone. This tonal contrast between divisions naturally lent itself also to echo effects and to works with strong sectional contrasts.

Buxtehude's organ music is almost evenly divided between freely composed pieces and those based on a chorale cantus firmus. Apart from three ostinato pieces, all the free works contain fugues and most of them preludes as well, but the usual designation 'prelude and fugue' is found in none of the original sources and would apply correctly to only five works, those with only one fugue (BUXWV138, 144–5, 147, 157). His more typical procedure in the praeludia is to alternate in a highly dramatic manner toccata-like free sections with two or three fugues, the subjects of which are related in the manner of the variation canzonas of Frescobaldi and Froberger. The free sections are improvisatory in character, concentrating on virtuoso passage-work for both hands and feet, dramatic gestures with bold harmonic progressions, and an occasional fugato to contrast with a prevailing texture that is basically chordal, albeit highly decorated. The opening section is usually the most developed, while the remaining free sections often take on a transitional or cadential character. Krummacher (1980) has demonstrated that these free sections represent excellent examples of the *stylus phantasticus* discussed by Kircher and Mattheson.

The fugues show a firm handling of contrapuntal technique, including the frequent appearance of double counterpoint and stretto. The subjects are decidedly instrumental in character: repeated notes, wide leaps and rests occur frequently. Each fugue consists of a series of expositions, usually confined to entrances in the tonic and dominant, with tonal answers predominating. There is very little episodic material or real modulation, these functions being fulfilled by the free sections between the fugues. The majority of these works call for obbligato pedal; this is most obvious in the virtuoso passages of the toccata sections but is equally important to the fugues. Buxtehude's fugues can indeed be considered the first body of such works to be really idiomatically conceived for the organ; the subjects are all suitable for the pedals, and the addition of a pedal part permits the extension and thickening of the contrapuntal texture, often making the works unplayable on manuals alone. Buxtehude's canzonas, on the other hand, all lack the participation of the pedal as well as the free sections. In their fugal technique they are similar to the fugues in the praeludia, if somewhat thinner in texture.

The three ostinato pieces (BUXWV159–61) are among Buxtehude's best-known works and exerted their influence on Brahms as well as on Bach. Here he took a form which was popular in Italy and south Germany but not in north Germany and made it into truly idiomatic organ music.

The organ chorale settings fall into three groups, each showing a distinctive approach to the chorale. Sets of chorale variations had been cultivated extensively by Sweelinck and Scheidt but do not figure very prominently in Buxtehude's oeuvre. Consisting of only three or four verses, they are often restricted to the manuals alone and sometimes to only two voices (as in the traditional bicinium) and the cantus firmus frequently appears unornamented (e.g. BUXWV213). In terms of variety and keyboard technique they do not match the variations of Pachelbel and Böhm. The variations on *Auf meinen lieben Gott* (BUXWV179) form an exception; as a dance suite on a chorale tune, however, they were more likely intended for performance on the harpsichord.

The chorale fantasia was deeply rooted in the north German improvisatory tradition, and Buxtehude's pieces in this form are his only virtuoso organ music based on the chorale. They are analogous in compositional method to vocal concertos: each phrase of the chorale is developed separately and extensively to form a highly sectionalized piece full of dramatic contrast (e.g. BUXWV210 and 223). The four 'liturgical' pieces (BUXWV203–5 and 218) belong to this group as well.

The majority of Buxtehude's surviving chorale settings are chorale preludes, highly unified settings of one stanza of the chorale (e.g. BUXWV184 and 219). Most of them are very similar in outward appearance: clearly intended for two manuals and pedal, the cantus firmus is set apart in the upper voice in a richly ornamented version, accompanied by three parts which are contrapuntal but not necessarily imitative. These short pieces, which Buxtehude probably used at Lübeck as an introduction to the congregational singing, are the most lyrical, subjective and devotional of his organ pieces; in this field they correspond quite closely to the arias in his vocal output.

II Harpsichord or clavichord music

Both Walther and Mattheson bemoaned the fact that Buxtehude had never published any of his keyboard music. Mattheson (1739, p.130) specifically mentioned seven keyboard suites depicting the nature of the planets. These have never come to light, but there is one manuscript tablature (Kongelige Bibliotek, Copenhagen, MS mu 6806, 1399) which contains 19 suites and six sets of secular variations ascribed to Buxtehude. The suites are nearly all in the standard order allemande–courante–sarabande–gigue, with an occasional *double*. The courantes always begin as variations of the allemandes, as do some of the sarabandes; the gigues go their own way, often in a loose fugal style that is not nearly as structured as that of the numerous fugues in gigue rhythm found in the praeludia and canzonas. The arpeggiation and continuous motion of the *style brisé* cultivated by the French harpsichord school is apparent throughout. The suites are more conventional than most of Buxtehude's organ music and do not match the individualized expression attained by Froberger. The secular variations, on the other hand – particularly the set of 32 on *La Capricciosa* (BUXWV250) – show a much greater interest in the variation process than can be seen in the organ music.

III Chamber music

Buxtehude's only major publications during his lifetime were collections of chamber music. One collection of sonatas for two or three violins, viola da gamba and continuo announced for publication in 1684 is either lost or never appeared; it is listed in the same catalogue as the two Abendmusiken that have never been found. Two extant prints, from ?1694 and 1696, each contain seven sonatas for violin, viola da gamba and harpsichord continuo, a scoring found in Germany, Austria and England but noticeably absent in Italy. These are sonatas *a due*, based on virtuoso and integrated writing for the violin and gamba. Although real trio textures sometimes occur, the continuo line is more often a simplification of the gamba part. The structure of the sonatas is based on an alternation of tempo and texture, but this can take place either by means of tonally closed, independent movements or with sections which flow together; there is no standardization of their number, which ranges from three to 13. Half the sonatas have at least one ostinato movement, with the pattern remaining unvaried in the continuo bass; the procedure is similar to that found in the passacaglia for organ (BUXWV161). The gamba part consists sometimes of divisions on this bass, sometimes of an independent part above it. The contrapuntal movements are fugal in style but are usually in only two real parts; as a result there is much more episodic writing than in the organ fugues. The continuo bass is more likely to be independent in the slow, homophonic sections, many of which are transitional in nature; in their harmonic intensity these sections are often reminiscent of the transitions in his organ praeludia and of Rosenmüller's sonatas of 1682.

The Sonatas preserved in manuscript all appear to be earlier than those of opp.1 and 2. BUXWV266, 269 and 271, scored for two violins, viola da gamba and continuo, may have belonged to the 1684 collection. Buxtehude's reworking of BUXWV273 (from the 1680s) as op.1 no.4 shows that he had experimented with the sonata-suite combination cultivated by Reincken, Becker and Erlebach but abandoned it in favour of the sonata alone, perhaps because of the greater opportunity it offered for dramatic contrast. Linfield has demonstrated that Buxtehude's sonatas represent the *stylus phantasticus* to an even greater degree than his organ praeludia; they mark the culmination of 17th-century style and have little in common with the more balanced and predictable sonatas of Corelli.

CHAPTER FOUR

Sources, chronology, literature

Since Buxtehude published no keyboard music and only a few occasional vocal works, the survival of the bulk of his works has depended on manuscript transmission, and it must be assumed that a considerable amount has been lost. The two principal sources for his vocal music were both compiled during his lifetime and with his knowledge. Gustaf Düben's collection, which his son gave to Uppsala University Library in 1732, contains 99 vocal works by Buxtehude in manuscript. These include five autographs in German organ tablature (BUXWV75, 78–9, 85, 88), one in score (BUXWV31) and numerous other manuscripts that appear to have been copied by Düben and his assistants from loaned autographs. The other important source is the Lübeck tablature A 373 (now in the Deutsche Staatsbibliothek, East Berlin), a large volume comprising 20 vocal works and one fragment. The first nine pieces contain autograph insertions, and the source appears to have been prepared under Buxtehude's direction, perhaps at the end of his life. There is a remarkable absence of his vocal music from all the important central German manuscript collections and inventories of the period.

The situation is completely reversed with regard to the organ music. Here the sources are widespread, many of them being of central German provenance, yet there is not a single manuscript that can be closely identified with Buxtehude himself. He undoubtedly wrote his organ music in tablature, but most of these manuscripts are in score; there is only one group of north German sources in tablature (in the Wenster Collection, Universitetsbiblioteket, Lund), dated 1713–14. Owing to the scattered nature of the sources, the variants between concordances are much greater than is the case with the vocal music.

The most striking aspect in all the organ manuscripts is the selectivity with which they were compiled. Walther was interested only in Buxtehude's chorale settings, and most of these works owe their preservation to seven separate manuscripts in his hand. The rest of the larger manuscripts, however, show a decided preference for free organ works, especially praeludia. The oldest of these is probably the 'codex E.B. 1688' at Yale University (MS LM 5056), which may date from as early as the year 1683. A different repertory of praeludia survives in a family of manuscripts circulated among pupils of Bach; one of these (in the Conservatoire Royal de Musique, Brussels, MS U 26659) was copied by J. F. Agricola and another belonged to Kirnberger. These manuscripts contain only pieces with obbligato pedal and were largely extracted from an earlier collection which also contained preludes and canzonas for manuals alone; one manuscript (Staatsbibliothek Preussischer Kulturbesitz, West Berlin, MS Mus 2681) survives from this more extensive collection, and it may have belonged to C. P. E. Bach. A final example of the selective manuscript compiler is seen in Johann Christoph Bach (1671–1721), J. S. Bach's elder brother. Among his unique copies of pieces by Buxtehude (in the Musikbibliothek der Stadt, Leipzig, MS III.8.4 and

Staatsbibliothek Preussischer Kulturbesitz, MS Mus 40644) are the three ostinato pieces and the famous Praeludium in C with its chaconne (BUXWV137).

There is as yet no complete chronology of Buxtehude's works. Research on the manuscripts of the Düben collection by Grusnick, Rudén and Snyder has established dates for the copying of many of his vocal works, giving a *terminus ante quem* which may be quite close to the composition date. Most of them came into the collection during 1680–87. Those copied before 1680 show a greater preference for concertos and arias with Latin texts; most of the cantatas appear after 1680. Very little vocal music survives from the 20 years of Buxtehude's life after Düben ceased collecting in 1687. Although the Lübeck tablature A 373 may have been copied later, not all of its repertory is more recent, for there are a number of concordances with the Düben collection. This manuscript seems, moreover, to have been compiled with the intention of providing a representative selection of Buxtehude's music; there is a higher proportion of cantatas here, however, including three of the five 'older mixed cantatas'. A chronology for the organ music is much more difficult, since most of the existing manuscripts were copied after Buxtehude's death. Archbold has proposed a chronology for the *pedaliter* praeludia (BUXWV136–58), based mainly on stylistic characteristics, whereby BUXWV152 and 158 would be the earliest praeludia, BUXWV140, 141, 142, 146, 149 and 155 the latest. Linfield has demonstrated that Buxtehude's composition of chamber music extended from the early 1660s (BUXWV270) to his op.2 of 1696.

It is clear from the state of the sources that Buxtehude

was regarded in the 18th century as a composer of keyboard music, and the scanty biographical notices reinforce this picture. Buxtehude scholarship began with Spitta's Bach biography in 1873, and he too was interested primarily in the organ music, with a decided preference for the free works over the chorale settings. He also published the first comprehensive edition of Buxtehude's organ music (1875–6), and Seiffert later revised and enlarged it. The only vocal music known to Spitta was the Lübeck tablature, but following Stiehl's discovery in 1889 of the works by Buxtehude at Uppsala there was a definite shift of interest to the vocal music, which is evident in Pirro's monograph of 1913. Publication of a complete edition of Buxtehude's works was begun in Germany in 1925 but reached only the eighth volume of vocal music. A new international edition, the *Collected Works* (1985–), will complete the publication of the vocal works and include the keyboard and chamber music as well.

Blume published the first systematic survey of Buxtehude's vocal music in 1940, suggesting that it would not have been performed within the context of a regular church service in Lübeck and was probably composed specifically for Düben. This thesis has been disputed in more recent studies on the vocal music. Kilian's dissertation (1956) concentrated on a study of the sources, and he offered evidence that some works have a clear *de tempore* character and that others were designated to be performed during Communion. Geck pushed this argument further in 1965. He presented as evidence a libretto for the Christmas season services at the Marienkirche in 1682; while Buxtehude's name is not mentioned in it, the types of text and scoring are very similar to those found in his extant vocal music. Blume's thesis was partly based on the rather high number of autographs supposed to be in the Düben collection, but Grusnick's study of the entire collection has disproved this, and Karstädt's discussion of the Lübeck tablature (1971) has confirmed that these pieces were performed at Lübeck. Some of the church account books kept by Buxtehude have recently become available again, and they too confirm Buxtehude's performances of vocal music in the church at times other than the Abendmusiken. Sørensen's dissertation (1958) did not address the liturgical issue but concentrated on formal analysis, arriving at the conclusion that Buxtehude's vocal music represented more the culmination of the preceding era than the beginning of the 18th-century cantata. This was followed by a revaluation of Buxtehude's historical importance by Krummacher (1966–7), who concluded that it lay in his organ music and not in his vocal music. A concurrent reawakening of interest in the keyboard music is evident in the work of Riedel, Pauly, Mortensen, Beckmann (see his editions of 1972 and 1980), Krummacher and Archbold. The influence of Buxtehude's organ music on Bach is undeniable, and it was in this context that Spitta first turned his attention to him. It appears that his vocal music, though of equal quality, did not exert such an important influence. Nevertheless, the continued interest which Buxtehude's vocal and instrumental music have evoked since Spitta's time demonstrates that they are both eminently worthy of performance and study on their own merits.

WORKS

Editions: *D. Buxtehude: Werke für Orgel*, ed. P. Spitta and M. Seiffert (Leipzig, 1903–4); suppl. ed. M. Seiffert (Leipzig, 1939) [SS] — 198
D. Buxtehude: Abendmusiken en und Kirchenkantaten, ed. M. Seiffert, DDT, xiv (1903, rev. 2/1957 by H. J. Moser) [S]
D. Buxtehude: Sonaten für Violine, Gambe und Cembalo, ed. C. Stiehl, DDT, xi (1903, rev. 2/1957 by H. J. Moser) [St]
D. Buxtehude: Werke, ed. W. Gurlitt and others, i–viii (Klecken and Hamburg, 1925–58/*R*1977) [G] [continued as Sn, see below] — 198
D. Buxtehude: Klavervaerker, ed. E. Bangert (Copenhagen, 1942) [Ba]
D. Buxtehude: Orgelwerke, ed. J. Hedar (Copenhagen and London, 1952) [H]
D. Buxtehude: Sämtliche Orgelwerke, ed. K. Beckmann (Wiesbaden, 1972) [B]
D. Buxtehude: Sämtliche Suiten und Variationen für Klavier/Cembalo, ed. K. Beckmann (Wiesbaden, 1980) [BK]
D. Buxtehude: the Collected Works, ed. K. Snyder and others, ix– (New York, 1986) [Sn] [continuation of G] — 198

Catalogue: *Thematisch-systematisches Verzeichnis der musikalischen Werke von Dietrich Buxtehude: Buxtehude-Werke-Verzeichnis*, ed. G. Karstädt (Wiesbaden, 1974) [BUXWV]

* = authenticity questionable; see also Geck (1961, 1963) and Blankenburg (1968)

Numbers in the right-hand column denote references in the text.

SACRED VOCAL

179–88, 195, 197, 198, 199

BuxWV	*Title*	*Text*	*Scoring*	*Source, edn., remarks*	
1*	Accedite gentes, accurite populi	Bible	SSATB, 2 vn, bc	*S-Uu*; ed. S. Sørenson, *Fire latinske kantater* (Copenhagen, 1957), 37	
2	Afferte Domino gloriam honorem	Bible	SSB, bc	G v, 10	
3	All solch dein Güt wir preisen	chorale	SSATB, 5 str, bc	*Uu*; ed. B. Grusnick (Kassel, 1956)	182
4	Alles, was ihr tut	Bible, poetry, chorale	SATB, 5 str, bc	S, 39; Sn ix, 3	186
5	Also hat Gott die Welt geliebet	Bible	S, 2 vn, va da gamba, bc	G i, 10	
6	An filius non est Dei	poetry	ATB, 3 va da gamba/trbn, bc	G vii, 49	182, 184
7	Aperite mihi portas justitiae	Bible	ATB, 2 vn, bc	G vii, 62	
8	Att du, Jesu, will mig hôra	poetry	S, 2 vn, bc	*L*; ed. J. Hedar (Copenhagen, 1944)	182
9	Bedenke, Mensch, das Ende	poetry	SSB, 3 vn, vle, bc	G v, 14	
10	Befiehl dem Engel, dass er komm	chorale	SATB, 2 vn, vle, bc	G viii, 73	
11	Canite Jesu nostro	prose	SSB, 2 vn, vle, bc	G v, 21	
12	Cantate Domino canticum novum	Bible	SSB, bc	G v, 29	181
13	Das neugeborne Kindelein	poetry	SATB, 3 vn, vle, bc	G viii, 121	
14	Dein edles Herz, der Liebe Thron	poetry	SATB, 5 str, bc	Sn ix, 35	
15	Der Herr ist mit mir	Bible	SATB, 2 vn, vle, bc	G viii, 85	
16	Dies ist der Tag	Bible	—	*D-Bds*, frag.; ed. in Pirro, p.437	
17	Dixit Dominus Domino meo	Bible	S/T, 2 vn, 2 va, bc	G ii, 27	181

18	Domine, salvum fac regem	Bible	SSATB, 5 str, bc	*S-Uu*; ed. S. Sørensen, *Fire latinske kantater* (Copenhagen, 1957), 51	
19	Drei schöne Dinge sind	Bible, poetry	SB, 2 vn, vle, bc	G iii, 10	
20	Du Frieden-Fürst, Herr Jesu Christ	chorale	SSATB, 2 vn, vle, bc	*Uu*; ed. B. Grusnick (Kassel, 1957)	
21	Du Frieden-Fürst, Herr Jesu Christ	chorale	SSB, 4 str, bn, bc	G v, 35	182
22	Du Lebensfürst, Herr Jesu Christ	poetry	SATB, 5 str, bc	Sn ix, 61	
23	Ecce nunc benedicite Domino	Bible	ATTB, 2 vn, bc	G viii, 105	
24	Eins bitte ich vom Herrn	Bible, poetry	SSATB, 5 str, bc	S, 15	
25	Entreisst euch, meine Sinnen	poetry	S, 2 vn, bc	G i, 15	
26	Erfreue dich, Erde!	poetry	SSAB, 2 tpt, 3 str, timp, bc	(parody of BUXWV122), *Uu*; ed. D. Kilian, *37 Kantaten von Dietrich Buxtehude*, xxvi (Berlin, 1958)	
27	Erhalt uns, Herr, bei deinem Wort	chorale	SATB, 2 vn, vle/bn, bc	G viii, 47	
28	Fallax mundus, ornat vultus	poetry	S, 2 vn, bc	G i, 17	182
29	Frohlocket mit Händen	Bible, chorale	SSATB, 2 tpt, 5 str, bc	*Uu*; ed. S. Sørensen (Copenhagen, 1972)	186
30	Fürchtet euch nicht	Bible, poetry	SB, 2 vn, bc	G iii, 18	
31	Fürwahr, er trug unsere Krankheit	Bible	SSATB, 5 str, bc	*Uu* (autograph); ed. B. Grusnick (Kassel, 1937)	195
32	Gen Himmel zu dem Vater mein	chorale	S, vn, va da gamba, bc	G i, 23	181
33	Gott fähret auf mit Jauchzen	Bible, poetry	SSB, 4 str, 6 brass, bn, bc	G v, 44	
34	Gott hilf mir	Bible, poetry, chorale	SSATB, 5 str, bc	S, 57	186
35	Herr, auf dich traue ich	Bible, poetry	S, 2 vn, bc	G i, 29	
36	Herr, ich lasse dich nicht	Bible	TB, 5 str, bc	G iii, 21	
37	Herr, nun lässt du deinen Diener	Bible	T, 2 vn, bc	G ii, 39	
38	Herr, wenn ich nur dich hab'	Bible	S, 2 vn, bc	G i, 35	180
39	Herr, wenn ich nur dich habe	Bible, poetry	S, 2 vn, vle, bc	G i, 38	
40	Herren vår Gud	chorale	SATB, 2 vn, vle, bc	G viii, 64	
41	Herzlich lieb, hab ich dich o Herr	chorale	SSATB, 5 str, bc	*D-Bds*; ed. B. Grusnick (Kassel, 1956)	182
42	Herzlich tut mich verlangen	chorale	S, 2 vn, bc	*S-L*; ed. J. Hedar (Copenhagen, 1943)	181
43*	Heut triumphieret Gottes Sohn	poetry, chorale	SSATB, 5 str, 2 tpt, bc	*D-B*; ed. T. Fedtke (Kassel, 1957), Alleluia in S, 167	186

BuxWV	Title	Text	Scoring	Source, edn., remarks	
44	Ich bin die Auferstehung	Bible	B, 4 str, 4 brass, bn, bc	G ii, 60	
45	Ich bin eine Blume zu Saron	Bible	B, 2 vn, vle, bc	G ii, 66	
46	Ich habe Lust abzuscheiden	Bible, poetry	SSB, 2 vn, vle/bn, bc	G v, 56	
47	Ich habe Lust abzuscheiden	Bible, poetry	SSB, 2 vn, vle, bc	G v, 62	
48	Ich halte es dafür	Bible, poetry	SB, vn, va, vle, bc	G iii, 30	
49	Ich sprach in meinem Herzen	Bible	S, 3 vn, bn, bc	G i, 47	
50	Ich suchte des Nachts	Bible, poetry	TB, 2 vn, 2 ob, vle, bc	G iii, 41	
51	Ihr lieben Christen, freut euch nun	Bible, poetry, chorale	SSATB, 6 str, 3 cornetts, 5 brass, bc	S, 107	186
52	In dulci jubilo	chorale	SSB, 2 vn, bc	G v, 69	
53	In te, Domine, speravi	Bible	SAB, bc	G vii, 8	
54	Ist es recht	Bible, poetry	SSATB, 5 str, bc	*S-Uu*; ed. B. Grusnick (Kassel, 1959)	
55	Je höher du bist	Bible, poetry	SSB, 2 vn, vle, bc	G v, 76	
56	Jesu dulcis memoria	poetry	SS, 2 vn, vle, bc	G iii, 51	182
57	Jesu dulcis memoria	poetry	ATB, 2 vn, bc	G vii, 72	182, 184
58	Jesu komm mein Trost und Lachen	poetry	ATB, 4 str, bc	G vii, 81	
59	Jesu meine Freud und Lust	poetry	A, 4 str, bc	G ii, 10	
60	Jesu meine Freude	chorale	SSB, 2 vn, bn, bc	G v, 87	182
61	Jesu, meiner Freuden Meister	poetry	SAB, 3 va, vle, bc	(Ratzeburg, 1677); ed. S. Sørensen (Copenhagen, 1977)	
62	Jesu, meines Lebens Leben	poetry	SATB, 5 str, bc	Sn ix, 91; early version, Sn ix, 247	184
63	Jesulein, du Tausendschön	poetry	ATB, 2 vn, vle/bn, bc	G vii, 89	184
64	Jubilate Domino, omnis terra	Bible	A, va da gamba, bc	G ii, 19	
65	Klinget mit Freuden	poetry	SSB, 3 str, 2 tpt, bc	(parody of BUXWV119); G v, 96	
66	Kommst du, Licht der Heiden	poetry	SSB, 5 str, bc	G vi, 14	
67	Lauda anima mea Dominum	Bible	S, 2 vn, vle, bc	G i, 57	
68	Lauda Sion Salvatorem	poetry	SSB, 2 vn, [va da gamba], bc	G vi, 24	182
69	Laudate pueri Dominum	Bible	SS, 5 va da gamba, vle, bc	G iii, 59	181
70	Liebster, meine Seele saget	poetry	SS, 2 vn, bc	G iii, 65	184
71	Lobe den Herrn, meine Seele	Bible	T, 3 vn, 2 va, bc	G ii, 44	
72	Mein Gemüt erfreuet sich	poetry	SAB, 4 vn, 2 fl, 9 brass, 3 bn, bc	G vii, 10	

73	Mein Herz ist bereit	Bible	B, 3 vn, vle, bc	G ii, 74	
74	Meine Seele, willtu ruhn	poetry	SSB, 2 vn, vle, bc	G vi, 30	
75	Membra Jesu	Bible, poetry		*Uu* (autograph, dated 1680); ed. B. Grusnick (Kassel, 1963)	185–6, 195
	1. Ecce super montes		SSATB, 2 vn, vle, bc		
	2. Ad ubera portabimini		SSATB, 2 vn, vle, bc		
	3. Quid sunt plagae istae		SSATB, 2 vn, vle, bc		
	4. Surge, amica mea		SSATB, 2 vn, vle, bc		
	5. Sicut modo geniti infantes		ATB, 2 vn, vle, bc		
	6. Vulnerasti cor meum		SSB, 5 va da gamba, bc		
	7. Illustra faciem tuam		SSATB, 2 vn, vle, bc		
76	Fried- und Freudenreiche Hinfarth			(Lübeck, 1674); facs. ed. M. Seiffert (Lübeck, 1937); G ii, 85	186
	1. Mit Fried und Freud	chorale	4vv		
	2. Klag-Lied	poetry	S, 2 str, bc		
77	Nichts soll uns scheiden	Bible, poetry	SAB, 2 vn, vle, bc	G vii, 20	
78	Nimm von uns, Herr	chorale	SATB, 4 str, bn, bc	*Uu* (autograph); Sn ix, 109	182, *183*, 195
79	Nun danket alle Gott	Bible	SSATB, 3 str, 4 brass, bn, bc	*D-Bds*, *S-Uu* (autograph); ed. S. Sørensen (Copenhagen, 1975)	195
80	Nun freut euch, ihr Frommen	poetry	SS, 2 vn, bc	G iii, 69	
81	Nun lasst uns Gott dem Herren	chorale	SATB, 2 vn, vle, bc	G viii, 9	
82	O clemens, o mitis	prose	S, 4 str, bc	G i, 65	181
83	O dulcis Jesu	prose	S, 2 vn, bc	G i, 71	181
84	O fröhliche Stunden	poetry	S, 2 vn, vle, bc	G i, 77	
85	O fröhliche Stunden	poetry	SSAB, 5 str, bc	*Uu* (autograph); Sn ix, 151	195
86	O Gott, wir danken deiner Güt'	poetry, chorale	SSATB, 2 vn, vle, bc	*Uu*; ed. D. Kilian, *37 Kantaten von Dietrich Buxtehude*, xii (Berlin, 1965)	186
87	O Gottes Stadt	poetry	S, 2 vn, va, vle, bc	G i, 84	184
88	O Jesu mi dulcissime	poetry	SSB, 2 vn, vle, bc	*Uu* (autograph); G vi, 39	182, 195
89	O lux beata trinitas	poetry	SS, 3 vn, bn, bc	G iii, 76	182
90	O wie selig sind	poetry	AB, 2 vn, vle, bc	G iii, 83	
91	Pange lingua gloriosi	poetry	SSAB, 5 str, bc	Sn ix, 183	182
92	Quemadmodum desiderat cervus	prose	T, 2 vn, bc	G ii, 54	181
93	Salve desiderium	poetry	SSB, 2 vn, vle/bn, bc	G vi, 46	182, 184
94	Salve, Jesu, Patris gnate unigenite	prose	SS, 2 vn, bc	G iii, 86	181
95	Schaffe in mir, Gott, ein rein Herz	Bible	S, 2 vn, vle, bc	G i, 96	

BuxWV	Title	Text	Scoring	Source, edn., remarks	
96	Schwinget euch himmelan	poetry	SSATB, 3 vn, vle, bc	*Uu*; ed. B. Grusnick (Kassel, 1959)	
97	Sicut Moses exaltavit serpentem	Bible	S, 2 vn, va da gamba, bc	G i, 101	
98	Singet dem Herrn ein neues Lied	Bible	S, vn, bc	G i, 108	180
99	Surrexit Christus hodie	poetry	SSB, 3 vn, bn, bc	G vi, 51	182
100	Wachet auf, ruft uns die Stimme	chorale	SSB, 4 vn, bn, bc	G vi, 60	182
101*	Wachet auf	chorale	ATB, 2 vn, bc	(without c.f.); G vii, 100	184
102	Wär Gott nicht mit uns diese Zeit	chorale	SATB, 2 vn, bc	G viii, 22	
103	Walts Gott mein Werk ich lasse	chorale	SATB, 2 vn, vle, bc	G viii, 31	
104	Was frag' ich nach der Welt	poetry	SAB, 2 vn, vle, bc	G vii, 29	
105	Was mich auf dieser Welt betrübt	poetry	S, 2 vn, bc	G i, 113	
106	Welte, packe dich	poetry	SSB, 2 vn, vle, bc	G vi, 75	
107	Wenn ich, Herr Jesu, habe dich	poetry	A, 2 vn, bc	G ii, 25	
108	Wie schmeckt es so lieblich	poetry	SAB, 2 vn, vle, bc	G vii, 39	184
109	Wie soll ich dich empfangen	poetry	SSB, 2 vn, bn, bc	G vi, 84	
110	Wie wird erneuet, wie wird erfreuet	poetry	SSATTB, 5 str, 6 brass, bc	*D-Bds*; ed. S. Sørensen (Copenhagen, 1977)	
111	Wo ist doch mein Freund geblieben	poetry	SB, 2 vn, bn, bc	G iii, 93	
112	Wo soll ich fliehen hin	Bible, poetry, chorale	SATB, 5 str, bc	S, 85; Sn ix, 211	186
113	Benedicam Dominum	Bible	6 choirs	G iv, 23	187
114*	Missa alla brevis	prose	SSATB, bc	G iv, 12	186

SECULAR VOCAL

179, 182, 195, 198

BuxWV	Title	Text	Scoring	Source, edn., remarks
115	Auf, Saiten, auf!	poetry	S, 2 vn, 2 va da gamba, bc	(Lübeck, 1673), lost; ed. W. Stahl (Kassel, n.d.)
116	Auf, stimmet die Saiten	poetry	AAB, 2 tpt, 2 trbn, bn, bc	(Lübeck, [1672]); G vii, 115
117	Deh credete il vostro vanto	poetry	S, 2 vn, bc	(Lübeck, 1695), lost; copy in *D-LÜh*
118	Gestreuet mit Blumen	poetry	A, 5 str, bc	(Lübeck, 1675)
119	Klinget für Freuden	poetry	SSB, 3 str, 2 tpt, bc	*S-Uu*[1680] [autograph; MS copy dated 1680]; G v, 96
120	O fröhliche Stunden, o herrlicher Tag	poetry	S, 2 vn, ob, bc	(Lübeck, 1705), lost; ed. W. Stahl (Kassel, n.d.)

121	Opachi boschetti	poetry	S/T, 2 vn, bc	(Lübeck, 1698), lost; extracts ed. in Pirro, p.473	
122	Schlagt, Künstler, die Pauken	poetry	SSAB, 2 tpt, 3 str, timp, bc	*Uu* [1681] (text Lübeck, 1681); ed. D. Kilian, *37 Kantaten von Dietrich Buxtehude*, xxvi (Berlin, 1958)	
			Canons		
123	Canon duplex per augmentationem		4vv	1674, MS presumably lost; facs. and solution in Snyder (1980)	186
124	Divertisons nous aujourd'hui	prose	3vv	1670, MS lost; facs. in Stahl (1926), 36; *MGG*, ii, cols. 553–4; solution in Snyder (1980)	186
124a	Canon quadruplex	poetry	5vv	1685 print, lost; ed. J. C. M. van Riemsdijk, *TVNM*, ii (1887), 91; see Beckmann (1980)	

LOST VOCAL WORKS — 175, 187–8, 193 199

125 Christum lieb haben ist viel besser, several vv, insts, listed in Lüneburg inventory

126 Music for ded. of Fredenhagen altar, 1697, 3 choirs, tpts, timp, text unknown

127 Pallidi salvete, several vv, insts, listed in Ansbach inventory

128 Die Hochzeit des Lamms, Abendmusik, 1678; lib, *B-Bc*, *S-Uu*, pr. in Pirro, 173ff — 187

129 Das Allerschröcklichste und Allererfreulichste, Abendmusik, listed in catalogue, 1684 — 187–8

130 Himmlische Seelenlust auf Erden über die Menschwerdung . . . Jesu Christi, Abendmusik, listed in catalogue, 1684 — 187

131 Der verlorene Sohn, Abendmusik, 1688, mentioned in Buxtehude letter — 187

132 Hundertjähriges Gedicht, mentioned in *Nova literaria*

133 Abendmusik, 1700; lib lost, extracts pr. in Stahl (1937), 18f — 187

134 Castrum doloris, Abendmusik, 1705; lib, *D-LÜh*, facs. in Karstädt (1962) — 177, 187

135 Templum honoris, Abendmusik, 1705; lib, *LÜh*, facs. in Karstädt (1962) — 177, 182

ORGAN — 181–92, 195–6, 197, 198, 199

136 Praeludium, C – SS suppl., 4; H ii, 2; B i, 1

137 Praeludium (Prelude, Fugue and Chaconne), C – SS i, 4; H ii, 1; B i, 2 — 197

138 Praeludium, C – B i, 3 — 190

139 Praeludium, D – SS i, 11 (suppl., 61); H ii, 11; B i, 4

140 Praeludium, d – SS i, 10; H ii, 19; B i, 5 — 197

BuxWV		
141	Praeludium, E – SS i, 8; H ii, 14; B i, 6	197
142	Praeludium, e – SS i, 6 (suppl., 54); H ii, 9; B i, 7	197
143	Praeludium, e – SS i, 13; H ii, 10; B i, 8	
144	Praeludium, F – SS suppl., 3; H ii, 16; B i, 9	190
145	Praeludium, F – SS i, 15; H ii, 15; B i, 10	190
146	Praeludium, f♯ – SS i, 12; H ii, 13; B i, 11	197
147	Praeludium, G – H ii, 7; B i, 12	190
148	Praeludium, g – SS i, 5 (suppl., 48); H ii, 22; B i, 13	
149	Praeludium, g – SS i, 14; H ii, 24; B i, 14	197
150	Praeludium, g – SS i, 7; H ii, 23; B i, 15	
151	Praeludium, A – SS suppl., 5; H ii, 12; B i, 16	
152	Praeludium, Phrygian – SS suppl., 1; H ii, 6; B i, 17	197
153	Praeludium, a – SS i, 9; H ii, 4; B i, 18	
154	Praeludium, B♭ (frag.) – H ii, 21; B i, suppl. 3	
155	Toccata, d – SS suppl., 7; H ii, 20; B i, 19	197
156	Toccata, F – SS i, 20; H ii, 17; B i, 20	
157	Toccata, F – SS i, 21; H ii, 18; B i, 21	190
158	Praeambulum, a – SS suppl., 2; H ii, 5; B i, 22	197
159	Ciacona, c – SS i, 2; H i, 3; B i, 23	191
160	Ciacona, e – SS i, 3; H i, 2; B i, 24	191
161	Passacaglia, d – SS i, 1; H i, 1; B i, 25	191, 194
162	Praeludium, G – H ii, 8; B i, 26	
163	Praeludium, g – SS i, 16; H ii, 25; B i, 27	
164	Toccata, G – SS i, 23 (suppl., 70); H ii, 27; B i, 28	
165	Toccata, G – SS i, 22 (suppl., 66); H ii, 26; B i, 29	
166	Canzona, C – SS suppl., 8; H i, 4; B i, 30	
167	Canzonetta, C – H i, 5; B i, 31	
168	Canzonetta, d – SS i, 25 (suppl., 72); H i, 10; B i, 32	
169	Canzonetta, e – H i, 9; B i, 33	
170	Canzona, G – H i, 6; B i, 34	
171	Canzonetta, G – SS i, 24; H i, 7; B i, 35	
172	Canzonetta, G – B i, 36	
173	Canzonetta, g – H i, 12; B i, 37	
174	Fuga, C – SS i, 17; H ii, 3; B i, 38	
175	Fuga, G – SS suppl., 6; H i, 8; B i, 39	
176	Fuga, B♭ – SS i, 18; H i, 11; B i, 40	
177	Ach Gott und Herr, d – SS suppl., 9; H iii/1, 1; B ii, 1	
178	Ach Herr, mich armen Sünder, Phrygian – SS ii/2, 1; H iv/3, 1; B ii, 2	

BuxWV		
179	Auf meinen lieben Gott, e – SS ii/2, 31; H iii/1, 7; B ii, suppl., 1	191–2
180	Christ unser Herr zum Jordan kam, Dorian – SS ii/2, 2; H iv/3, 2; B ii, 3	
181	Danket dem Herrn, g – SS ii/1, 1; H iii/1, 2; B ii, 4	
182	Der Tag, der ist so freudenreich, G – SS ii/2, 3; H iv/3, 3; B ii, 5	
183	Durch Adams Fall ist ganz verderbt, Dorian – SS ii/2, 4; H iv/3, 4; B ii, 6	
184	Ein feste Burg ist unser Gott, C – SS ii/2, 5; H iv/3, 5; B ii, 7	192
185	Erhalt uns, Herr, bei deinem Wort, g – SS ii/2, 7; H iv/3, 6; B ii, 8	
186	Es ist das Heil uns kommen her, C – SS ii/2, 8; H iv/3, 7; B ii, 9	
187	Es spricht der Unweisen Mund wohl, G – SS ii/2, 9; H iv/3, 8; B ii, 10	
188	Gelobet seist du, Jesu Christ, G – SS ii/1, 2; H iii/2, 1; B ii, 11	
189	Gelobet seist du, Jesu Christ, G – SS ii/2, 10; H iv/3, 9; B ii, 12	
190	Gott der Vater wohn uns bei, C – SS ii/2, 11; H iv/3, 10; B ii, 13	
191	Herr Christ, der einig Gottes Sohn, G – SS ii/2, 13; H iv/3, 11b; B ii, 14	
192	Herr Christ, der einig Gottes Sohn, G – SS ii/2, 12; H iv/3, 11a; B ii, 15	
193	Herr Jesu Christ, ich weiss gar wohl, a – SS ii/2, 14; H iv/3, 12; B ii, 16	
194	Ich dank dir, lieber Herre, F – SS ii/1, 3; H iii/2, 2; B ii, 17	
195	Ich dank dir schon durch deinen Sohn, F – SS ii/1, 4; H iii/2, 3; B ii, 18	
196	Ich ruf zu dir, Herr Jesu Christ, d – SS ii/2, 15; H iii/2, 4; B ii, 19	
197	In dulci jubilo, G – SS ii/2, 17; H iv/3, 14; B ii, 20	
198	Jesus Christus, unser Heiland, g (Dorian) – SS ii/2, 16; H iv/3, 13; B ii, 21	
199	Komm heiliger Geist, Herre Gott, F – SS ii/2, 18; H iv/3, 15a; B ii, 22	
200	Komm heiliger Geist, Herre Gott, F – SS ii/2, 19; H iv/3, 15b; B ii, 23	
201	Kommt her zu mir, spricht Gottes Sohn, g – SS ii/2, 20; H iv/3, 16; B ii, 24	
202	Lobt Gott, ihr Christen, allzugleich, G – SS ii/2, 21; H iv/3, 17; B ii, 25	

203 Magnificat primi toni, Dorian – SS ii/1, 5a; H iii/2, 5; B ii, 26 — 192
204 Magnificat primi toni, Dorian – SS ii/1, 5b (suppl., 10a); H iii/1, 3a; B ii, 27 — 192
205 Magnificat noni toni, d – SS suppl., 10b–c; H iii/1, 3b; B ii, 28 — 192
206 Mensch, willt du leben seliglich, Phrygian – SS ii/2, 22; H iv/3, 18; B ii, 29
207 Nimm von uns, Herr, du treuer Gott (Vater unser in Himmelreich), d – SS ii/1, 9a; H iii/1, 6; B ii, 31
208 Nun bitten wir den heiligen Geist, G – SS ii/2, 24; H iv/3, 19b; B ii, 32
209 Nun bitten wir den heiligen Geist, G – SS ii/2, 23; H iv/3, 19a; B ii, 33
210 Nun freut euch lieben Christen g'mein, G – SS ii/1, 6; H iii/2, 6; B ii, 34 — 192
211 Nun komm, der Heiden Heiland, g – SS ii/2, 25; H iv/3, 20; B ii, 35
212 Nun lob mein Seel' den Herren, C – SS suppl., 11; H iii/1, 5; B ii, 36
213 Nun lob mein Seel' den Herren, G – SS ii/1, 7a; H iii/1, 4a; B ii, 37 — 191
214 Nun lob mein Seel' den Herren, G – SS ii/1, 7b[1]; H iii/1, 4b[1]; B ii, 38
215 Nun lob mein Seel' den Herren, G – SS ii/1, 7b[2]; H iii/1, 4b[2]; B ii, 39
216 O lux beata Trinitas (frag.) – B ii, suppl. 2
217 Puer natus in Bethlehem, a – SS ii/2, 26; H iv/3, 21; B ii,40
218 Te Deum laudamus, Phrygian – SS ii/1, 8; H iii/2, 7; B ii, 41 — 192
219 Vater unser in Himmelreich, d – SS ii/1, 9b; H iv/3, 22; B ii, 42 — 192
220 Von Gott will ich nicht lassen, a – SS ii/2, 27; H iv/3, 23a; B ii, 43
221 Von Gott will ich nicht lassen, a – SS ii/2, 28; H iv/3, 23b; B ii, 44
222 Wär Gott nicht mit uns diese Zeit, a – SS ii/2, 29; H iv/3, 24; B ii, 45
223 Wie schön leuchtet der Morgenstern, G – SS ii/1, 10; H iii/2, 8; B ii, 46 — 192
224 Wir danken dir, Herr Jesu Christ, Dorian – SS ii/2, 30; H iv/3, 25; B ii, 47
225 Canzonetta, a – B i, 190

OTHER KEYBOARD — 192–3, 198, 199
(*all MSS in DK-Kk*)

226 Suite, C; Ba 3; BK4
227 Suite, C; Ba 7; BK 8
228 Suite, C; Ba 10; BK 12
229* Suite, C; Ba 13; BK 112
230 Suite, C; Ba 15; BK 16
231 Suite, C; BK 20
232 Suite, D; Ba 29; BK 23
233 Suite, d; Ba 18; BK 25
234 Suite, d; Ba 21, BK 29
235 Suite, e; Ba 31; BK 34
236 Suite, e; Ba 35; BK 38
237 Suite, e; Ba 38; BK 42
238 Suite, F; Ba 42; BK 46
239 Suite, F; BK 49
240 Suite, G; Ba 54; BK 52
241 Suite, g; Ba 45; BK 55
242 Suite, g; Ba 48; BK 58
243 Suite, A; Ba 61; BK 62
244 Suite, a; Ba 57; BK 66
245 Courant zimble, 8 variations, a; Ba 80; BK 106
246 Aria and 10 variations, C; Ba 64; BK 70
247 More Palatino, aria and 12 variations, C; Ba 72; BK 78
248 Rofilis, aria and 3 variations, d; Ba 78, BK 86
249 Aria and 3 variations, a; Ba 84; BK 101
250 La Capricciosa, 32 variations, G; Ba 88; BK 88 — 193
251 7 suites, lost, mentioned in Mattheson (1739), 130

OTHER INSTRUMENTAL — 193–4, 197, 198

252–8 VII suonate, vn, va da gamba, hpd, op.1 (Hamburg, ?1694): Sonata I, F, St 3; Sonata II, G, St 13; Sonata III, a, St 22; Sonata IV, B♭, St 33; Sonata V, C, St 44; Sonata VI, d, St 55; Sonata VII, e, St 66 — 193
259–65 VII suonate, vn, va da gamba, hpd, op.2 (Hamburg, 1696): Sonata I, B♭, St 79; Sonata II, D, St 90; Sonata III, g, St 103; Sonata IV, c, St 116; Sonata V, A, St 126; Sonata VI, E, St 139; Sonata VII, F, St 150 — 193–4, 197
266 Sonata, C, 2 vn, va da gamba, bc; St 164 — 194

BuxWV

267 Sonata, D, va da gamba, vle, bc; St 176

268 Sonata, D, va da gamba, bc, *GB-Ob*; ed. F. Längin (Mainz, 1956)

269 Sonata, F, 2 vn, va da gamba, bc, *S-Uu*; ed. in Einstein, 103ff — 194

270 Sonata, F, 2 vn, bc, *Uu* (bc only); ed. in Linfield, 345 — 197

271 Sonata, G, 2 vn, va da gamba, bc, *Uu*; ed in Linfield, 347 — 194

272 Sonata, a, vn, va da gamba, bc, *Uu*; ed. in Linfield, 359

273 Sonata, B♭, vn, va da gamba, bc (early version of BUXWV255 plus suite); St 160 (suite only); ed. (complete) in Linfield, 369 — 194

DOUBTFUL AND MISATTRIBUTED WORKS

(*for further information see Geck, 1961, 1963*)

BuxWV suppl.

1 Magnificat, *S-Uu* (anon.); ed. B. Grusnick (Kassel, 1931)

2 Man singet mit Freuden vom Sieg, *D-B*; ed. T. Fedtke (Stuttgart, 1964), ? by J. Schelle, see Kilian

BuxWV

3 Das jüngste Gericht, *S-Uu* (anon.); ed. W. Maxton (Kassel, 1939), see Maxton (1927–8, 1962), Blankenburg (1968 Krummacher (1978), Karstädt (1979), Snyder (1981) and Ruhle — 188

5 Trio sonata, org, *US-NH*; B i, suppl. 4; see Linfield

6 Courante, d, kbd, *DK-Kk* (anon.); Ba 101; BK 111

7 Courante, G, kbd, *Kk* (anon.); Ba 102

8 Simphonie, G, kbd, *Kk* (anon.); Ba 103; BK 114

9 Erbarm dich mein, O Herre Gott, *S-Uu*; ed. B. Grusnick (Kassel, 1937), attrib. L. Busbetzky

10 Laudate Dominum omnes gentes, *Uu*; ed. B. Grusnick (Kassel, 1937), attrib. L. Busbetzky

11 Erhalt uns Herr, bei deinem Wort, *D-Bds*, *NL-DHgm*; ed. in SS ii/2, 6 and H iv, 3, 26, ? by J. Pachelbel, see B ii, vi

12 Suite, d, kbd, *DK-Kk* (entitled 'di D.B.H.'); Ba 26, attrib. N.-A. Lebègue, Second livre de clavessin (Paris, ?1687)

13 Suite, g, kbd, *Kk* (entitled 'di D.B.H.'); Ba 51, attrib. N.-A. Lebègue, Second livre de clavessin (Paris, ?1687)

BIBLIOGRAPHY

CATALOGUES, BIBLIOGRAPHIES

C. Stiehl: 'Die Familie Düben und die Buxtehudeschen Manuscripte auf der Bibliothek zu Uppsala', *MMg*, xxi (1889), 2

A. Göhler: *Verzeichnis der in den Frankfurter und Leipziger Messkatalogen der Jahre 1564 bis 1759 angezeigten Musikalien* (Leipzig, 1902/*R*1965)

F. Lindberg: *Katalog över Dübensamlingen i Uppsala Universitets Bibliotek* (MS, 1946, *S-Uu*)

B. Grusnick: 'Zur Chronologie von Dietrich Buxtehudes Vokalwerken', *Mf*, x (1957), 75

——: 'Die Dübensammlung: ein Versuch ihrer chronologischen Ordnung', *STMf*, xlvi (1964), 27–82; xlviii (1966), 63–186

W. Blankenburg: 'Neue Forschungen über das geistliche Vokalschaffen Dietrich Buxtehudes', *AcM*, xl (1968), 130

J. O. Rudén: *Vattenmärken och Musikforskning: Presentation och Tillämpning av en Dateringsmetod på musikalier i handskrift i Uppsala Universitetsbiblioteks Dübensamling* (Licentiatavhandling i musikforskning, Uppsala U., 1968)

H. Kümmerling: *Katalog der Sammlung Bokemeyer* (Kassel, 1970)

G. Karstädt: *Thematisch-systematisches Verzeichnis der musikalischen Werke von Dietrich Buxtehude: Buxtehude-Werke-Verzeichnis* (Wiesbaden, 1974)

H. Wettstein: *Dietrich Buxtehude (1637–1707): eine Bibliographie* (Freiburg, 1979)

HISTORICAL STUDIES, BIOGRAPHY

WaltherML

M. H. Schacht: *Musicus Danicus eller Danske Sangmester* (MS, 1687, *DK-Kk*); ed. G. Skjerne (Copenhagen, 1928)

H. E. Elmenhorst: *Dramatologia antiquo-hodierna* (Hamburg, 1688)

M. H. Fuhrmann: *Musicalischer-Trichter* (Berlin [Frankfurt an der Spree], 1706)

Nova literaria Maris Balthici et Septentrionis (Hamburg, July 1707)

J. Mattheson: *Der vollkommene Capellmeister* (Hamburg, 1739/*R*1954)

——: *Grundlage einer Ehren-Pforte* (Hamburg, 1740); ed. M. Schneider (Berlin, 1910/*R*1969)

J. Moller: *Cimbria literata*, ii (Kiel, 1744)

C. Ruetz: *Widerlegte Vorurtheile von der Beschaffenheit der heutigen Kirchenmusic und von der Lebens-Art einiger Musicorum* (Lübeck, 1752)

H. Jimmerthal: *Zur Geschichte der St. Marien Kirche in Lübeck* (MS, 1857, Nordelbisches Ev.-Luth. Kirchenarchiv, Lübeck)

A. Hagedorn: 'Briefe von Dietrich Buxtehude', *Mitteilungen des Vereins für Lübeckische Geschichte und Altertumskunde*, iii (1887–8), 192
S. A. E. Hagen: *Diderik Buxtehude (o. 1637–1707), hans Familie og lidet kjendte Ungdom, inden han kom til Lübeck 1668* (Copenhagen, 1920)
W. Stahl: *Franz Tunder und Dietrich Buxtehude* (Leipzig, 1926)
——: *Die Lübecker Abendmusiken im 17. und 18. Jahrhundert* (Lübeck, 1937)
L. Pedersen: 'Fra Didrik Hansen Buxtehudes barndom og ungdom 1636–37 til l. maj 1668', *Medlemsblad for Dansk Organist- og Kantorsamfund af 1905*, iii (1937), 25
W. Stahl: 'Dietrich Buxtehudes Geburtsort', *Mf*, iv (1951), 382
A. von Brandt: *Geist und Politik in der Lübeckischen Geschichte* (Lübeck, 1954)
N. Friis: *Buxtehude: hans By og hans Orgel* (Helsingør, 1960)
——: *Diderik Buxtehude* (Copenhagen, 1960)
G. Jaacks: ' "Häusliche Musikszene" von Johannes Voorhout: zu einem neu erworbenen Gemälde im Museum für Hamburgische Geschichte', *Beiträge zur deutschen Volks- und Altertumskunde*, xvii (1978), 56 [see fig. 7 above, p.176]
K. Beckmann: 'Reincken und Buxtehude: zu einem wiederentdeckten Gemälde in Hamburg', *Der Kirchenmusiker*, xxxi (1980), 172
W.-D. Hauschild: *Kirchengeschichte Lübecks: Christentum und Bürgertum in neun Jahrhunderten* (Lübeck, 1981)
G. Karstädt: 'Buxtehude und die Neuordnung der Abendmusiken', *Festschrift für Bruno Grusnick zum 80. Geburtstag* (Neuhausen-Stuttgart, 1981), 119
H.-B. Spies: 'Vier neuentdeckte Briefe Dietrich Buxtehudes', *Zeitschrift des Vereins für Lübeckische Geschichte und Altertumskunde*, lxi (1981), 81
A. Elder: *Der nordelbische Organist: Studien zu Sozialstatus, Funktion und kompositorischer Produktion eines Musikerberufes von der Reformation bis zum 20. Jahrhundert* (Kassel, 1982)
C. Wolff: 'Das Hamburger Buxtehude-Bild', *800 Jahre Musik in Lübeck*, i: *zur Ausstellung im Museum am Dom aus Anlass des Lübecker Musikfestes 1982*, ed. A. Grassmann and W. Neugebauer (Lübeck, 1982), 64
H.-B. Spies: 'Buxtehude und die finanzielle Musikförderung in Lübeck', *Musik und Kirche*, liii (1983), 5

DISCUSSIONS OF WORKS, SOME WITH BIOGRAPHY

P. Spitta: *Johann Sebastian Bach*, i (Leipzig, 1873, 5/1962; Eng. trans., 1884/*R*1951), 256–310

A. Einstein: *Zur deutschen Literatur für Viola da Gamba im 16. und 17. Jahrhundert* (Leipzig, 1905/*R*1972), 58f, 103ff
A. Pirro: *Dietrich Buxtehude* (Paris, 1913/*R*1976)
W. Maxton: 'Mitteilungen über eine vollständige Abendmusik Dietrich Buxtehudes', *ZMw*, x (1927–8), 387
W. Buszin: 'Dietrich Buxtehude, 1637–1707, on the Tercentenary of his Birth', *MQ*, xxiii (1937), 465
F. Blume: 'Das Kantatenwerk Dietrich Buxtehudes', *JbMP 1940*, 10–39; repr. in *Syntagma musicologicum*, i (see below: 1963), 320–51
J. Hedar: *Dietrich Buxtehudes Orgelwerke: zur Geschichte des norddeutschen Orgelstils* (Stockholm, 1951)
F. Blume: 'Buxtehude, Dietrich', *MGG*; repr. in *Syntagma musicologicum*, i (see below: 1963), 302
H. Lorenz: 'Die Klaviermusik Dietrich Buxtehudes', *AMf*, xi (1954), 238
F. K. Hutchins: *Dietrich Buxtehude: the Man, his Music, his Era* (Paterson, NJ, 1955)
H. D. Kilian: *Das Vokalwerk Dietrich Buxtehudes: Quellenstudien zu seiner Überlieferung und Verwendung* (diss., Free U. of Berlin, 1956)
S. Sørensen: *Diderich Buxtehudes vokale kirkemusik: studier til den evangeliske kirkekantates udviklingshistorie* (Copenhagen, 1958)
W. S. Newman: *The Sonata in the Baroque Era* (Chapel Hill, 1959, rev. 4/1983), 251ff
M. Geck: 'Quellenkritische Bemerkungen zu Dietrich Buxtehudes Missa brevis', *Mf*, xiii (1960), 47
F. W. Riedel: *Quellenkundliche Beiträge zur Geschichte der Musik für Tasteninstrumente in der 2. Hälfte des 17. Jahrhunderts* (Kassel, 1960), 194ff
M. Geck: 'Die Authentizität des Vokalwerks Dietrich Buxtehudes in quellenkritischer Sicht', *Mf*, xiv (1961), 393; xvi (1963), 175
W. Maxton: 'Die Authentizität des "Jüngsten Gerichts" von Dietrich Buxtehude', *Mf*, xv (1962), 382
G. Karstädt: *Die 'extraordinairen' Abendmusiken Dietrich Buxtehudes* (Lübeck, 1962)
F. Blume: 'Dietrich Buxtehude in Geschichte und Gegenwart', *Syntagma musicologicum: gesammelte Reden und Schriften*, i, ed. M. Ruhnke and A. A. Abert (Kassel, 1963), 351
S. Sørensen: 'Monteverdi–Förster–Buxtehude: Entwurf zu einer entwicklungsgeschichtlichen Untersuchung', *DAM*, [iii] (1963), 87; It. trans. as 'L'eredità monteverdiana nella musica sacra del nord: Monteverdi–Foerster–Buxtehude', *RIM*, ii (1967), 341
H. J. Pauly: *Die Fuge in den Orgelwerken Dietrich Buxtehudes* (Regensburg, 1964)
M. Geck: *Die Vokalmusik Dietrich Buxtehudes und der frühe Pietismus* (Kassel, 1965)

F. Krummacher: *Die Überlieferung der Choralbearbeitungen in der frühen evangelischen Kantate* (Berlin, 1965)
——: 'Orgel- und Vokalmusik im Oeuvre norddeutscher Organisten um Buxtehude', *DAM*, [v] (1966–7), 63
W. Apel: *Geschiche der Orgel- und Klaviermusik bis 1700* (Kassel, 1967; Eng. trans., rev. 1972), 610ff
O. Mortensen: 'Über Typologisierung der Couranten und Sarabanden Buxtehudes', *DAM*, vi (1968–72), 5–51
G. Karstädt: *Der Lübecker Kantatenband Dietrich Buxtehudes: eine Studie über die Tabulatur Mus. A373* (Lübeck, 1971)
S. Sørensen: *Das Buxtehudebild im Wandel der Zeit* (Lübeck, 1972)
R. Roberts: *The Sacred Cantatas of Dietrich Buxtehude: their Texts and Music* (diss., U. of Birmingham, 1974)
H. E. Smither: *A History of the Oratorio*, ii (Chapel Hill, 1977)
F. Krummacher: *Die Choralbearbeitung in der protestantischen Figuralmusik zwischen Praetorius und Bach* (Kassel, 1978)
G. Karstädt: 'Die Instrumente in den Kantaten und Abendmusiken Dietrich Buxtehudes', *Festschrift Kurt Gudewill zum 65. Geburtstag* (Wolfenbüttel and Zurich, 1978), 111
——: 'Richtiges und Zweifelhaftes in Leben und Werk Dietrich Buxtehudes', *Musik und Kirche*, xlix (1979), 163
K. J. Snyder: 'Buxtehude's Organ Music: Drama without Words', *MT*, cxx (1979), 517
F. Krummacher: 'Stylus phantasticus und phantastische Musik: kompositorische Verfahren in Toccaten von Frescobaldi und Buxtehude', *SJb*, ii (1980), 7–77
K. J. Snyder: 'Dietrich Buxtehude's Studies in Learned Counterpoint', *JAMS*, xxxiii (1980), 544
L. L. Archbold: *Style and Structure in the Praeludia of Dietrich Buxtehude* (diss., U. of California, Berkeley, 1981)
K. J. Snyder: 'Buxtehude and "Das jüngste Gericht": a New Look at an Old Problem', *Festschrift für Bruno Grusnick zum 80. Geburtstag* (1981), 128
S. C. Ruhle: *An Anonymous Seventeenth-century German Oratorio in the Düben Collection* (*Uppsala University Library vok. mus. i hskr. 71*) (diss., U. of North Carolina, 1982)
W. Breig: 'Der "Stylus Phantasticus" in der Lübecker Orgelmusik', *800 Jahre Musik in Lübeck*, ii: *Dokumentation zum Lübecker Musikfest 1982*, ed. A. Edler, W. Neugebauer and H. W. Schwab (Lübeck, 1983), 43
N. M. Jensen: 'Buxtehudes vokale und instrumentale Ensemblemusik', *800 Jahre Musik in Lübeck*, ii: *Dokumentation zum Lübecker Musikfest 1982*, ed. A. Edler, W. Neugebauer and H. W. Schwab (Lübeck, 1983), 53

H. W. Schwab: 'Ratsmusik und Hausmusik – Offizielles und privates Musizieren aus der Zeit zwischen 1600 und 1800', *800 Jahre Musik in Lübeck*, ii: *Dokumentation zum Lübecker Musikfest 1982*, ed. A. Edler, W. Neugebauer and H. W. Schwab (1983), 31
K. J. Snyder: 'Lübecker Abendmusiken', *800 Jahre Musik in Lübeck*, ii: *Dokumentation zum Lübecker Musikfest 1982*, ed. A. Edler, W. Neugebauer and H. W. Schwab (1983), 63
E. Linfield: *Dietrich Buxtehude's Sonatas: a Historical and Analytical Study* (diss., Brandeis U., 1984)
K. J. Snyder: *Dietrich Buxtehude* (in preparation)

HENRY PURCELL

Jack Westrup

CHAPTER ONE

Life

Henry Purcell was born, probably at Westminster, London, on an unknown date in 1659. He came of an active family of musicians who were possibly of Buckinghamshire origin. Other spellings of the name found in contemporary records are Persill, Purcel, Purcill, Pursal, Pursall, Pursel, Pursell and Pursill. On Henry's parentage, the only contemporary evidence is a letter, dated 8 February 1679, from Thomas Purcell, a tenor singer and composer, to the singer John Gostling (facsimile in Westrup, *Purcell*, rev. 4/1980), which refers to 'my sonne Henry' and mentions that 'my sonne is composing: wherin you will be cheifly consern'd'. John Hingeston (see below) in his will (12 Dec 1683) left £5 'to my Godson Henry Pursall (son of Elizabeth Pursall)', probably a son of an older Henry Purcell, a singer who died in 1664 and who was probably Thomas's brother; but he did not identify him as a musician or mention him as his assistant, and there is no certainty that this was the composer. (For an alternative view of the composer's parentage, strongly advancing the claims of the elder Henry as his father, see Zimmerman, 1967, rev. 2/1983, 1f, 331ff.) The composer is known to have had three brothers, of whom one, Daniel, achieved some recognition as a composer.

As a boy Purcell was a chorister in the Chapel Royal. He was obviously prodigiously gifted. John Playford's

Catch that Catch Can, or The Musical Companion (1667) includes a three-part song, *Sweet tyranness*, attributed to him. Even though a version for solo voice was published with his name 11 years later in *New Ayres and Dialogues*, it may well be by the elder Henry Purcell. But there is nothing inherently improbable in supposing that the younger Henry was writing music at the age of eight; we also know that boys in the Chapel Royal choir were encouraged to compose.

In 1673 Purcell's voice broke at what was then an unusually early age, and he was appointed an unpaid assistant to John Hingeston, who had charge of the king's keyboard and wind instruments, with the prospect of eventually succeeding him as keeper. Having acquired the necessary experience he was engaged from 1674 to 1678 to tune the organ at Westminster Abbey, where in 1675–6 he was also paid for copying two books of organ parts. In 1677 he was appointed composer-in-ordinary for the violins in succession to Matthew Locke and in 1679 succeeded John Blow as organist of Westminster Abbey – a post which not only entitled him to a salary but also provided the rent of a house. It was probably in 1680 or 1681 that he married. In 1680 he wrote the first of his welcome songs for Charles II and also contributed music to the theatre for the first time. On 14 July 1682 he succeeded Edward Lowe as one of the organists of the Chapel Royal, which meant that he was also a singer in the choir. In February he took the sacrament at St Margaret's, Westminster, before witnesses – perhaps as a result of the recent enforcement of the laws against Dissenters.

In June 1683 he published his *Sonnata's of III Parts*, dedicated to Charles II, and announced that he could

9. Henry Purcell: portrait in black chalk heightened with white, attributed to Godfrey Kneller (1646–1723); it is believed to be the only surviving portrait of Purcell drawn from life

supply copies from his house in St Ann's Lane, Westminster. In December that year, on the death of Hingeston, he was appointed organ maker and keeper of the king's instruments. His court appointments were renewed in 1685 by James II, for whose coronation he provided not only music but a second organ. The post of keeper of the instruments was not a sinecure: in 1687 he drew up a detailed statement showing that the cost of cleaning and repairing the Chapel Royal organ and tuning the harpsichords, plus his own salary, would come to £81. On the accession of William III in 1689 he was confirmed in his court appointments and once again had the duty of providing a second organ for the coronation. On this occasion he was involved in a dispute with the dean and chapter of Westminster over the money that he had received for places in the main organ loft: it should have been handed over, after reasonable deductions, more promptly.

Apart from his increased activity as a composer, particularly for the theatre, there is little information about the remaining years of his life. The last royal event for which he provided music was the Duke of Gloucester's birthday in July 1695. He made his will on the day of his death at Westminster, 21 November 1695, when he was so feeble that he could hardly hold the pen. His funeral took place in Westminster Abbey five days later. The whole chapter attended in their vestments, and the music was sung by the united choirs of the Abbey and the Chapel Royal. Purcell was buried, at no cost to his widow, in the north aisle, adjoining the organ. Lady Elizabeth Howard paid for the erection of a marble tablet, fixed to a pillar above the grave. By his wife Frances, who died in February 1706, Purcell

Westmr Novr ye 2d. 1686

Sr/

I have wrote severall times to Mr Webber concerning
what was due to me on Hodg's account and recd no
answer, which has occasion'd this presumption in
giving you the trouble of a few Lines relating to the
matter; It is ever since ye beginning of June last that
the Money has been due: the Sum is 27l for
for half a Years teaching & boarding the other a
Bill of 7l for nessecarys wch I laid out for him,
the Bill Mr Webber has; Compassion Moves
me to acquaint you of a great many debts Mr
Hodg contracted whilst in London and to some who
are so poor 'twere an act of Charity as well as
Justice to pay 'em I hope you will be so kind to
take it into Your consideration and also pardon
this boldness from

Sr Yor most obliged
humble Servt
Henry Purcell

10. Purcell's autograph letter of 1686 to the Dean of Exeter had several children, of whom only two are known to have survived him: a daughter, Frances (*b* May 1688), and a son, Edward (baptized 6 Sept 1689; *d* 1 July 1740).

CHAPTER TWO

Style and development

Purcell's earliest surviving works, about a dozen anthems of variable quality and a handful of songs, date from the later 1670s. The works of 1680, however – the string fantasias, the music for *Theodosius* and the first welcome song for Charles II – show such a complete mastery of the craft of composition that they must be the result of constant practice over a number of years. They also illustrate two aspects of his work. The music for *Theodosius* and the welcome song were written to order; the fantasias, we may assume, were written either for his own satisfaction or for private performance by friends or colleagues. As the years went on he found himself more and more in demand as a composer. His music for the theatre in particular must have made his name familiar to many people who knew nothing of his church music and had no opportunity of hearing his court odes. He might have been tempted to take some of his duties lightly; but in fact everything he wrote shows the same professional skill and scrupulous attention to detail. His music is sometimes dull, but it is never slovenly. He clearly learnt much from Locke, Humfrey and Blow and quickly became familiar with the Italian music of his time. But whatever he absorbed from others became transformed in his hands into something that was peculiarly his own. Of all the English composers of his time he was the most individual.

When Henry Playford published the first volume of *Orpheus Britannicus* – an anthology of Purcell's songs – in 1698, he wrote: 'The Author's extraordinary Talent in all sorts of Musick is sufficiently known, but he was especially admir'd for the *Vocal*, having a peculiar Genius to express the Energy of *English* Words, whereby he mov'd the Passions of all his Auditors'. This is a just comment, but it is not the whole truth. In Purcell's early work the vocal music is less remarkable than the instrumental. The songs in *Theodosius* and the writing for voices in the earliest welcome songs have a certain stiffness, which disappeared in his later work as he acquired the art of combining just accentuation with a flexible melodic line. The string fantasias, on the other hand, show a mastery of instrumental writing and an intensity of imagination which he never surpassed. There is a further difference between his songs and his instrumental music. The songs rely for their effect primarily, though not wholly, on melody: the bass, even when figured, provides only a general outline of the harmonic progressions. On the other hand the instrumental writing, including movements in vocal works, is marked by a harmonic subtlety which in vocal music appears only in ensembles. This is very noticeable when Purcell provides an instrumental version of a song, e.g. 'They tell us that you mighty powers above' in *The Indian Queen*, where the instrumental version introduces dissonances to which there is no clue in the original song. The same thing happens when a song is repeated by a chorus, e.g. 'If love's a sweet passion' in *The Fairy Queen*, where there is a further variant in the instrumental prelude.

Purcell's technical mastery needs no better illustration than his use of canon and ground bass. He employed both with such effortless skill that one can easily listen without being aware that they exist. The examples of canon which he contributed to the 12th edition of John Playford's *Introduction to the Skill of Musick* (1694) do not explain to the reader how to write one: they merely exhibit the finished article. One has the impression that to him such technical devices were instinctive. Perhaps he underestimated the capacity of less gifted practitioners. He described 'Composing upon a *Ground*' as 'a very easie thing to do, and requires but little judgment', with no hint of the subtlety with which he ensured that the phrases of the melody did not coincide with the repetitions of the bass. He did, however, mention that when, as often, the ground is simply four notes descending, 'to maintain *Fuges* upon it would be difficult', forgetting perhaps how cleverly he had done this in *Dioclesian*.

Purcell's early music, like Monteverdi's, tends to be conservative. The chromaticisms in the string fantasias and in the early anthems, so far from being 'prophetic' or 'curiously modern', are a logical extension of the practice of his immediate predecessors, particularly Locke. He never abandoned chromaticism, but as he became acquainted with the directness and relative simplicity of Italian music he tended to use it much more as a melodic feature within the framework of diatonic tonality. His contribution to Playford's book does not mention chromaticism, and there are scarcely any instances of it in his examples. Instead he drew particular attention to the Italians' use of the diminished 7th and the Neapolitan 6th, both of which are basically modifications of diatonic harmony. He was attracted also by the sensuous effect of passages in 3rds, particularly in three-part harmony. He criticized an example from Christopher Simpson's *Compendium of Practical Musick* (1667) on the ground that the alto part moves awkwardly:

Now in my opinion the *Alt* or *Second Part* should move gradually in *Thirds* with the *Treble*; though the other be fuller, this is the smoothest, and carries more Air and Form in it, and I'm sure 'tis the constant Practice of the *Italians* in all their Musick, either Vocal or Instrumental, which I presume ought to be a Guide to us; the way I would have, is thus: [ex.1].

He recommended that a second treble should normally be kept below the first but at the same time pointed out that in sonatas 'one *Treble* has as much Predominancy as the other; and you are not tied to such a strict Rule, but one may interfere with the other'. Both principles may be illustrated from his own sonatas, e.g. no.9 of the *Ten Sonata's in Four Parts*, where the first and second violins move in 3rds through practically the whole of the second movement. Purcell's vocal writing is also frequently Italianate in character, in both recitative and aria. His recitative, both in church music and in secular works, is often marked by the use of pathetic inflections of the kind found earlier in Locke and Humfrey. Florid writing occurs here, as well as in arias. Its use in arias is not confined to extended pieces like 'Hark, the ech'ing air' in *The Fairy Queen* but is found also in such relatively simple songs as 'Dear pretty youth' in *The Tempest*.

In spite of Purcell's emphasis on Italian practice and his publisher's declaration, in the preface to the *Sonnata's of III Parts*, that Englishmen 'should begin to loath the levity, and balladry of our neighbours', he was not unaffected by French music. He may be presumed to have agreed with Dryden's observation, in the dedication to the score of *Dioclesian* (1691), that English music was 'now learning *Italian*, which is its best Master, and studying a little of the *French* Air, to give it more of Gayety and Fashion'. There are plenty of examples of gaiety and fashion in Purcell's dances and entr'actes. But French influence appears also in more solemn passages, notably the introductions to his overtures and the ritual scenes in his dramatic music, as well as in some of the airs of a more popular cast, e.g. 'I come to sing great Zempoalla's story' in *The Indian Queen*. Besides all this there is a strong English element, both in instrumental writing which preserves an older tradition and in songs in the popular idiom of the day. None of these influences accounts for the personal element in Purcell's music, which won him a reputation in his day and has maintained it ever since. This personal element, though often difficult to pin down exactly, has several recognizable features, among them a vein of nostalgia which is not to be explained simply by the use of chromaticism and pathetic inflections, a wide-ranging vocal line and continuity without repetition. Much of his simpler, more direct music does not differ in character from the work of his contemporaries: it merely happens to be better.

The range of Purcell's invention is remarkable. If there were not positive evidence it might be difficult to believe that the same person was the composer of *Ah! few and full of sorrow* and the duet 'Hark! how the songsters' in *Timon of Athens*, or that the man who wrote the chaconne published in *Ten Sonata's in Four Parts* could also produce a series of vulgar catches. If this diversity is not always appreciated, the reason is that although all the surviving music is now in print relatively little of it is performed. The court odes are virtually unknown, the dramatic music, apart from *Dido and Aeneas*, is mainly represented by a handful of songs and a few instrumental suites, and the anthems with strings are heard only at festivals when the necessary resources can be assembled. Only a few of the songs and duets are regularly sung, and then often to the accompaniment of 'realizations' which, when not inept or even inaccurate, distract attention from the vocal line by misplaced ingenuity. On the whole the chamber works have fared best, but even here the works with continuo accompaniment have suffered from over-editing. The survival of Purcell's music depends not on factitious additions to his text but on performance by singers and instrumentalists who have the technique and the imagination to do it justice. Recently, more such performers have been doing so.

CHAPTER THREE

Dramatic music

The music that Purcell composed for the stage was conditioned by the circumstances of the time. The reopening of the London theatres at the Restoration created a desire to see plays. Songs and instrumental music could be introduced, as they had been in Shakespeare's time, but opera as it was understood on the Continent was not welcomed, nor was it feasible without a substantial subsidy and facilities for training singers. In the 1660s there were various plans for introducing Italian opera to London, but they were not successful. In consequence most of Purcell's dramatic music consisted of overtures, entr'actes, dances and songs. There were, however, a limited number of productions which provided more generous opportunities for music by including masques and scenes of a sacrificial or ceremonial character. Purcell wrote music for five productions of this kind. Four of them – *Dioclesian*, *The Fairy Queen*, *The Tempest* and *The Indian Queen* – were adaptations of existing plays; serious doubts as to the authenticity of the *Tempest* music except for 'Dear pretty youth' have recently been expressed (e.g. 'Purcell did not set *The Tempest* . . .', Price, 1984, p.20). The fifth, *King Arthur*, was adapted by John Dryden from his first version specifically with Purcell's music in mind. It was no doubt Dryden's patronage that stimulated Purcell's work for the theatre in the last five years of his life. Not all his dramatic music can be dated exactly, but it is certain that the greater part of it was written between 1690 and 1695, in contrast to his church music, hardly any of which dates from these years.

The most substantial music is naturally to be found in what Roger North called 'semi-operas' – the five works mentioned above. Some of it is on grandiose lines – for instance, the chaconne at the end of the masque in *Dioclesian*, built on a pompous ground bass with brilliant antiphony between trumpets, oboes and strings, or the setting of 'How happy the lover' in *King Arthur*, where every possibility of the ground bass is exploited by voices and instruments. Particularly impressive are the ceremonial scenes (e.g. the preparations for the sacrifice in Act 5 of *The Indian Queen*), which are not confined to the 'semi-operas'. They occur also in the incidental music written for other plays: examples are the scene for the druids in *Bonduca* and the incantation in *Oedipus*, which includes the song 'Music for a while'. But since many of the plays for which Purcell wrote music were comedies the opportunities for such extended scenes were limited.

Although the settings of some of the plays are exotic, there is no attempt at local colour in the music: any such attempt would have been foreign to the ideas not only of Purcell but of his contemporaries. There is, however, plenty of atmosphere and often lively characterization. The scene in *King Arthur* where the Cold Genius, with chattering teeth, rises slowly from the ground is made more vivid by a throbbing accompaniment for strings and a chorus which follows a little later in the same vein. Equally imaginative is the appearance of Night, with muted strings, in *The Fairy Queen* and the hissing of Envy and his followers in *The Indian Queen.* Purcell was as much at home with comic characters as with allegorical figures: the dialogue for Corydon and Mopsa in *The Fairy Queen* may sound a little too sophisticated for a representation of country life, but it is richly humorous in conception. Some of the instrumental music is clearly designed to illustrate action on the stage, e.g. the Dance of the Furies and the Chair Dance in *Dioclesian,* and the 'Symphony while the Swans come forward' and the Monkeys' Dance in *The Fairy Queen.* But the greater part of the instrumental music has no dramatic significance: for this reason it is included in the general discussion of instrumental music in chapter 7. Similarly many of the songs have no direct relevance to the plays in which they occur. They were not in any case sung by the principal actors, though they might be assigned to priests, spirits, allegorical figures and subordinate characters generally: for a discussion see chapter 6.

Dido and Aeneas is an exception to the general run of Purcell's music for the stage in having a libretto which is set to music throughout. It was written for performance by 'Young Gentlewomen' at Josias Priest's boarding-school at Chelsea in 1689; though divided into three acts it lasts little more than an hour. Since Aeneas is a tenor, at least in the copies that have survived, and there are parts for tenors and basses in the chorus, the work cannot have been performed entirely by Priest's pupils; and although the music does not make any extravagant demands, some of it calls for a professional standard of technique. Though the opera is a miniature it covers a wide range of emotional expression, from Dido's soliloquies and Aeneas's passionate regret to the sombre utterances of the sorceress and the breezy heartiness of the sailor. Since Priest was a dancing-master there are opportunities for ballet – particularly a Triumphing Dance in chaconne rhythm for the members of Dido's court and a fantastic dance for the witches; there may well have been more in the original production. The work suffers a little from its brevity: the drama moves too rapidly for all the episodes to make their full effect, and Aeneas hardly has time to establish himself as a character. But against this must be set the expressive quality of the recitative, the inspired treatment of ground basses, and the imaginative portrayal of the witches – not least in the use of an echo behind the scenes. Everywhere the music triumphs over a prosaic libretto.

CHAPTER FOUR

Odes and welcome songs

The practice of honouring the royal family with music was established early in the reign of Charles II. The regular occasions were New Year's Day, a royal birthday and the king's return to the capital after a visit elsewhere – generally to Windsor or Newmarket. The only occasion when Purcell set an ode for New Year's Day seems to have been in 1694, when according to the author of the words, Matthew Prior, *Light of the world* was sung; but the music has not survived. During the reigns of Charles II and James II Purcell set nine welcome songs, as well as an ode for the marriage of Prince George of Denmark to Princess Anne. After the accession of William III welcome songs were virtually abandoned, but Purcell wrote six birthday odes for Queen Mary and one for the birthday of Princess Anne's son, the Duke of Gloucester.

The court odes and welcome songs cover almost the entire period of Purcell's known activity as a composer, from 1680 to 1695. Although they do not represent all the facets of his art they give a good general idea of the way in which he was developing. They might properly be described as cantatas for solo voices, chorus and orchestra. Those for Charles II and James II employ mainly strings; however, recorders are used in three of them and in *Swifter, Isis, swifter flow* (1681) an oboe is introduced for the first time. Some of the odes written during the reign of William III are more elaborately orchestrated: five out of the seven have oboes and trumpets. The timpani part found in the only surviving copy (1765) of the last of the Queen Mary odes, *Come, ye sons of art, away*, is probably not authentic. Like the verse anthems the odes include sections for solo voices – in some cases as many as seven – and a limited amount of recitative. By a quaint conceit *Why, why are all the Muses mute?*, the first welcome song for James II, begins with recitative and the overture is not heard until after the chorus has sung 'Awake, awake'. The overtures, as in Purcell's works in general, are mostly modelled on the French opera overture, with a pompous introduction, a canzona in fugal style and sometimes an adagio to conclude. Two of the odes introduce popular tunes. In *Ye tuneful Muses* (1686) the melody of *Hey then, up go we* is ingeniously used as the bass of a countertenor solo, then as a counterpoint to a chorus and finally as the bass of an orchestral ritornello. In *Love's goddess sure was blind* (1692) the Scots tune *Cold and raw* is introduced as a bass to the soprano solo 'May her blest example'; according to a tradition reported by John Hawkins, it was one of Queen Mary's favourite tunes.

The texts of the odes, most of which are anonymous, are obsequious in the manner of the period. Since Purcell was thoroughly familiar with this conventional attitude to royalty there is no reason to suppose that he was hampered by the words that he had to set. At the same time his imagination was clearly stirred more vigorously by any suggestion of imagery. Examples are the solos 'Welcome, more welcome does he come' in *From those serene and rapturous joys* (1684) and 'See how the glitt'ring ruler of the day' in *Arise, my Muse*

11. Henry Purcell: portrait (1695) by John Closterman

(1690), both on a ground bass. The first of these, sombre in tone, refers to Lazarus emerging in a winding-sheet 'from his drowsy tomb'; the second, in minuet rhythm, represents the sun summoning the planets to 'dance in a solemn ball'. For obvious reasons the texts only rarely offer opportunities for a pathetic style. Where they do Purcell clearly welcomed them, e.g. in the countertenor solo 'But ah, I see Eusebia drowned in tears' in *Arise, my Muse*. Here there is genuine emotion, in spite of the fact that 'Eusebia' is the Anglican Church, regretting that William III is compelled to champion her cause in Ireland. Much of the music in these works, however, is in a simple, extrovert style, which nevertheless does not exclude contrapuntal treatment in the choruses. How successful this style could be is best illustrated by *Come, ye sons of art, away*, where there are splendid, swinging tunes which invite the cooperation of trumpets. The same ode also includes one of the most brilliant of Purcell's songs on a ground bass, the duet 'Sound the trumpet' for two countertenors, as well as an elaborate soprano solo, 'Bid the virtues', with an equally elaborate oboe obbligato.

Purcell's odes for other occasions are planned on similar lines. Some of them have rather better texts than the court odes – but not all: Nahum Tate's *Great parent, hail*, for the centenary of Trinity College, Dublin (1694), is a monument of banality, which clearly failed to inspire the composer. Nor did D'Urfey's *Of old when heroes thought it base*, written for the annual reunion of Yorkshire gentlemen (1690), evoke more than a conventional response. Brady's ode for St Cecilia's Day, *Hail, bright Cecilia* (1692), is superior to these mundane offerings. The text, in praise of music, offers particular opportunities in its references to individual instruments. The passage beginning

> Hark, each tree its silence breaks,
> The box and fir to talk begin

refers to the recorder and the violin – which led Purcell to accompany his duet setting with an antiphonal treatment of strings and three recorders (one a bass). 'The fife and all the harmony of war' has a martial setting for two trumpets and timpani. The organ, extolled as 'Wondrous machine' over a ground bass, is not specifically illustrated, though it is not impossible that one was present in Stationers' Hall, where the work was first performed. According to the *Gentleman's Journal* the declamatory solo for countertenor, ' 'Tis Nature's voice', was 'sung with incredible Graces by Mr. *Purcell* himself'; this may not mean that he actually sang the piece but merely that he had written out the graces, preserved in the score, instead of leaving them to the singer. Though the music for the other soloists is on the whole less arduous it all demands a very proficient technique. The choral writing in this work is more elaborate than in many of Purcell's odes, particularly in the six-part writing in the final chorus, which includes a canon by double augmentation. An earlier St Cecilia ode, *Welcome to all the pleasures* (1683), is much slighter but has a kind of youthful freshness which is very agreeable. It includes a particularly effective song on a ground bass, 'Here the deities approve', which seems to have been popular, since a keyboard transcription was published in *The Second Part of Musick's Hand-maid* (1689) under the title 'A New Ground'.

CHAPTER FIVE

Sacred music

The repertory of church music at the time of the Restoration was inevitably based on the work of pre-Commonwealth composers, whose anthems and services continued in use, particularly in cathedrals. Purcell himself began at an early age to make copies of anthems by Tallis, Byrd, Orlando Gibbons and other composers of earlier generations. In the Chapel Royal the need for new music was supplied by composers like Child, Locke and Cooke, who was Master of the Choristers. Cooke was fortunate in having a number of talented boys in the choir, and some of these – notably Humfrey, Blow and Wise – contributed anthems in a style which probably owed something to Cooke's Italian training. A feature of the new style was the extensive use of verse sections for solo voices. Another, found only in music for the Chapel Royal, was the use of a string orchestra in works to be performed when the king was present. Purcell thus became familiar as a boy both with the older tradition of the full anthem and with the newer style, in which solo voices and instruments played an important part. The full anthem could also include sections for a group of solo voices by way of contrast, but they did not call for an independent accompaniment.

Thomas Purcell's letter to Gostling cited on p.217 above shows that Purcell was writing anthems for the Chapel Royal as early as 1679. In fact it has been shown that the earliest anthem that can be dated (approximately), *Lord, who can tell how oft he offendeth?*, was probably written before 1678. Until Purcell was appointed an organist of the Chapel Royal he seems to have divided his attention between full anthems and verse anthems. Some of these must have been sung in Westminster Abbey, since the records for 1681–2 mention a payment 'for writing Mr. Purcell's Service & Anthems'. Once he became a member of the Chapel Royal in 1682 he abandoned the full anthem, either because of the obligations of his new post or because he was inclined to move with the times. His full anthems follow a well-established tradition of vocal polyphony, though not without signs of his individual attitude to harmony, e.g. in the chromatic progressions which illustrate the 'bitter pains of eternal death' in the first version of *In the midst of life*. The early verse anthems with organ are in some respects not unlike the full anthems. Here too there are several instances of an imaginative treatment of the text. These characteristics appear equally in the sacred music designed for private use, which is discussed later in this chapter. In the Latin psalm *Jehovah, quam multi*, which appears to be an early work, the most remarkable details of the harmony are produced by diatonic clashes between the parts. This work, and *Beati omnes*, may have been written for the queen's chapel at Somerset House, where G. B. Draghi was organist.

Purcell never abandoned his personal response to the texts that he set, but in the larger verse anthems it tended to find its principal expression in recitative. Most of his anthems with strings appear to date from his appointment at the Chapel Royal, and he continued to write works of this kind until the accession of William III. Though James II, as a Roman Catholic, had his own chapel, the Chapel Royal was maintained, but with a declining number of singers, and the string players were ordered to attend when Princess Anne was present. After the accession of William III the use of strings in the Chapel Royal seems to have been abandoned, and only three of Purcell's anthems, all with organ, are known to have been written in his reign.

The main function of the string orchestra was to provide an overture and ritornellos, though it might also on occasion join the voices. Purcell's overtures generally begin in the pompous style of the French opera overture, usually followed not by a fugal canzona but by a brisk, rhythmical movement in triple time – a style which recurs in the ritornellos. The sections for solo voices also have a strongly marked rhythm, but not the symmetrical structure to be found in the odes, no doubt because the words are not in verse. The ground bass, which is a constant feature of the odes, occurs rarely: one example is in the anthem *In thee, O Lord, do I put my trust*, and there it is used not for the vocal sections but for the introductory symphony. The writing, both for voices and for instruments, is often florid, with a profusion of dotted notes. This style has sometimes been described as 'secular' – a misuse of the term, since music can be sacred or secular only by association. No doubt the Restoration composers, including Purcell, were endeavouring to satisfy Charles II's desire for a lively, rhythmical style. But this kind of church music had been current on the Continent for a good many years before the Restoration, and in adopting it the English composers were merely showing that they were

12. Autograph MS of part of Purcell's anthem 'My heart is inditing', composed in 1685

up to date. Nor was liveliness the only characteristic of the verse anthem. Much of the music is dignified and pathetic in expression, especially in recitative. Like Italian opera composers Purcell obviously wrote for individual singers. He had in John Gostling the services of an unusually accomplished bass soloist. Few singers even at the present day would care to tackle a passage like that in ex.2, from *I will give thanks unto thee, O Lord.*

Ex.2

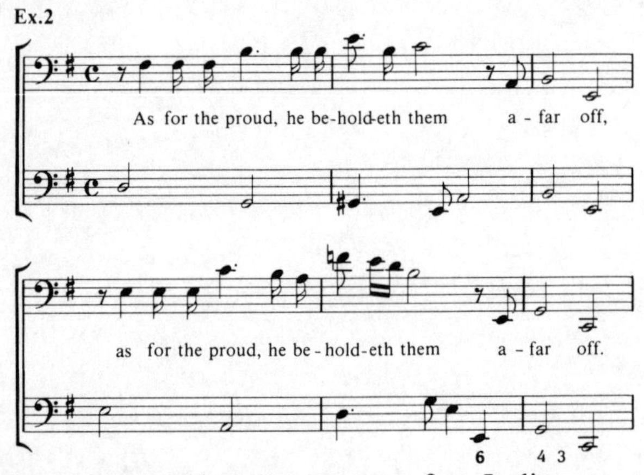

The most complete acceptance of an Italianate style is to be found in *O sing unto the Lord*, with its vigorous antiphony between instruments and chorus. The longest of Purcell's anthems, and certainly the most imposing, is *My heart is inditing*, written for the coronation of James II (see fig.12). Here he had the advantage of two choirs, which made possible extensive subdivision of the voices, and he took the opportunity to produce something both more solid and less episodic than the general run of his anthems. The choral antiphony at 'Hearken, O daughter' is no less impressive than the massive 'Alleluia' with which the work ends. Purcell seems to be completely involved. This is not always the case in the anthems in general, where the music, however well tailored, sometimes seems to lack conviction and to be marked more by mannerisms than by ideas.

The works written for private performance, several of which date from early in Purcell's career, are rather different in character. For one thing they are more intimate: there is no question of impressing a Sunday congregation in the Chapel Royal. Here one finds more than once a manipulation of harmony which carries even further the ambiguity to be found in many of Purcell's earlier works. The opening of *Plung'd in the confines of despair* is a characteristic example (see ex.3). A number of solo settings establish a similar mood. A penitential attitude to God seems to have been favoured for private devotion. Not all the settings, however, are of this character. Here too, as in the verse anthems, there are jaunty 'Alleluias' and sometimes even a symmetrical tune, as in the four-part *Early, O Lord, my fainting soul*, though it falls a good deal short of the individuality shown in many of Purcell's secular songs. One piece, *In guilty night*, is in effect a dramatic cantata on the subject of Saul's visit to the Witch of Endor. Of the solo songs the *Evening Hymn* ('Now that the sun hath veil'd his light') deserves its high reputation. Founded on a simple descending ground bass, it exhibits Purcell's customary skill in handling this form. The vocal line is adjusted to disguise the repetition of the bass, which modulates in the middle section of the piece. The seemingly endless 'Alleluia' that fills the last 45 bars is sufficient to make one forget more jogtrot settings of the word in other contexts.

Ex.3

Since Purcell was an organist for 18 years of his life it is curious that he appears to have written relatively little for the Anglican service. What we have is a complete service in B♭ (including the alternative canticles, Kyrie and the Nicene Creed), an Evening Service in G minor, and a *Te Deum* and *Jubilate* in D. The last is a special case, as it was written for St Cecilia's Day 1694 and is scored for trumpets and strings. Both the services are 'full' settings, but they include many verse sections. They are relatively simple works, even though they include several ingenious canons, and look as if they might have been written for ordinary cathedral choirs rather than for the Chapel Royal: the fact that the B♭ service occurs, either complete or in part, in such a large number of manuscripts suggests that it was welcomed in cathedrals. The verse setting of the *Te Deum* and *Jubilate*, which makes much play with trumpets, was admired at the time and for some years after Purcell's death. But its superficial brilliance does not bear comparison with the best of his anthems: only the more intimate moments are rewarding.

CHAPTER SIX

Secular vocal music

The sheer quantity of Purcell's secular music for one or more voices is almost embarrassing. In the dramatic music alone there are nearly 150 solo songs, and there are more than 100 others, most of which were published in the songbooks of the time. There are also a large number of duets and catches, not to mention a handful of cantatas (a term not actually used by Purcell) for two or more voices, some of which have parts for violins or recorders or both. Some of the solo songs are simple ditties, others are themselves cantatas, with a sequence of recitative and aria. Purcell's treatment of recitative was not restricted to a simple enunciation of the words. It was often florid, incorporating in its notation what singers might otherwise have improvised from a simple vocal line. It could more properly be described as arioso, which does not imply that it should be sung in strict time. The opening of *The fatal hour*, published in the second volume of *Orpheus Britannicus*, will serve as an example (see ex.4).

The cantata pattern was particularly suitable for 'mad songs', popular at the time, in which the various stages of frenzy, from dejection to wild fantasy, could be illustrated by contrasted sections of the music. A characteristic example is 'From rosy bowers' in the third part of *Don Quixote*, described when it was published posthumously as the last song that Purcell wrote. D'Urfey,

the author of the play, listed the five sections as follows:

1. Sullenly Mad ('From rosy bowers, where sleeps the God of Love')
2. Mirthfully Mad. A Swift Movement ('Or if more influencing, Is to be brisk and airy')
3. Melancholy Madness ('Ah, 'tis in vain, 'tis all, 'tis all in vain')
4. Fantastically Mad ('Or, say ye powers, my peace to crown, Shall I thaw myself or drown')
5. Stark Mad ('No, no, no, no, I'll straight run mad').

Purcell's music faithfully follows these divisions. Sections 1 and 3 are set as a slow arioso, section 2 in brisk duple time, section 4 in triple time, while section 5, marked 'Quick', develops into a miniature aria with an increasingly elaborate bass. Cantata settings other than mad songs did not call for such extravagant gestures. There was, however, no standard pattern. 'Sweeter than roses' in *Pausanias*, for instance, has only two sections: a melting arioso which blazes briefly into fire, and a splendid song of triumph in triple time, 'What magic has victorious love'. Some other examples, particularly dialogues, have more.

The songs which consist of a single movement show a considerable variety of structure. Some are in a simple da capo form, e.g. Belinda's song 'Pursue thy conquest, love' in *Dido and Aeneas*. Some are in the rondeau form familiar from French music, e.g. 'I attempt from love's sickness to fly' in *The Indian Queen*. Others exploit variety over a ground bass, often of the simplest possible form. What is most remarkable is Purcell's power of extending a melody so that one phrase leads effortlessly to the next. He may repeat the first phrase, with or without a different ending; but after that the song frequently goes its own way without looking back, as if propelled by some hidden energy. The popular 'Nymphs and shepherds' in *The Libertine* returns briefly to its opening phrase at the end of the song, but most of it is a continuous and apparently spontaneous flow of new invention. Even a da capo aria like 'Hark how all things in one sound rejoice' in *The Fairy Queen* shows this capacity for uninhibited continuity. Another striking characteristic of many of the songs is the wide range of intervals employed. The traditional precepts of good vocal writing are often flouted with considerable

13. *The beginning of Purcell's song 'Celia has a thousand charms': from 'Orpheus Britannicus'* (*London: Pearson, 1706 edition*)

Ex.5

When I have of - ten heard young maids com - plain - ing, That when men pro - mise most they most __ de - ceive,
Then I thought none of them wor - thy my gain - ing, And what they swore, I would nev - er be - lieve.
But when so hum - bly one made his ad - dress - es, With looks so soft, and with lan - guage so kind,
I thought it sin to re - fuse his ca - ress - es, Na - ture o'er - came and I soon changed my mind.

success. An outstanding example is the song 'When I have often heard young maids complaining' in *The Fairy Queen*, which is quoted complete (as ex.5) because it also illustrates the principle of continuity just mentioned. Except possibly at the cadences there are no intervals in the melody that can be predicted, yet they all sound unforced and natural. The modulation in the second half springs no surprises, but the details of its progress and its return to the tonic key are equally unpredictable.

Songs built on ground basses had been used by Italian composers for a good many years before Purcell was born. He was obviously attracted by the challenge that the form presented, since he used it constantly. The difficulty with a short diatonic ground is that the music may be too firmly tied to the tonic key. Purcell did not always avoid this difficulty: the song *O solitude, my sweetest choice* hardly ever escapes from the key of C minor. No doubt it was awareness of the problem that prompted him often to modulate and present the bass at one or more different levels, as in Dido's song 'Ah! Belinda', where a temporary switch from C minor to G minor avoids monotony. At the same time the very flexible vocal line distracts attention from any sameness of key. Where the bass has a less definite tonality there are more opportunities for divergence. The ambiguous ground bass of 'Music for a while' in *Oedipus* allows greater freedom, and Purcell enlarged this by skilful modulation and extension. A particularly striking example of his adroitness in handling the form is the short section of *Gentle shepherds* (an elegy on the death of John Playford) beginning 'Muses, bring your roses hither'. Here the bass is a simple descending scale of C in quavers; but, far from being tied to this key, Purcell managed to make brief allusions to F, E minor and G. In *Cease, anxious world* he not only modulated but used the same formula for the bass of two different sections of the song – one in triple, the other in duple time.

Purcell's increasing preoccupation with Italian music led him to write a good many songs in a florid style, often with quickly moving basses. Examples are the opening sections of *Love arms himself in Celia's eyes* and *Anacreon's Defeat* ('This poet sings the Trojan wars'), both in a heroic vein. These characteristics are strongly marked in the big da capo arias in *The Tempest*, though (see chapter 3) these may well be by some other composer. Though many of his simpler songs can be comfortably performed by amateurs, and no doubt were at the time, it is clear that Purcell was writing primarily for professional singers who, on the evidence of his music, must have had a highly developed technique. He was careful, as the Italians were, to put his runs on an accented syllable and generally choose a long vowel. In passages of this kind music takes precedence over speech; elsewhere he was punctilious in preserving the natural rhythms of English words – over and over again the music fits the text like a glove.

In duets there is the added attraction of two strands of melody woven together, most beautifully in the Latin elegy on Queen Mary, *O dive custos Auriacae domus*, and, in a different vein, 'My dearest, my fairest' in *Pausanias*. There are others less serious in intention, including comic dialogues. The catches, nearly all for three voices, range from the merely convivial to the frankly obscene. The music, though expert, is of minor importance and would hardly be worth mentioning except as an indication of the composer's versatility.

CHAPTER SEVEN

Instrumental music

The music Purcell composed for instrumental forces, whether written in association with choral works or for the theatre or for independent performance, is the most convincing evidence of his genius. In his overtures he favoured the French type, though occasionally inclining to one more nearly related to the Italian. Where there are three movements, the third is normally a solemn adagio, often of a particularly poignant character. In the case of the dramatic music it would be tempting to see some relation between such passages and the play that follows; but in fact there is no obvious distinction between overtures to tragedies and those for comedies. What is common to all of them is a scrupulous attention to details. The same is true of the act tunes (i.e. entr'-actes) and dances. Quite apart from their rhythmical exuberance there are many subtleties of harmony which must have passed unnoticed by many of those who attended the performances. The resources available were limited. In the Chapel Royal Purcell was restricted to strings: the same is true of practically all his incidental music for the theatre. For the odes, recorders, oboes, trumpets and, exceptionally, timpani were available, and he used these resources in all his major works for the stage, with the exception of *Dido and Aeneas*, which is for strings only. *Dioclesian* is unique in having a part for tenor oboe in the masque, and the symphony in Act 4 of *The Fairy Queen* ('A Sonata plays while the Sun rises') must be one of the earliest instrumental pieces to begin with a timpani solo.

In the dramatic music and the odes Purcell wrote effective obbligatos for wind instruments – for recorders, for example, in the ode *What, what shall be done in behalf of the man?*, in the duet 'Hark! how the songsters' in *Timon of Athens*, and in the girl's song 'Why should men quarrel?' in *The Indian Queen*; for solo oboe in the ode *Come, ye sons of art, away*; and for trumpet in many pieces of a heroic character. He was also adept at the antiphonal treatment of strings and wind. But since the bulk of his work was for string orchestra it is here that his gift for instrumental writing is most evident – in its dignity, its vivacity and the illumination created by the inner parts.

All this was public music, in which intimacy was incidental. In his chamber music he could give freer expression to his more personal invention. It divides into two obvious categories: works for a viol consort, a medium that was virtually obsolete in his day, and sonatas for the more up-to-date combination of two violins, bass viol and keyboard continuo. The two categories are near to each other in date. Most of the fantasias for viols are exactly dated in the year 1680; and several of the *Ten Sonata's in Four Parts*, published by Purcell's widow in 1697, occur in the same manuscript and in a similar handwriting. The *Sonnata's of III Parts*, published by Purcell himself in 1683, were probably written about the same time, though the autograph does not survive to prove this. Though there are obvious differences between the fantasias and the sonatas, the two categories overlap to some extent, since there are details in the sonatas which are nearer to older

English methods than to the current Italian style. It is perhaps not surprising that Corelli was said to have had a 'mean opinion' of Purcell's sonatas: to him the Italian characteristics to which Purcell's publisher drew attention would have seemed less convincing. A similar view is implied by Roger North's reference to Purcell's 'noble set of sonnatas' as 'clog'd with somewhat of an English vein'.

The fantasias, which owe something to the example of Locke, are not only masterpieces of contrapuntal writing but are also passionate revelations of the composer's most secret thoughts. Every possible device of imitation, inversion and augmentation is employed, but the result, far from being an academic exercise, is appealing to the ear and might even be termed romantic in expression. In contrast to the composer's later music the tonality is often ambiguous, not least in some of the homophonic passages which occur by way of contrast. Chromatic movement appears frequently, and virtuosity is not excluded. In the Fantasia upon One Note, where the fourth part repeats middle C throughout, Purcell set himself a severe problem but solved it by exploiting all the possible harmonic progressions and by rhythmic vivacity. The two In Nomines look back firmly to the old English tradition of weaving counterpoint around the plainsong *Gloria tibi Trinitas* as it appears in the Benedictus of Taverner's mass with that title.

The *Sonnata's of III Parts* are described on the title-page as for 'Two Viollins and Basse: to the Organ or Harpsecord'. We learn from the preface that Purcell had not originally intended to publish a basso continuo part – a curious observation, since the bass viol part is not always identical with the basso continuo, and in one

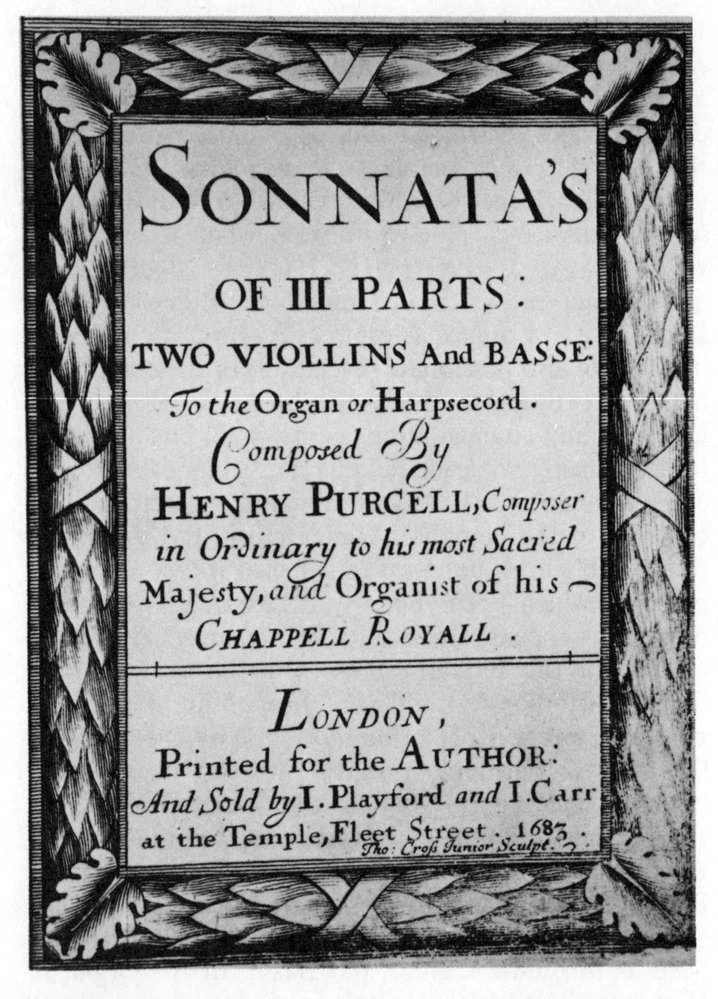

SONNATA'S

OF III PARTS:

TWO VIOLLINS And BASSE:

To the Organ or Harpsecord.

Composed By

HENRY PURCELL, Composer in Ordinary to his most Sacred Majesty, and Organist of his CHAPPELL ROYALL.

LONDON,

Printed for the AUTHOR:

And Sold by I. Playford and I. Carr at the Temple, Fleet Street. 1683.

Tho: Cross Junior Sculpt.

14. Title-page of Purcell's 'Sonnata's of III Parts' (*London: Author and John Playford, 1683*) movement of the fifth sonata the two are completely independent for about 30 bars. If the sonatas were in fact string trios in their original form, a good deal of rewriting must have taken place before they were published. Here, as in the fantasias, there is masterly counterpoint, although the extremes of ingenuity occur less frequently. There are some particularly happy examples of imitation by augmentation. The brisk canzona movements are nearer to the Italian style than the stately introductions, which in general are closer to the older English style. Even though the violin parts are largely confined within a modest range, they show a thorough understanding of the technique of the instrument and, particularly in the *largo* movements, an appreciation of its sensuous qualities.

The *Ten Sonata's in Four Parts* include a single movement in the form of a chaconne which is not a sonata at all. However, no one is likely to regret its inclusion, as it is one of the very finest, as well as one of the longest, of Purcell's instrumental works. Every possible device of harmonization, figuration and imitation is employed to produce a piece which impresses both by the grandeur of the conception, the wealth of invention and the intimacy of expression. The fact that it was not published until after Purcell's death probably reflects a lack of public support for instrumental music of this kind. This lack of interest would explain why there are no contemporary manuscript copies; and since the autograph has disappeared it is difficult to say when it was written.

Purcell's keyboard music forms only a minor part of his work. Apart from the eight suites dedicated to Princess Anne, which are agreeable miniatures, there are a number of little pieces presumably designed for amateurs, as well as transcriptions from other works. It is all tasteful but of no great significance when compared with the rest of Purcell's music. Nor do the organ pieces, whether authentic or not, call for more than a passing mention.

APPENDIX

Autographs

A detailed list of Purcell's autographs is printed in Holst, pp.106ff. The principal ones are:

(i) Fitzwilliam Museum, Cambridge, MS 88, ?1677–82. In addition to anthems by Purcell (some incomplete) it includes his copies of anthems by other composers, including Blow, Locke and Humfrey.

(ii) British Library, London, MS Add.30930, *c*1680–83. Sacred music of various kinds and instrumental ensemble music, including 13 fantasias and eight trio sonatas (some incomplete), the latter printed in *Ten Sonata's in Four Parts* (1697).

(iii) British Library, London, MS RM.20.h.8, *c*1681–90. Anthems, welcome songs, odes, cantatas and songs (some incomplete). A few items in a copyist's hand.

(iv) Gresham College (Guildhall Library), London, Safe 3, 1690–95. Songs (some incomplete), mostly from the dramatic music.

Publications

Apart from contributions to anthologies and to collective works, and single issues of songs, only a few works by Purcell were published in his lifetime, all of them in London:

Theodosius: or, The Force of Love . . . with the Musick betwixt the Acts (1680) [music anonymous]
Sonnata's of III Parts: Two Viollins and Basse: to the Organ or Harpsecord (1683, 2/1684)
A Musical Entertainment perform'd on November XXII, 1683 it being the Festival of St. Cecilia . . . (1684) [the ode Welcome to all the pleasures]
A Pastoral Elegy on the Death of Mr. John Playford (1687) [Gentle shepherds]
New Songs sung in The Fool's Preferment, or, The Three Dukes of Dunstable (1688)
The Songs of Amphitryon (1690)
The Vocal and Instrumental Musick of the Prophetess, or The History of Dioclesian (1691)
Some Select Songs as they are Sung in the Fairy Queen (1692)
The Songs in the Tragedy of Bonduca (?1695)
The Songs in The Indian Queen: as it is now Compos'd into an Opera (1695)

The following works were published posthumously, also in London:

A Choice Collection of Lessons for the Harpsichord or Spinnet (1696, 3/1699)
A Collection of Ayres, compos'd for the Theatre, and upon other Occasions (1697)
Ten Sonata's in Four Parts (1697)
Te Deum & Jubilate . . . made for St. Caecilia's Day, 1694 (1697)
Orpheus Britannicus: a Collection of all The Choicest Songs for One, Two, and Three Voices (1698–1702/ *R*1967, 2/1706–11, 3/1721)

Portraits

The only authentic portraits of Purcell known today are the following:

(i) Portrait in black chalk heightened with white, head, attributed to Godfrey Kneller (1646–1723); it belonged to William Seward (1749–99) and then to Charles Burney, and was acquired by the National Portrait Gallery, London, in 1974. It is believed to be the only surviving portrait of Purcell from life (fig.9).

(ii) Portrait in oils, head and shoulders, formerly attributed to Kneller. Depicts Purcell as a young man. Inscribed: 'Mr. Henry Purcell'. National Portrait Gallery.

(iii) Engraving, head and shoulders, printed as the frontispiece to the first violin part of *Sonnata's of III Parts* (1683). Presumably taken from a painting now lost. Description: 'Vera Effigies Henrici Purcell, Aetat: Suae 24'. The engraving in John Hawkins's *A General History of the Science and Practice of Music* (London, 1776, repr. 1875/*R*1969) is an inferior copy and like all the portraits in that work is reversed.

(iv) Crayon sketch, head, formerly attributed to Kneller. Department of Prints and Drawings, British Museum, London.

(v) Portrait in oils, head and shoulders, by John Closterman (1656–1713) (fig.11). Traditionally said to have been painted after Purcell's death. National Portrait Gallery. The engraving printed as a frontispiece to *Orpheus Britannicus* (1698) is a copy of this, reversed. Description: 'Henricus Purcell. Aetat: Suae 37.95.'.

WORKS

(*see also Appendix above*)

Detailed list, including doubtful and spurious attributions, in F. B. Zimmerman: *Henry Purcell, 1659–1695: an Analytic Catalogue of his Music* (London, 1963) [Z]

Edition: *The Works of Henry Purcell*, The Purcell Society (London, 1878–1965; newly ed. or rev. 2/1961–) [PS]

Numbers in the right-hand column denote references in the text.

OPERAS AND SEMI-OPERAS

220, 223, 227, 228, 253–4, 229–32

Z	Title	Text	Type	First performance	Date	PS	
626	Dido and Aeneas	N. Tate	opera	London, Josias Priest's School for Young Ladies, Chelsea	1689	iii	228, 231–2, 248, 251, 253
627	The Prophetess, or The History of Dioclesian	T. Betterton, after J. Fletcher, P. Massinger	semi-opera	London, Dorset Garden	1690	ix	225, 227, 229, 230, 231, 253, 259
628	King Arthur, or The British Worthy	J. Dryden	semi-opera	London, Dorset Garden	1691	xxvi	229, 230
629	The Fairy Queen	?E. Settle, after Shakespeare: A Midsummer Night's Dream	semi-opera	London, Dorset Garden	1692	xii	224, 227, 229, 231, 248, 251, 253–4, 259
630	The Indian Queen (final masque by D. Purcell)	Dryden, R. Howard	semi-opera	London, Drury Lane	1695	xix, 1	224, 227, 229, 230, 231, 248, 254, 259
631	The Tempest, or The Enchanted Island [?only 'Dear pretty youth']	T. Shadwell, after Shakespeare	semi-opera	—	*c*1695	xix, 111	227, 229, 252

PLAYS WITH INCIDENTAL MUSIC AND SONGS

218, 220, 223, 227, 228, 230, 253–4

Z	Title	Text	Date	Purcell's contribution	PS	
606	Theodosius, or The Force of Love	N. Lee	1680	songs, ensembles, choruses	xxi, 115	223, 224, 259
589	Sir Barnaby Whigg, or No Wit like a Woman's	T. D'Urfey	1681	1 song with chorus	xxi, 103	
581	The History of King Richard II (The Sicilian Usurper)	Tate, after Shakespeare	1681	1 song	xx, 43	
590	Sophonisba, or Hannibal's Overthrow	Lee	?1685	1 song	xxi, 109	
571	A Fool's Preferment, or The Three Dukes of Dunstable	D'Urfey, after Fletcher: The Noble Gentleman	1688	7 songs, 1 duet	xx, 11	259
572	Amphitryon, or The Two Sosias	Dryden	1690	2 songs, 1 duet, inst music	xvi, 21	259
577	Distressed Innocence, or The Princess of Persia	E. Settle	1690	inst music	xvi, 122	

Z	Title	Text	Date	Purcell's contribution	PS	
588	Sir Anthony Love, or The Rambling Lady	T. Southerne	1690	2 songs, 1 duet, inst music	xxi, 87	
604	The Massacre of Paris	Lee	1690	1 song (2 settings, the 2nd for a 1695 revival)	xx, 106	
575	Circe	C. D'Avenant	?1690	2 songs, ensembles, choruses	xvi, 95	
597	The Gordian Knot Unty'd	—	1691	inst music	xx, 23	
598	The Indian Emperor, or The Conquest of Mexico	Dryden, R. Howard	1691	1 song	xx, 41	
612	The Wives' Excuse, or Cuckolds make Themselves	Southerne	1691	4 songs	xxi, 162	
576	Cleomenes, the Spartan Hero	Dryden	1692	1 song	xvi, 120	
580	Henry the Second, King of England	?W. Mountfort, J. Bancroft	1692	1 song	xx, 38	
586	Regulus	J. Crowne	1692	1 song	xxi, 51	
602	The Marriage-hater Match'd	D'Urfey	1692	1 duet	xx, 84	
573	Aureng-Zebe	Dryden	?1692	1 song	xvi, 42	
583	Oedipus	Dryden, Lee	?1692	2 songs, 2 trios, choruses	xxi, 1	230, 251
579	Epsom Wells	Shadwell	1693	1 duet	xvi, 221	
587	Rule a Wife and Have a Wife	Fletcher	1693	1 song	xxi, 85	
592	The Double Dealer	W. Congreve	1693	1 song, inst music	xvi, 194	
596	The Female Vertuosos	T. Wright, after Molière: Les femmes savantes	1693	1 duet	xx, 7	
601	The Maid's Last Prayer, or Any rather than Fail	Southerne	1693	2 songs, 1 duet	xx, 72	
607	The Old Bachelor	Congreve	1693	1 song, 1 duet, inst music	xxi, 19	
608	The Richmond Heiress, or A Woman once in the Right	D'Urfey	1693	1 duet (dialogue)	xxi, 53	
582	Love Triumphant, or Nature will Prevail	Dryden	1694	1 song	xx, 70	
591	The Canterbury Guests, or A Bargain Broken	E. Ravenscroft	1694	1 quartet	xvi, 87	
595	The Fatal Marriage, or The Innocent Adultery	Southerne	1694	2 songs	xx, 1	
603	The Married Beau, or The Curious Impertinent	Crowne	1694	1 song, inst music	xx, 89	
613	Tyrannic Love, or The Royal Martyr	Dryden	1694	1 song, 1 duet	xxi, 135	
578	The Comical History of Don Quixote, parts i–iii	D'Urfey	1694–5	songs, duets, ensembles, choruses	xvi, 132	246–8
570	Abdelazer, or The Moor's Revenge	A. Behn	1695	1 song, inst music	xvi, 1	

574	Bonduca, or The British Heroine	after Beaumont, Fletcher	1695	3 songs, 2 duets, ensembles, choruses, inst music	xvi, 45	230, 259
584	Oroonoko	Southerne	1695	1 duet	xxi, 38	
585	Pausanias, the Betrayer of his Country	R. Norton	1695	1 song, 1 duet	xxi, 44	248, 252
605	The Mock Marriage	T. Scott	1695	3 songs	xx, 113	
609	The Rival Sisters, or The Violence of Love	R. Gould	1695	3 songs, inst music	xxi, 63	
600	The Libertine	Shadwell	?1695	2 songs, ensembles, choruses	xx, 45	248
610	The Spanish Friar, or The Double Discovery	Dryden	?1695	1 song	xxi, 112	
611	The Virtuous Wife, or Good Luck at Last	D'Urfey	?1695	inst music	xxi, 148	
632	Timon of Athens	Shadwell, after Shakespeare	?1695	songs, duets, inst music, choruses	ii	228, 254

Spurious: z593 The Double Marriage (Beaumont, Fletcher), inst music, PS xvi, 211, by L. Grabu

ANTHEMS AND SERVICES

220, 223, 225, 228, 230, 238–45, 259

(*all with bc (org); with chorus 4vv unless otherwise stated*)

Z	*First line or Title*	*Type*	*Date*	*Forces*	*PS*
1	Awake, put on thy strength	verse	*c*1682–5	inc., 2 A, B, ?vv, 2 vn, va	xiv, 41
2	Behold, I bring you glad tidings	verse	1687	A, T, B, 2 vn, va	xxviii, 1
3	Behold now, praise the Lord	verse	*c*1680	A, T, B, 2 vn, va	xiii, 49
4	Be merciful unto me	verse	? before 1683	A, T, B	xxviii, 28
5	Blessed are they that fear the Lord	verse	1688	2 S, A, B, 2 vn, va	xxviii, 42
6	Blessed be the Lord my strength	verse	by Feb 1679	A, T, B	xxviii, 60
7	Blessed is he that considereth the poor	verse	?*c*1688	A, T, B	xxviii, 71
8	Blessed is he whose unrighteousness is forgiven	verse	*c*1678	2 S, A, 2 T, B	xiii, 71
9	Blessed is the man that feareth the Lord	verse	?*c*1688	A, T, B	xxviii, 83
10	Blow up the trumpet in Sion	full	by Feb 1679	cantoris (S, 2 A, T, B, 5vv), decani (2 S, A, T, B, 5vv)	xxviii, 96
11	Bow down thine ear, O Lord	verse	*c*1678	S, A, T, B	xiii, 103
12	Give sentence with me, O God	verse	by Nov 1681	A/T, T, B	xxxii, 117 (inc.)
13A	Hear me, O Lord, and that soon	verse	*c*1680	S, A, T, B	xiii, 87
13B	Hear me, O Lord, and that soon	verse	*c*1680	S, A, T, B, 5vv	xiii, 90

Z	First line or Title	Type	Date	Forces	PS	
14	Hear my prayer, O God	verse	before 1683	A, T, B	xxviii, 125	
15	Hear my prayer, O Lord	full	*c*1680–82	inc., 2 S, 2 A, 2 T, 2 B, ?vv	xxviii, 135	
16	In thee, O Lord, do I put my trust	verse	*c*1682	A, T, B, 2 vn, va	xiv, 53	240
17A	In the midst of life	full	before 1682	S, A, T, B	xiii, 1	239
17B	In the midst of life	full	before 1682	A, T, B, 2 vn, va	xxviii, 215	
18	It is a good thing to give thanks	verse	*c*1682–5	A, T, B, 2 vn, va	xiv, 1	
19	I was glad when they said unto me	verse	1682–3	A, T, B, 2 vn, va	xiv, 97	
—	I was glad when they said unto me (incorrectly attrib. Blow in *GB-EL* 6)	full	1685	2 S, A, T, B	ed. B. Wood (London, 1977)	
20	I will give thanks unto thee, O Lord	verse	*c*1682–5	2 S, A, T, B, 9vv, 2 vn, va	xvii, 47	242
21	I will give thanks unto the Lord	verse	1685	T, 2 B, 2 vn	xxviii, 139	
N67	I will love thee, O Lord	verse		B	xxviii, 157	
22	I will sing unto the Lord	full	by Feb 1679	2 S, A, T, B, 5vv	xxviii, 165	
23	Let God arise	verse	by Feb 1679	2 T	xxviii, 173	
24	Let mine eyes run down with tears	verse	*c*1682	2 S, A, T, B	xxix, 1	
25	Lord, how long wilt thou be angry?	full	*c*1680–82	A, T, B, 5vv	xxix, 19	
26	Lord, who can tell how oft he offendeth?	verse	*c*1677	2 T, B	xxix, 28	239
27	Man that is born of a woman	full	*c*1680–82	S, A, T, B	xxix, 36	
28	My beloved spake	verse	before 1678	A, T, 2 B, 2 vn, va	xiii, 24	
29	My heart is fixed, O God	verse	*c*1682–5	A, T, B, 2 vn, va	xiv, 112	
30	My heart is inditing	verse	1685	2 S, 2 A, 2 T, 2 B, 8vv, 2 vn, va	xvii, 69	220, *241*, 242–3
31	My song shall be alway	verse	probably 1690	B/S, 2 vn, va	xxix, 51	
32	O consider my adversity	verse		A, T, B	xxix, 68	
33	O give thanks unto the Lord	verse	1693	S, A, T, B, 2 vn	xxix, 88	
34	O God, the king of glory	full	by Feb 1679	4vv only	xxix, 108	
D4	O God, they that love thy name	—		inc.	xxxii, 120	
35	O God, thou art my God	full	*c*1680–82	2 S, A, T, B, 8vv	xxix, 111	
36	O God, thou hast cast us out	full	*c*1680–82	2 S, 2 A, T, B, 6vv	xxix, 120	
37	O Lord God of hosts	full	*c*1680–82	2 S, 2 A, T, B, 8vv	xxix, 130	
38	O Lord, grant the king a long life	verse	1685	A, T, B, 2 vn	xxix, 141	
39	O Lord our governor	verse	before 1683	3 S, 2 B	xxix, 152	
40	O Lord, rebuke me not	verse	by Feb 1679	2 S	xxix, 168	
41	O Lord, thou art my God	verse	*c*1680–82	A, T, B	xxix, 179	
42	O praise God in his holiness	verse	*c*1682–5	A, T, 2 B, 2 vn, va	xiv, 21	
43	O praise the Lord, all ye heathen	verse	by Nov 1681	2 T	xxxii, 1	
44	O sing unto the Lord	verse	1688	S, A, T, 2 B, 2 vn, va	xvii, 119	242
45	Out of the deep have I called	verse	?*c*1680	S, A, B	xxxii, 8	

46	Praise the Lord, O Jerusalem	verse	probably 1689	2 S, A, T, B, 5vv, 2 vn, va	xvii, 146	
47	Praise the Lord, O my soul, and all that is within me	verse	*c*1682–5	2 S, 2 T, 2 B, 2 vn, va	xiv, 131	
48	Praise the Lord, O my soul, O Lord my God	verse	1687	A, B, 2 vn	xvii, 166	
49	Rejoice in the Lord alway	verse	*c*1682–5	A, T, B, 2 vn, va	xiv, 155	
50	Remember not, Lord, our offences	full	*c*1680–82	5vv	xxxii, 19	
51	Save me, O God	full	by Nov 1681	2 S, A, T, B, 6vv	xiii, 64	
52	Sing unto God	verse	1687	B	xxxii, 23	
53	The Lord is king, be the people never so impatient	verse		2 S	xxxii, 36	
54	The Lord is king, the earth may be glad	verse	1688	B	xxxii, 44	
55	The Lord is my light	verse	*c*1682–5	A, T, B, 2 vn, va	xiv, 78	
56	The way of God is an undefiled way	verse	1694	2 A, B, 6vv	xxxii, 58	
57	They that go down to the sea in ships	verse	1685	A, B, 2 vn	xxxii, 71	
58A	Thou know'st, Lord, the secrets of our hearts	verse	before 1683	no solo vv	xiii, 6	
58B	Thou know'st, Lord, the secrets of out hearts	verse	before 1683	S, A, T, B	xxix, 46	
58C	Thou know'st, Lord, the secrets of our hearts	full	1695	flatt tpts (slide tpts)	xxxii, 88	
59	Thy righteousness, O God, is very high	full		inc., ?4vv	xxxii, 124	
60	Thy way, O God, is holy	verse	1687	A, B, 2 vn	xxxii, 91 (inc.)	
61	Thy word is a lantern unto my feet	verse		A, T, B	xxxii, 101	
62	Turn thou us, O good Lord	verse		A, T, B	xxxii, 111	
63	Unto thee will I cry	verse	*c*1682–5	A, T, B, 2 vn, va	xvii, 20	
64	Who hath believed our report?	verse	*c*1677–8	A, 2 T, B	xiii, 11	
65	Why do the heathen so furiously rage together?	verse	*c*1682–5	A, T, B, 2 vn, va	xvii, 1	
230	Morning and Evening Service, B♭	full	before Oct 1682	cantoris (S, A, T, B), decani (S, A, T, B)	xxiii, 1	244–5
231	Magnificat and Nunc dimittis, g	full		cantoris (A, T, B), decani (2 S, A)	xxiii, 80	244–5
232	Te Deum and Jubilate, D	verse	1694	2 S, 2 A, T, B, 5vv, 2 tpt, 2 vn, va	xxiii, 90	244, 245, 259

OTHER SACRED

(*all with bc; firm dates are those of publication*) 230, 243, 259

Z	*First line or Title*	*Text*	*Voices*	*Date*	*PS*	
130	Ah! few and full of sorrow	G. Sandys	T, B, SATB	*c*1680	xxx, 109	228
181	Awake and with attention hear	A. Cowley	B	1688	xxx, 1	
182	Awake, ye dead	N. Tate	2 B	1693	xxx, 98	
131	Beati omnes qui timent Dominum	Ps cxxviii	S, B, SSAB	*c*1680	xxxii, 137	239
183	Begin the song, and strike the living lyre	Cowley	B	1693	xxx, 18	
184	Close thine eyes and sleep secure	F. Quarles	S, B	1688	xxx, 105	
102	Domine, non est exaltatum	Ps cxxxi	inc.	*c*1680		

Z	First line or Title	Text	Voices	Date	PS	
132	Early, O Lord, my fainting soul	J. Patrick	S, B, SSAB	c1680	xxx, 117	243
103–6	4 doxologies		canons, 3–4vv	1, in c, c1680	xxxii, 159, 161, 163, 168	
107	God is gone up	Ps xlvii. 5	canon, 7vv		xxxii, 170	
186	Great God and just	J. Taylor	S, SSB	1688	xxx, 33	
133	Hear me, O Lord, the great support	Patrick	A, T, ATB	1680–82	xxx, 127	
187	Hosanna to the highest	—	B, AB		xxx, 38	
188	How have I strayed	W. Fuller	S, SB	1688	xxx, 44	
189	How long, great God	J. Norris	S	1688	xxx, 48	
134	In guilty night	—	S, A, B, SAB	1693	xxxii, 128	243
190	In the black dismal dungeon of despair	Fuller	S	1688	xxx, 53	
135	Jehovah, quam multi sunt	Ps iii	T, B, SSATB	c1680	xxxii, 147	239
108	Laudate Dominum	Ps cxvii	canon, 3vv		xxxii, 170	
191	Let the night perish (Job's Curse)	Taylor	S, SB	1688	xxx, 57	
136	Lord, I can suffer thy rebukes	Patrick	S, S, A, B, SSAB	c1680	xxx, 136	
137	Lord, not to us but to thy name	Patrick	ATB	c1680	xxx, 146	
192	Lord, what is man?	Fuller	S	1693	xxx, 62	
109	Miserere mei	—	canon, 4vv	1687	xxxii, 171	
193	Now that the sun hath veil'd his light (An Evening Hymn on a Ground)	Fuller	S	1688	xxx, 70	243
138	O all ye people, clap your hands	Patrick	SSTB	c1680	xxx, 148	
139	O happy man that fears the Lord	Patrick	SSAB		xxx, 157	
140	O, I'm sick of life	Sandys	A, T, ATB	c1680	xxx, 160	
141	O Lord our governor	Patrick	B, SSAB	c1680	xxx, 167	
142	Plung'd in the confines of despair	Patrick	T, B, TTB	c1680	xxx, 180	243, 244
143	Since God so tender a regard	Patrick	T, B, TTB	c1680	xxx, 187	
195	Sleep, Adam, sleep and take thy rest	—	S	1683	xxx, 75	
196	Tell me, some pitying angel (The Blessed Virgin's Expostulation)	Tate	S	1693	xxx, 77	
197	The earth trembled	Quarles	S/B	1688	xxx, 85	
198	Thou wakeful shepherd (A Morning Hymn)	Fuller	S	1688	xxx, 88	
199	We sing to him whose wisdom form'd the ear	N. Ingelo	S, SB	1688	xxx, 91	
144	When on my sick bed I languish	T. Flatman	T, B, TTB	c1680	xxx, 194	
200	With sick and famish'd eyes	G. Herbert	S	1688	xxx, 94	
120–5	5 chants, Burford psalm-tune, all doubtful	—	SATB		xxxii, 173f	

Spurious: z185 Full of wrath, his threat'ning breath (J. Taylor), S, PS xxx, 28, by H. Brailsford

ODES AND WELCOME SONGS

(*with 4vv, 2 vn, va, bc, unless otherwise stated*)

220, 223, 228, 233–7, 240, 253, 254, 259

Z	First line	Occasion	Date	Forces	PS	
340	Welcome, vicegerent of the mighty king	welcome, Charles II	1680	2 S, A, T, B	xv, 1	218, 223
336	Swifter, Isis, swifter flow	welcome, Charles II	1681	2 S, A, T, B, 2 rec, ob, 3 vn	xv, 24	233
341	What, what shall be done in behalf of the man?	welcome, Duke of York	1682	2 S, A, T, B, 2 rec	xv, 52	254
337	The summer's absence unconcerned we bear	welcome, Charles II	1682	2 S, 2 A, T, 2 B	xv, 83	
325	From hardy climes and dangerous toils of war	wedding of Prince George of Denmark and Princess Anne	1683	B	xxvii, 1	
324	Fly, bold rebellion	welcome, Charles II	1683	2 S, 2 A, T, 2 B, 8vv	xv, 116	
339	Welcome to all the pleasures (C. Fishburn)	St Cecilia's Day	1683	2 S, A, T, B	x, 1	259
329	Laudate Ceciliam	St Cecilia's Day	1683	A, T, B, ?no chorus, no va	x, 44	
326	From those serene and rapturous joys (T. Flatman)	welcome, Charles II	1684	2 S, A, T, B	xviii, 1	234
343	Why, why are all the Muses mute?	welcome, James II	1685	2 S, A, T, 2 B, 5vv	xviii, 37	234
334	Raise, raise the voice	St Cecilia's Day	c1685	S, B, 3vv, no va	x, 26	
344	Ye tuneful Muses	welcome, James II	1686	2 S, A, T, 2 B, 2 rec	xviii, 80	234
335	Sound the trumpet	welcome, James II	1687	2 A, 2 T, 2 B	xviii, 121	
332	Now does the glorious day appear (T. Shadwell)	birthday, Mary II	1689	S, A, T, 2 B, 3 vn	xi, 1	
322	Celestial music did the gods inspire	perf. at Mr Maidwell's school	1689	S, A, T, B, 2 rec	xxvii, 29	
333	Of old when heroes thought it base (T. D'Urfey)	Yorkshire Feast	1690	2 A, T, 2 B, 5vv, 2 rec, 2 ob, 2 tpt	i	236
320	Arise, my Muse (D'Urfey)	birthday, Mary II	1690	2 A, T, B, 2 rec, 2 ob, 2 tpt, 2 va	xi, 36	234, 236
338	Welcome, welcome, glorious morn	birthday, Mary II	1691	S, A, T, 2 B, 2 ob, 2 tpt	xi, 72	237
331	Love's goddess sure was blind (C. Sedley)	birthday, Mary II	1692	S, 2 A, T, B, 2 rec	xxiv, 1	234
328	Hail, bright Cecilia (N. Brady)	St Cecilia's Day	1692	S, 2 A, T, 2 B, 2 rec, b rec, 2 ob, 2 tpt, timp	viii	236
321	Celebrate this festival (N. Tate)	birthday, Mary II	1693	2 S, A, T, B, 8vv, 2 ob, 2 tpt	xxiv, 36	
327	Great parent, hail (Tate)	centenary of Trinity College, Dublin	1694	S, A, T, B, 2 rec	xxvii, 59	236–7
323	Come, ye sons of art, away (?Tate)	birthday, Mary II	1694	S, 2 A, B, 2 ob, 2 tpt, ?timp	xxiv, 87	234, 236, 254
342	Who can from joy refrain? (?Tate)	birthday, Duke of Gloucester	1695	2 S, 2 A, T, B, 5vv, 2 ob, tpt	iv	220

Z	SECULAR SONGS FOR SOLO VOICE AND CONTINUO	223, 228, 246, 251–2, 259
	(*dates are those of composition as given in PS xxv, rev.1985*)	
352	Ah! cruel nymph, you give despair	
353	Ah! how pleasant 'tis to love, 1687–8	
354	Ah! what pains, what racking thoughts (Congreve), 1695, inc., bc lost or never written	
355	Amidst the shades and cool refreshing streams, 1683	
356	Amintas, to my grief I see, 1678	
357	Amintor, heedless of his flocks, 1680	
358	Ask me to love no more (A. Hammond), 1694	
359	A thousand sev'ral ways I tried, 1683–4	
360	Bacchus is a pow'r divine	
461	Beneath a dark and melancholy grove, ? by 1683, with SB	
361	Beware, poor shepherds, 1683–4	
362	Cease, anxious world (G. Etherege), 1684	252
364	Celia's fond, too long I've loved her (A. Motteux), 1964	
365	Corinna is divinely fair, 1692	
367	Cupid, the slyest rogue alive, 1685	
462	Draw near, you lovers (T. Stanley), 1683, with SB	
368	Farewell, all joys, 1684	
463	Farewell, ye rocks (D'Urfey), 1684, with SB	
369	Fly swift, ye hours, 1691	
370	From silent shades, 1682, also known as Bess of Bedlam	
464	Gentle shepherds, you that know (Tate), on the death of John Playford, 1686–7, with SB	251–2, 259
371	Hears not my Phillis (C. Sedley), 1684, also known as The Knotting Song	
372	He himself courts his own ruin, 1683–4	
465	High on a throne of glitt'ring ore (D'Urfey), 1689, ode to the Queen, with SB	
—	Honours may crown, *c*1691, mock-song: music from 4th act trumpet tune in Dioclesian	
373	How delightful's the life of an innocent swain, with SB	
374	How I sigh when I think of the charms, 1680	
375	I came, I saw, and was undone (A. Cowley), 1685	
376	I envy not a monarch's fate, 1693, mock-song: music same as I envy not in ode z321	
377	I fain would be free, 1694, bc lost or never written	
378	If grief has any power to kill, 1685	
379	If music be the food of love (H. Heveningham), 2 settings, 1692–5, 2 versions of the first	
380	If pray'rs and tears, 1685, on the death of Charles II	
381	I lov'd fair Celia (B. Howard), by 1693	
382	I love and I must, 1693, also known as Bell Barr	
383	Incassum, Lesbia, rogas (G. Herbert), 1695, also known as The Queen's Epicedium	
384	In Cloris all soft charms (J. Howe), 1683–4	
385	In vain we dissemble, 1684	
386	I resolve against cringing, 1678	
388	I take no pleasure in the sun's bright beams (Chamberlaine), 1680	
389	Leave these useless arts in loving, solo version of duet in Epsom Wells	
390	Let each gallant heart (J. Turner), 1682	
391	Let formal lovers still pursue, 1687	
466	Let us, kind Lesbia, give away, 1683–4, with SB	
392	Love arms himself in Celia's eyes (M. Prior)	252
393	Love is now become a trade, 1684	
394	Lovely Albina's come ashore, 1695	
395	Love's pow'r in my heart shall find no compliance, 1687–8	
396	Love, thou canst hear, tho' thou art blind (R. Howard), 1695	
467	Musing on cares of human fate (D'Urfey), 1685, with SB	
399	My heart, whenever you appear, 1684	
400	Not all my torments can your pity move, 1693	
468	No, to what purpose should I speak (Cowley), 1683, with SB	
401	No watch, dear Celia, just is found (?Motteux), 1693, mock-song: music same as Thou tun'st this world in ode z328	
402	Oh! fair Cedaria, hide those eyes	
403	Oh! how happy's he (W. Mountfort), bc lost, mock-song: melody from 1st act tune of Dioclesian	
404	Olinda in the shades unseen, 1694	
405	On the brow of Richmond Hill (D'Urfey), 1691	
406	O solitude, my sweetest choice (K. Philips), 1684–5	251
407	Pastora's beauties, when unblown, 1680	
408	Phillis, I can ne'er forgive it, 1687–8	

409 Phillis, talk no more of passion, 1685
410 Pious Celinda goes to prayers (Congreve), 1695
411 Rashly I swore I would disown, 1682
412 Sawney is a bonny lad (Motteux), 1694
469 Scarce had the rising sun appear'd, 1678
470 See how the fading glories of the year, 1689, with SB
413 She loves and she confesses too (Cowley), 1680
414 She that would gain a faithful lover (Lady E. M.), 1695
415 She who my poor heart possesses, 1682
416 Since one poor view has drawn my heart, 1680
471 Since the pox or the plague, 1678, with SB
417 Spite of the godhead, powerful love (A. Wharton), 1687
444 Stripp'd of their green our groves appear (Motteux), 1691
420 Sylvia, now your scorn give over, 1687–8
512 Sylvia, 'tis true you're fair, 1685, with SB; also in PS xxii as duet
421 The fatal hour comes on apace — 246–7
422 They say you're angry (Cowley), 1684
423 This poet sings the Trojan wars, 1687, also known as Anacron's Defeat — 252
424 Through mournful shades and solitary groves (R. Duke), 1683–4
425 Turn then thine eyes, solo version of duet in The Fairy Queen
426 Urge me no more
427 We now, my Thyrsis, never find (Motteux), 1693, mock-song: music same as I lov'd fair Celia
428 What a sad fate is mine, 2 settings, 2nd 1693–4
429 What can we poor females do?, 1693–4, solo version of duet
472 What hope for us remains now he is gone?, on the death of Matthew Locke, 1678
430 When first Amintas sued for a kiss (D'Urfey), 1686, mock-song
431 When first my shepherdess and I, 1686
432 When her languishing eyes said 'love', 1680
434 When my Aemelia smiles
435 When Strephon found his passion vain, 1682, ?duet
437 While Thirsis, wrapp'd in downy sleep, 1685
438 Whilst Cynthia sang, all angry winds lay still, 1685
440 Who but a slave can well express
441 Who can behold Florella's charms?, 1695
443 Ye happy swains, whose nymphs are kind, 1684
473 Young Thirsis' fate, ye hills and groves, deplore, on the death of Thomas Farmer, 1689, with SB

(doubtful)

Dates are of publication; also in PS xxv, rev. 1985

D130 A choir of bright beauties
363 Cease, O my sad soul (C. Webbe) (1678), ? by the elder Henry Purcell
D133 How peaceful the days are (1678–9)
387 I saw that you were grown so high (1678), ? by the elder Henry Purcell
397 More love or more disdain I crave (Webbe) (1678), ? by the elder Henry Purcell
418 Sweet, be no longer sad (Webbe) (1678), ? by the elder Henry Purcell
S70 Sweet tyranness, I now resign (1678), solo version of trio, ? by the elder Henry Purcell
433 When I a lover pale do see (1678), ? by the elder Henry Purcell
D201 When night her purple veil has softly spread, B, 2 vn, PS xxi; probably by D. Purcell
436 When Thirsis did the splendid eye (1675), ? by the elder Henry Purcell
442 Why so serious, why so grave? (T. Flatman), bc lost

SONGS FOR SOLO VOICE AND CONTINUO IN STAGE WORKS — 231, 246, 259

Some with chorus; Z no. of stage work in parentheses; see note to The Tempest, Z631, above.

Acolus, you must appear [see While these pass o'er]; Ah! Belinda (626); Ah! cruel, bloody fate (606); Ah! how sweet it is to love (613); Ah! me to many deaths (586); All our days and out nights (627); A thousand, thousand ways (629); Begone, curst fiends of hell (630); Beneath the poplar's shadow (590); Blow, Boreas, blow (589); Britons, strike home (574) — 251

Celia has a thousand charms (609); Celia, that I once was blest (572); Charon the peaceful shade invites (627); Come all to me (632); Come, all ye songsters of the sky (629); Come away, do not stay (583); Come away, fellow sailors (626); Come ev'ry demon who o'ersees (575); Come if you dare (628); Corinna, I excuse thy face (612); Cynthia frowns whene'er I woo her (592) *249*

Dear, pretty youth (631); Dream no more of pleasures past (606); Fairest isle (628); Fled is my love (571); For Iris I sigh (572); From rosy bowers (578 III); Genius of England (578 II); Great Diocles the boar has killed (627); Great Love, I know thee now (628) 227, 229 246–8

Hail to the myrtle shade (606); Hang this whining way of wooing (612); Hark, behold the heavenly choir (606); Hark how all things (629); Hark, the ech'ing air (629); Hear, ye gods of Britain (574); Hence with your trifling deity (632); Here's the summer, sprightly, gay (629); Hither this way (628); How blest are shepherds (628); How happy is she (609); How happy's the husband (582); Hush, no more (629) 248 227

I am come to lock all fast (629); I attempt from love's sickness to fly (630); I call you all to Woden's hall (628); I come to sing great Zempoalla's story (630); If love's a sweet passion (629); If thou wilt give me back my love (571); I'll mount to yon blue coelum (571); I'll sail upon the dog-star (571); I look'd and saw within the book of fate (598); Ingrateful love (612); In vain, Clemene, you bestow (588); In vain 'gainst love I strove (580); I see she flies me (573); I sigh'd and I pin'd (571); I sigh'd and owned my love (595) 248 227 224

Lads and lasses, blithe and gay (578 II); Let monarchs fight (627); Let not a moon-born elf mislead ye (628); Let the dreadful engines (578 I); Let the graces and pleasures repair (627); Let the soldiers rejoice (627); Let us dance, let us sing (627); Love in their little veins (632); Love quickly is pall'd (632); Lucinda is bewitching fair (570)

Man is for the woman made (605); Music for a while (583); Next winter comes slowly (629); No, no, poor suffering heart (576); Now, now the fight's done (606); Now the night is chas'd away (629); Nymphs and shepherds, come away (600); Oft she visits this lone mountain (626); Oh, how you protest (605); O lead me to some peaceful gloom (574); O let me weep (629); One charming night (629) 230, 251 248

Pluto, arise (575); Pursue thy conquest, love (626); Pursuing beauty (588); Retir'd from any mortal's sight (581); Return, revolting rebels (632); Sad as death at dead of night (606); Say, cruel Amoret (612); See, even night herself is here (629); See, I obey (629); Seek not to know (630); See my many colour'd fields (629) 248

See where repenting Celia lies (603); Shake the cloud from off your brow (626); Since from my dear (627); Since the toils and the hazards (627); Sing while we trip it (629); Sound, fame (627); Still I'm wishing (627); Sweeter than roses (585) 248

Take not a woman's anger ill (609); Tell me no more I am deceived (601); Thanks to these lonesome vales (626); The air with music gently wound (575); The cares of lovers (632); The danger is over (595); The gate to bliss does open stand (606); There's not a swain (587); There's nothing so fatal as woman (571); They tell us that you mighty powers above (630); Thou doting fool (628) 224

Though you make no return to my passion (601); Thrice happy lovers (629); Thus happy and free (629); Thus the ever grateful spring (629); Thus the gloomy world (629); Thus to a ripe consenting maid (607); Thy genius, lo! (2 settings) (604); 'Tis death alone can give me ease (571); 'Tis I that have warmed ye (628); To arms, heroic prince (600); Trip it, trip it in a ring (629), also as duet; 'Twas within a furlong (605)

Wake, Quivera (630); What power art thou? (628); What shall I do to show? (627); When a cruel long winter (629); When first I saw the bright Aurelia's eyes (627); When I am laid in earth (626); When I have often heard (629); When the world first knew creation (578 I); While these pass o'er the deep (631); Whilst I with grief (610); Why should men quarrel? (630); Ye blustering brethren (628); Ye gentle spirits of the air (629); Yes Daphne [Xansi], in your looks I find (629); Ye twice ten hundred deities (630); Your hay it is mowed (628) 250,251 254

(*doubtful; see note to 'The Tempest', z631, above*)

Arise, ye subterranean winds (631); Come down, come down, my blusterers (631); Come unto these yellow sands (631); Dry those eyes (631); Fair and serene (631); Full fathom five (631); Halcyon days (631); Kind fortune smiles (631); Saint George, the patron of our isle (628); See, see the heavens smile (631)

SONGS FOR TWO OR MORE VOICES AND CONTINUO 228, 252, 259

Dates given only for those published in Purcell's lifetime; for S, B, and in PS xxii unless otherwise stated.

Z	
480	Above the tumults of a busy state
481	A grasshopper and a fly (D'Urfey) (1686); PS xxv (1928)
482	Alas, how barbarous are we (K. Philips)

D171 A poor blind woman, 2 S, B
483 Come, dear companions of th'Arcadian fields (1686)
484 Come, lay by all care (1685)
485 Dulcibella, whene'er I sue for a kiss (A. Henley) (1694)
486 Fair Cloe, my breast so alarms (J. Glanvill) (1692)
487 Fill the bowl with rosy wine (A. Cowley) (1687)
489 Go, tell Amynta, gentle swain (Dryden)
541 Hark, Damon, hark, 2 S, B, 2 rec, 2 vn; PS xxvii
542 Hark how the wild musicians sing, 2 T, B, 2 vn; PS xxvii
490 Haste, gentle Charon, 2 B
491 Has yet your breast no pity learn'd? (1688)
492 Hence, fond deceiver (1687)
493 Here's to thee, Dick (Cowley) (1688)
494 How great are the blessings (Tate) (1686), also known as A Health to King James
543 How pleasant is this flowery plain, S, T, 2 rec (1688)
495 How sweet is the air and refreshing (1687)
544 If ever I more riches did desire (Cowley), 2 S, T, B, 2 vn; PS xxvii
545 In a deep vision's intellectual scene (Cowley), 2 S, B, also known as The Complaint
497 In some kind dream (G. Etherege) (1687)
498 I saw fair Cloris all alone (W. Strode) (1686)
499 I spy Celia, Celia eyes me
500 Julia, your unjust disdain
501 Let Hector, Achilles and each brave commander (1689)
502 Lost is my quiet for ever (1691)
503 Nestor, who did to thrice man's age attain (1689)
504 O dive custos Auriacae domus (H. Parker), 2 S, on the death of Queen Mary (1695) 252
505 Oft am I by the women told (Cowley) (1687)
506 Oh! what a scene does entertain my sight, with rec/vn
507 Saccharissa's grown old (1686)
508 See where she sits (Cowley), with 2 vn
509 Sit down, my dear Sylvia (D'Urfey) (1685)
510 Soft notes and gently raised (J. Howe), with 2 rec (1685)
511 Sylvia, thou brighter eye of night
513 There ne'er was so wretched a lover as I (Congreve)
514 Though my mistress be fair (1685)
546 'Tis wine was made to rule the day, S, SSB
515 Trip it, trip it in a ring, duet version of solo song in The Fairy Queen
516 Underneath this myrtle shade (Cowley) (1692)
547 We reap all the pleasures, inc., STB, 2 rec; PS xxvii
517 Were I to choose the greatest bliss (1689)
518 What can we poor females do? (1694), also as solo song
519 When gay Philander left the plain (1684)
520 When, lovely Phyllis, thou art kind (1685)
521 When Myra sings (G. Granville) (1695)
435 When Strephon found his passion vain (1683), ? duet; in PS xxv, rev. 1985
522 When Teucer from his father fled (D. Kenrick) (1686)
D172 When the cock begins to crow, 2 S, B
523 While bolts and bars my day control
524 While you for me alone had charms (J. Oldham)
525 Why, my Daphne, why complaining? (1691)
512 Sylvia, 'tis true, see 'Secular songs for solo voice and continuo'

(doubtful or spurious)

496 In all our Cynthia's shining sphere (from E. Settle: The World in the Moon); by D. Purcell
s69 Sweet tyrannes, I now resign, 2 S, B (1667); ? by the elder Henry Purcell 218

SONGS FOR TWO OR MORE VOICES AND CONTINUO IN STAGE WORKS 231, 246, 259
(some with chorus; Z no. of stage work in parentheses)

Ah! how happy are we, A, T (630); Art all can do, why then will mortals, 2 S, B (578 I); As Amoret and Thyrsis, S, B (607); As soon as the chaos was made, S, B (602); Behold the man that with gigantic might (A Dialogue between a Mad Man and a Mad Woman), S, B (608); But ere we this perform, 2 S (626); Can'st thou, Marina, leave, A, T, B (606); Celemene, pray tell me, 2 S (584); Come away, no delay, 2 B (627); Come let us agree, S, B (632); Come, let us leave the town, S, B (629)

Fair Iris and her swain, S, B (572); Fear no danger, 2 S (626); For folded flocks, S, B (628); Good neighbour, why do you look awry?, 2 S, A, B (591); Hark! how the songsters, 2 S (632); Hark, my Damilcar, S, B (613); Hear, ye sullen powers below, A, T, B (583); Jenny, 'gin you can love, S, T (571); Laius, hear!, A, T, B (583); Leave these useless arts in loving, S, B (579), also as solo song; Let all mankind the pleasures share, S, B (627); Let the fifes and the clarions, 2 A (629); Love, thou art best of human joys, 2 S (596) 228, 254

Make room for the great god of wine, 2 B (627); May the god of wit inspire, A, T, B (629); My dearest, my fairest, S, T (585); No more, Sir, no more, S, B (588); No, no, resistance is but vain, 2 S (601); Now the maids and the men (Dialogue between Corydon and Mopsa), A, B (629); O, the sweet delights of love, 2 S (627); Shepherd, leave decoying, 2 S (628); Since times are so bad, S, B (578 II); Sing all ye muses, A, B (578 I); Sing ye Druids all, 2 S (574); Sound a parley, S, B (628) — 252

Tell me why, my charming fair, S, B (627); They shall be as happy, 2 S, B (629); To arms, your ensigns straight display, A, B (574); To Mars let 'em raise, A, T, B (627); Turn then thine eyes, 2 S (629), also as solo song; Two daughters of this aged stream, 2 S (628); What flatt'ring noise is this, A, T, B (630); With this sacred charming wand, 2 S, B (578 I); You say 'tis love, S, B (628) — 228, 246, 252

CATCHES

Publication dates given only for those published in Purcell's lifetime; for 3vv and in PS xxii unless otherwise stated.

Z	
240	A health to the nut-brown lass (J. Suckling), 4vv (1685)
241	An ape, a lion, a fox and an ass (1686)
242	As Roger last night to Jenny lay close
599	At the close of the evening, 3 B (1691), in Beaumont, Fletcher: The Knight of Malta; PS xx
243	Bring the bowl and cool Nantz (?1693–4)
244	Call for the reckoning
245	Come, let us drink (A. Brome), with bc
246	Come, my hearts, play your parts (1685)
247	Down, down with Bacchus (1693)
248	Drink on till night be spent (P. Ayres) (1686)
249	Full bags, a brisk bottle (J. Allestry) (1686)
250	God save our sovereign Charles (1685)
251	Great Apollo and Bacchus
252	Here's a health, pray let it pass
253	Here's that will challenge all the fair (1680), also known as Bartholomew Fair
254	He that drinks is immortal (1686)
255	If all be true that I do think (1689)
256	I gave her cakes and I gave her ale (1690)
257	Is Charleroy's siege come too? (?1693)
574	Jack, thou'rt a toper (1695), in Fletcher: Bonduca; PS xvi
101	Joy, mirth, triumphs I do defy, 4vv
258	Let the grave folks go preach (1685)
259	Let us drink to the blades (?1691)
260	My lady's coachman, John (1688)
594	My wife has a tongue (1685), in E. Ravenscroft: The English Lawyer; PS xvi
261	Now England's great council's assembled (1685)
262	Now, now we are met and humours agree (1688)
263	Of all the instruments that are (1693)
264	Once in our lives let us drink to our wives (1686)
265	Once, twice, thrice, I Julia tried
266	One industrious insect (?R. Thomlinson), also known as Insecta praecauta, alterius merda
267	Pale faces, stand by (Mr Taverner) (1688)
268	Pox on you for a fop
269	Prithee be n't so sad and serious (Brome)
270	Room for th'express, written on the fall of Limerick, July 1694
271	Since the duke is return'd (1685)
272	Since time so kind to us does prove
273	Sir Walter enjoying his damsel
274	Soldier, soldier, take off thy wine, 4vv
275	Sum up all the delights (1688)
276	The Macedon youth (Suckling), 4vv (1686)
277	The miller's daughter riding (1686)
278	The surrender of Limerick (?1691)
279	'Tis easy to force, 4 B (1685)
280	'Tis too late for a coach (1686)
281	'Tis women makes us love, 4vv (1685)
282	To all lovers of music (Carr) (1687)
283	To thee, to thee and to a maid (1685)
284	True Englishmen drink a good health, 'Song with music on the 7 Bishops' (*c*1689)
285	Under a green elm lies Luke Shepherd's helm, 4vv (1686)
286	Under this stone lies Gabriel John (1686)

287	When V and I together meet (1686)	
288	Who comes there? (1685)	
289	Wine in a morning makes us frolic and gay (T. Brown) (1686)	
290	Would you know how we meet (T. Otway) (1685)	
291	Young Colin cleaving of a beam (D'Urfey) (1691)	
292	Young John the gard'ner, 4vv (1683)	

STRINGS AND WIND

(*in PS xxxi unless otherwise stated*)

Z		225, 228, 253–7, 259
730	Chacony, g, 4 str	
—	Cibell, C, tr, str [? orig. of T678: see below]	
744	Fantasia, a, 2 vn, b viol, bc, 1683, inc.	
731	Fantasia upon a Ground, D/F, 3 vn/rec, bc, *c*1680	
745	Fantasia upon One Note, F, 5 viols, *c*1680	255
732–4	3 fantasias, d, F, g, 3 viols, *c*1680	
735–43	9 fantasias, g, B♭, F, c, d, a, e, G, d, 4 viols, 1680	223, 224, 254, 255
746	In Nomine, g, 6 viols, *c*1680	255
747	In Nomine, g Dorian, 7 viols, *c*1680	255
860	March and Canzona, c, 4 flatt (slide) tpt, 1692	
770	Overture, G [concert version of introduction to Swifter, Isis, swifter flow], 4 str, 1681	
771	Overture, d, 4 str	
772	Overture, g, 5 str	
752	Pavan, g, 3 vn, bc, *c*1680	
748–51	4 pavans, A, a, B♭, g, 2 vn, bc, *c*1680	
N773	Prelude, g/d, vn/rec	
850	Sonata, D, tpt, str, ?1694	
780	Sonata, g, vn, bc; reconstructed by T. Dart for vn, b viol, bc	
790–801	[12] Sonnata's of III Parts, 2 vn, b viol, bc (org/hpd), *c*1680 (London, 1683, 2/1684); PS v	218, 227, 254–7, 259, 260
802–11	Ten Sonata's in Four Parts, 2 vn, b viol, bc, *c*1680 (London, 1697); PS vii	226, 228, 254, 257, 259
770	Suite, G, 4 str, inner parts inc.	
—	The Staircase Overture, B♭, 4 str, ?*c*1676	
—	Bass parts of 2 overtures, C, 2 minuets, C, f, and 2 pavans, f	

(*doubtful*)

—	Bass parts of a prelude and alman, b	

HARPSICHORD — 257–8

(*source of transcriptions, with Z no., in square brackets*)

Editions: PS vi [unreliable]
H. Purcell: Complete Harpsichord Works, ed. H. Ferguson (London, 1964)
Old English Composers for the Virginal and Harpsichord, ed. E. Pauer (London, 1879)

* – *in Ferguson, not in PS*
† – *in PS, not in Ferguson*
†* – *in Pauer only*

		The Second Part of Musick's Hand-maid (London, 1689[7]):	237
T694		Song Tune, C [Ah! how pleasant 'tis to love, 353]	
T695	*	Song Tune, C [Sylvia, now your scorn give over, 420]	
647–8		2 marches, C	
T689		A New Minuet, d	
649–50		2 minuets, a, a	
T688		Minuet, d [Raise, raise the voice, 344/6]	
655		A New Scotch Tune, G	
T682		A New Ground, e [Welcome to all the pleasures, 339/3]	237
646		A New Irish Tune [Lilliburlero], G	
653		Rigadoon, C	
656		Sefauchi's Farewell, d	
665	*	Suite, C	
		A Choice Collection of Lessons, hpd/spinet (London, 1696):	259
660–63, 666–9		8 suites, G, g, G, a, C, D, d [Hornpipe from The Married Beau, 603/3], F [Minuet from The Double Dealer, 592/3]	257
T687		March, C [The Married Beau, 603/8]	
T698		Trumpet Tune, C [The Indian Queen, 630/4a]	
T680		Chaconne, g [Timon of Athens, 632/20]	
T686		Jig, g [Abdelazer, 570/7]	
T678		Trumpet Tune 'Cibell', C	
T697		Trumpet Tune, C [Dioclesian, 627/21]	

		8 airs:
641		G
T675		d [The Indian Queen, 630/22]
T676		d [The Double Dealer, 592/7]
T693(2)		g [Abdelazer, 570/6]
T696(1)	*	d [2nd version of T675]
T696(2)	*	d
—		d [The Fairy Queen], 629/17 bc, transposed]; *D-Ka*
—	*	F [The Indian Queen]
		5 grounds:
645		Ground in Gamut, G
T681		c [Ye tuneful Muses, 344/11]
D221		c, possibly by Croft
D222	*	d [Celebrate this festival, 321/2a]
—		a, inc.; *US-Cn*
		4 hornpipes:
T683	*	B♭ [Abdelazer, 570/8]
T684		d 'Round O' [Abdelazer, 570/2]
—		F [The Indian Queen, 630/2b], *Cn*
T685		c [The Old Bachelor, 607/4]
—		Jig, g [Abdelazer, 570/7], Cn
		4 overtures:
T690	†*	c [The Indian Queen, 630/3a]
T691	†*	D [Timon of Athens, 632/1]
T692	†*	D [The Fairy Queen, 629/3ab]
T693(1)	†	g [The Virtuous Wife, 611/1]
		Suite, a:
652	†	Prelude
642	†*	Almand, Corant
654		Saraband
664	†*	Suite, B♭: Almand, Corant, Saraband
644		Corant, G
651		Minuet, G
670	*	The Queen's Dolour, a
T677	*	Canary, B♭ [The Indian Queen, 630/18]
—		Prelude for the Fingering, C, attrib. Purcell in The Harpsichord Master, i (London, 1697), anon. in later vols.; ed. in Petre
—		Voluntary, no.9 of Ten Select Voluntaries (London, *c*1780), 1 movt of which may be by Purcell; see Cooper

ORGAN (all in *PS vi*) 258

716	Verse, F
717–20	4 voluntaries, C, d, d (double org), G
721	Voluntary on the 100th Psalm, A

THEORETICAL WORKS

'A Brief Introduction to the Art of Descant: or, Composing Musick in Parts', in J. Playford: *An Introduction to the Skill of Musick* (London, 12/1694/*R*1972) [partly rev. from earlier work of Campion, Simpson, Playford and others] 225

BIBLIOGRAPHY

W. H. Husk: *An Account of the Musical Celebrations on St Cecilia's Day* (London, 1857), 178, 180, 184–6

W. H. Cummings: *Purcell* (London, 1881)

A. Nicoll: *A History of Restoration Drama, 1660–1700* (Cambridge, 1923, 4/1952)

D. Arundell: *Henry Purcell* (London, 1927/*R*1971)

H. Dupré: *Purcell* (Paris, 1927; Eng. trans., 1928/*R*1975)

H. C. Colles: *Voice and Verse: a Study in English Song* (London, 1928)

E. J. Dent: *Foundations of English Opera* (Cambridge, 1928/*R*1965)

A. K. Holland: *Henry Purcell: the English Musical Tradition* (London, 1932, 2/1948)

F. de. Quervain: *Der Chorstil Henry Purcell's* (Berne, 1935)

J. A. Westrup: 'Fact and Fiction about Purcell', *PMA*, lxii (1935–6), 93

——: *Purcell* (London, 1937, rev. 4/1980)

C. L. Day and E. B. Murrie: *English Song-books, 1651–1702* (London, 1940)

H. M. Miller: 'Henry Purcell and the Ground Bass', *ML*, xxix (1948), 340

S. Favre-Lingorow: *Der Instrumentalstil von Purcell* (Berne, 1950)

W. Meinardus: *Die Technik des Basso Ostinato bei Henry Purcell* (diss., U. of Cologne, 1950)

S. Demarquez: *Purcell: la vie, l'oeuvre* (Paris, 1951)

G. van Ravenzwaaij: *Purcell* (Haarlem and Antwerp, 1954)

H. Wessely-Kropik: 'Henry Purcell als Instrumentalkomponist', *SMw*, xxii (1955), 85–141

R. Sietz: *Henry Purcell: Zeit, Leben, Werk* (Leipzig, 1956)

F. B. Zimmerman: 'Purcell and Monteverdi', *MT*, xcix (1958), 368

T. Dart: 'Purcell's Chamber Music', *PRMA*, lxxxv (1958–9), 81

I. Holst, ed.: *Henry Purcell (1659–1695): Essays on his Music* (London, 1959)

M. Tilmouth: 'The Technique and Forms of Purcell's Sonatas', *ML*, xl (1959), 109

J. Wilson, ed.: *Roger North on Music* (London, 1959)

R. E. Moore: *Henry Purcell and the Restoration Theatre* (London, 1961)

M. Laurie: *Purcell's Stage Works* (diss., U. of Cambridge, 1962)

D. Schjelderup-Ebbe: *Purcell's Cadences* (Oslo, 1962)

J. A. Westrup: 'Purcell's Music for "Timon of Athens" ', *Festschrift Karl Gustav Fellerer* (Regensburg, 1962), 573

F. B. Zimmerman: 'Purcell and the Dean of Westminster: some New Evidence', *ML*, xliii (1962), 7

——: *Henry Purcell, 1659–1695: an Analytical Catalogue of his Music* (London, 1963)

M. Laurie: 'Did Purcell Set *The Tempest*?', *PRMA*, xc (1963–4), 43
J. A. Westrup: 'Purcell's Parentage', *MR*, xxv (1964), 100
F. B. Zimmerman: 'Purcell's "Service Anthem" *O God, thou art my God* and the B-flat major Service', *MQ*, l (1964), 207
H. Ferguson: 'Purcell's Harpsichord Music', *PRMA*, xci (1964–5), 1
J. Buttrey: *The Evolution of English Opera between 1656 and 1695: a Reinvestigation* (diss., U. of Cambridge, 1967)
F. B. Zimmerman: *Henry Purcell, 1659–1695: his Life and Times* (London, 1967; rev. 2/1983)
——: 'Sound and Sense in Purcell's "Single Songs" ', in V. Duckles and F. B. Zimmerman: *Words to Music* (Los Angeles, 1967), 45–90
J. Buttrey: 'Dating Purcell's "Dido and Aeneas" ', *PRMA*, xciv (1967–8), 51
R. Covell: 'Seventeenth-century Music for The Tempest', *SMA*, ii (1968), 43
A. H. King: 'Benjamin Goodison and the First "Complete Edition" of Purcell', *Musik und Verlag: Karl Vötterle zum 65. Geburtstag* (Kassel, 1968), 391
G. Rose: 'Purcell, Michelangelo Rossi and J. S. Bach: Problems of Authorship', *AcM*, xl (1968), 203
H. D. Johnstone: 'English Solo Song, c. 1710–1760', *PRMA*, xcv (1968–9), 67
F. B. Zimmerman: 'Anthems of Purcell and Contemporaries in a Newly Rediscovered "Gostling Manuscript" ', *AcM*, xli (1969), 55
R. McGuinness: *English Court Odes, 1660–1820* (Oxford, 1971)
F. B. Zimmerman: *The Anthems of Henry Purcell* (New York, 1971)
R. E. Burkart: *The Trumpet in England in the Seventeenth Century with Emphasis on its Treatment in the Works of Henry Purcell* (diss., U. of Wisconsin, 1972)
G. Rose: 'A New Purcell Source', *JAMS*, xxv (1972), 230
R. Savage: 'The Shakespeare–Purcell *Fairy Queen*: a Defence and Recommendation', *Early Music*, i (1973), 200
D. L. Smithers: *The Music and History of the Baroque Trumpet before 1721* (London, 1973) [chap. on Purcell]
I. Spink: *English Song: Dowland to Purcell* (London, 1974)
P. Dennison: 'The Stylistic Origins of the Early Church Music [of Purcell]', *Essays on Opera and English Music in Honour of Sir Jack Westrup* (Oxford, 1975), 44
N. Fortune: 'The Domestic Sacred Music [of Purcell]', *Essays on Opera and English Music in Honour of Sir Jack Westrup* (Oxford, 1975), 62
F. B. Zimmerman: *Henry Purcell 1659–1695: Melodic and Intervallic Indexes to his Complete Works* (Philadelphia, 1975)

R. Savage: 'Producing Dido and Aeneas: an Investigation into Sixteen Problems', *Early Music*, iv (1976), 393
E. van Tassel: 'Two Purcell Discoveries – 1: Purcell's "Give Sentence" ', *MT*, cxviii (1977), 381 [with music suppl.]
B. Wood: 'Two Purcell Discoveries – 2: A Coronation Anthem Lost and Found', *MT*, cxviii (1977), 466
The Gostling Manuscript (Austin, Texas, and London, 1977) [facs. edn.]
B. Cooper: 'Did Purcell write a Trumpet Voluntary?', *MT*, cxix (1978), 791, 1073
J. Meffen: 'A Question of Temperament: Purcell and Croft', *MT*, cxix (1978), 504
R. Petre: 'A New Piece by Henry Purcell', *Early Music*, vi (1978), 374
H. Siedentopf: 'Eine Komposition Purcells im Klavierbuch einer württembergischen Prinzessin', *Mf*, xxxi (1978), 446
B. Wood: 'A Newly Identified Purcell Autograph', *ML*, lix (1978), 329
C. A. Price: *Music in the Restoration Theatre* (Ann Arbor, 1979)
A. Browning: 'Purcell's "Stairre Case Overture" ', *MT*, cxxi (1980), 768
K. T. Rohrer: *'The Energy of English Words': a Linguistic Approach to Henry Purcell's Methods of Setting Texts* (diss., Princeton U., 1980)
R. Charteris: 'Some Manuscript Discoveries of Henry Purcell and his Contemporaries in the Newberry Library, Chicago', *Notes*, xxxvii (1980–81), 7
R. Manning: 'Revisions and Reworkings in Purcell's Anthems', *Soundings*, ix (1982), 29
D. Charlton: ' "King Arthur": Dramatick Opera', *ML*, lxiv (1983), 183
A. D. Ford: 'A Purcell Service and its Sources', *MT*, cxxiv (1983), 121
R. Ford: 'A Sacred Song not by Purcell [Full of wrath, his threat'ning breath]', *MT*, cxxv (1984), 45
M. Laurie: 'Purcell's Extended Solo Songs', *MT*, cxxv (1984), 19
C. A. Price: *Henry Purcell and the London Stage* (Cambridge, 1984)

GEORG PHILIPP TELEMANN

Martin Ruhnke

CHAPTER ONE

Life

Georg Philipp Telemann was born at Magdeburg on 14 March 1681. The most prolific composer of his day, he was widely regarded as Germany's leading composer in the early and middle 18th century; his fluent command of melody and uncomplicated textures show him as an important link between the late Baroque and the new Classical style. He was also highly influential in concert organization, music education and theory. He died in Hamburg on 25 June 1767.

I Sources, ancestry, school years

Telemann left three autobiographies, in which he described his career, his artistic development and to some extent his attitude to music. The first was written in 1718, at the request of Mattheson, who had intended to publish biographies of the best-known musicians. The second is shorter and is in the form of a letter from Telemann to Walther in 1729, giving him information for inclusion in his *Lexicon*. The third, the most comprehensive, dates from 1739, and was published by Mattheson in his *Grundlage einer Ehren-Pforte* (1740). A biographical study, published in both German and French in *c*1745, draws heavily on the *Ehren-Pforte* but includes additional material and information that the editor, B. Schmid, can have obtained only from Telemann himself. Although the autobiographies and the biography provide a wealth of background for an evaluation of Telemann's character and his approach to music, they contain a number of contradictions.

Telemann's forebears belonged to the upper middle class; there were no musicians among them. His father's family came from the area of Nordhausen, near Erfurt. His grandfather was vicar of Cochstedt, near Aschersleben, and his grandmother was a clergyman's daughter. His father, Heinrich Telemann (1646–85), went to school in Halberstadt and Quedlinburg, studied at the University of Helmstedt from 1664, and in 1668 was appointed headmaster of a school before becoming a parish priest in 1669 and subsequently, in 1676, a deacon in Magdeburg. In 1669 he married Maria Haltmeier (1642–1711), daughter of a Protestant clergyman from Regensburg, who, having been dismissed from his living in Freistadt, near Linz, in 1624, had found a new appointment near Magdeburg. Although Telemann claimed that his musical talent was inherited from his mother, there is no evidence that her family showed any musical talent except for her nephew, Joachim Friedrich Haltmeier (1668–1720), who became Kantor at Verden after spending some time at university. His son Carl, author of a treatise on thoroughbass published by Telemann in 1737, was an organist in Hanover.

Almost all Telemann's ancestors had received a university education, and most of them had entered the church. When his father died in 1685 his mother was left with the task of guiding her two sons along the same path. The elder studied theology and became a clergyman. The younger, Georg Philipp, attended two schools in Magdeburg, the Altstädtisches Gymnasium and the

Domschule, where he received instruction in Latin, rhetoric and dialectic, and became interested in German poetry. Although he had no special coaching, by the time he was ten he had learnt to play the violin, the flute, the zither and keyboard instruments, studied the compositions of his music master the Kantor Benedikt Christiani, transcribed other compositions, and tried his hand at writing arias, motets and instrumental pieces. When at the age of 12 he embarked on the composition of an opera, *Sigismundus* (to a libretto by Postel), his mother and her advisers are said to have forbidden him any further involvement with music and taken away all his musical instruments. It was felt that in different surroundings he would find his way back to his true vocation; and to that end he was sent in late 1693 or early 1694 to school at Zellerfeld, where he was placed in the care of the superintendent Caspar Calvoer, who had apparently become acquainted with the family while studying at Helmstedt. Calvoer did more than guide Telemann's academic progress: an informed devotee of theoretical music studies, he taught his pupil the relationship between music and mathematics; with Calvoer's approval, Telemann continued to complement his general education by teaching himself composition and thoroughbass, and from time to time he seems to have composed symphonies for the local Stadtpfeifer. After four years he moved to Hildesheim, where he became a scholar at the famous Gymnasium Andreanum. The Rektor of the school, J. C. Losius, had been educated in Magdeburg and Helmstedt, and he too did more than simply supervise Telemann's general education, encouraging him to compose incidental songs for his numerous Latin school dramas (texts of four have survived in their entirety; six more are known by name). This music is no longer extant; but it is possible that it was also Telemann who wrote the anonymous songs for the collection *Singende Geographie*, in which Losius recorded his geography syllabus in verse form. Father Crispus, in charge of Roman Catholic church music, also made use of Telemann's talents in the Catholic Godehardi church, where Telemann and some of his Protestant fellow students gave performances of German cantatas. On visits to Hanover and Brunswick he had his first taste of French instrumental music and Italian opera, and in his private studies in composition he modelled his writing on the music of Steffani, Rosenmüller, Corelli and Caldara.

II Leipzig, Sorau

In autumn 1701 Telemann embarked on his university studies, not in Helmstedt, where his ancestors and teachers had been students, but in Leipzig. By his own account (in the autobiographies of 1718 and 1739) he had intended to study law, in accordance with his mother's wishes. He had allegedly left behind in Magdeburg all his instruments, compositions and notes, and contrived for a time to conceal his musical gifts from the music lovers among his fellow students. The story goes, however, that one day a room-mate happened to discover one of Telemann's compositions, which he arranged to have performed in the Thomaskirche. When as a result of this Telemann was commissioned by the mayor of Leipzig to write a cantata every two weeks for performance there, the stage was set for a musical career. Telemann had the gift of attracting musical students to himself and of engaging them in pleasurable activities. In 1702 he founded a student collegium musicum; the regular public concerts he organized began a new chapter in the history of the collegia musica. While in the 17th century student music-making was a casual, leisure-time activity, Telemann and his collegium musicum were orientated towards public performance. In 1702 he became musical director of the Leipzig Opera, whose founder, N. A. Strungk, had died two years before; here he was able to employ students as singers and instrumentalists. Within three years he had composed at least four operas, and he later continued to supply operas for Leipzig from Sorau, Eisenach and Frankfurt – more than 20, according to the 1739 autobiography, though evidence of no more than five is available. When in 1704 a new organ was installed in the Neukirche, which until 1710 was also the university church, Telemann applied for the post of organist, supporting his application with the promise that, with no increase in stipend, he would also act as musical director and that he and his collegium musicum would give concerts of sacred music in the church on feast days and fair days. On these terms he was offered the appointment.

Telemann's many activities offended against the existing order of Leipzig's musical life, and encroached on the territory of Kuhnau, then Kantor at the Thomaskirche. As the city director of music, Kuhnau was responsible for the music in all the Leipzig churches, and until this time he had been able to decide what was or was not possible with the available resources. The choristers and Stadtpfeifer could take it in turn to perform cantatas on alternate Sundays in the Thomaskirche and the Nicolaikirche, while the second choir, conducted by a prefect, sang traditional motets and German chorales in the Neukirche. More money would have been needed to augment this programme. But Telemann achieved what had hitherto seemed out of reach. In the Thomaskirche cantatas were now sung every Sunday instead of in alternate weeks, and university church services no longer had to forgo performances of fine church music. On several occasions Kuhnau petitioned against Telemann's infringement of his rights and tried to discredit him as an 'opera musician'; the only result was that the city fathers forbade Telemann to appear on the operatic stage. Kuhnau complained bitterly about the students' 'rush to opera', for they had flocked to join Telemann and no longer supported Kuhnau in providing music for the church services. Even after Telemann had left Leipzig, the leaders of the students' collegium musicum still held on to the organist's post at the Neukirche; and to the end of his life Kuhnau inveighed against what he considered to be the illegal activities of the students, trying in vain to reassert his original rights.

In 1705 Telemann was summoned to the court of Count Erdmann II of Promnitz at Sorau, Lower Lusatia (now Żary, in Poland), where he became Kapellmeister. According to the autobiographies of 1718 and 1739, the invitation had been issued in 1704; but Telemann was still in receipt of his organist's stipend in Leipzig on 22 April 1705, and the 1729 autobiography states that he spent four years at Leipzig University. Before taking up the reins of government in 1703, the Count of Promnitz had travelled through Italy and France, and had acquired a taste for excessive displays of courtly splendour. In particular he had become enamoured of French instrumental music, and his new Kapellmeister was required to provide French overtures in the style of Lully and Campra. When the court spent six months at Pless, one of the count's domains in Upper Silesia, Telemann came into contact with Polish folk music, and visits to Kraków helped to develop his admiration for this fascinating form of art, whose 'barbaric beauty' captivated him. He made several journeys from Sorau to Berlin, where he became familiar with the court instrumental music and court opera. Controversies with the Sorau Kantor and theorist W. C. Printz brought him to grips with problems relating to musical theory. It was also in Sorau that Telemann met the reformer Erdmann Neumeister, who wrote cantata texts and since 1706 had been superintendent and court chaplain. In 1711 Neumeister stood godfather at the baptism of Telemann's first daughter; ten years later, in Hamburg, he successfully supported Telemann's appointment to a post in that city.

III Eisenach, Frankfurt

The date of Telemann's move from Sorau to the court at Eisenach has long been the subject of dispute. In two of the autobiographies, 1718 and 1739, Telemann gave the year as 1708; the published biography (*c*1745) gives 1709. C. Freyse (*MGG*, 'Eisenach') suggested that the document recording Telemann's appointment as court Kapellmeister was drawn up as early as 11 March 1707; this would be corroborated by what Telemann said in his poem on the death of his first wife (DDT, xxviii) – that he and others fled before the Swedish troops sweeping through Saxony from the east. This document however is a ratification on 11 March 1717 of Telemann's appointment as visiting Kapellmeister;

15. Georg Philipp Telemann: engraving by G. Lichtensteger

and nowhere in the poem does Telemann say that it was to Eisenach that he fled. In fact he was initially appointed to take charge of the newly formed musical establishment at the court on 24 December 1708. Before that he had been Konzertmeister of the court orchestra under Pantaleon Hebenstreit, and eventually he had been given the task of recruiting singers for the proposed establishment. It seems that 1708 may have been the date of his actual move to Eisenach. For the orchestra, of which he evidently thought highly, Telemann began to compose overtures, concertos and chamber works. When the new Hofkapelle was established he was also required to compose church cantatas and music for special occasions. While he was at Eisenach he must have met J. S. Bach, whose cousin Johann Bernhard Bach was town organist there and was involved in the musical life at court; in 1714 Telemann was to stand godfather to C. P. E. Bach. In autumn 1709, on his solemn undertaking that he would return to Eisenach and refuse all other offers of employment, Telemann was granted leave of absence. He returned to Sorau, where he married Louise Eberlin, a lady-in-waiting to the Countess of Promnitz and daughter of the musician Daniel Eberlin – only to lose her in January 1711 after the birth of his first daughter. Looking back in 1718, he was to claim that in Eisenach he not only came of age musically but that as a Christian he became a different man.

In February 1712, a year after his wife's death, Telemann accepted an invitation to Frankfurt am Main, where he became the city director of music and Kapellmeister at the Barfüsserkirche. No school appointment was connected with this post; Telemann merely supervised such singing instruction as was given in the schools. His six to eight choristers, personally selected from the schoolboys, had to be trained privately by him. He composed at least five cycles of cantatas in Frankfurt, each of them spanning the liturgical year. In addition he was expected to write and arrange performances of special works for civic celebrations. Though at court he was relatively restricted as a musician and composer by the demands of his official duties, his civic appointment allowed him a much greater degree of freedom to influence and reshape the city's musical life. He assumed the directorship of the collegium musicum of the Frauenstein Society, an association of the aristocracy and the bourgeoisie, who immediately made him their secretary and administrator. In conjunction with the collegium musicum he organized weekly public concerts for which he composed chamber and orchestral music and oratorios. At one special concert he arranged a performance of his setting of the Brockes Passion in the Barfüsserkirche, in the presence of the Landgrave of Hesse, whose own court musicians took part (1716).

In 1714 he married Maria Katharina Textor, daughter of a Frankfurt council clerk: from this union were born eight sons (none became a musician) and two daughters. Through his marriage Telemann became a citizen of Frankfurt, a privilege that he retained even after he left for Hamburg, by promising to continue to write church cantatas for Frankfurt.

In September 1716 Telemann visited Eisenach, where he conducted a special concert in honour of the duchess's birthday. Shortly afterwards he was made visiting Kapellmeister at Eisenach. In 1717 he was offered the post of Kapellmeister in Gotha, and there were moves to have him appointed Kapellmeister to all the courts of Duke Ernst's line. Telemann, however, took advantage of the offer from Gotha to strengthen his position in Frankfurt, and he obtained not only better conditions for himself but also the employment of extra musicians. In his application Telemann stressed that he had been active as a singer and instrumentalist as well as a composer and conductor. In 1719 he went to Dresden for the celebrations on the marriage of the Prince Elector of Saxony, Friedrich August II. During this visit he had the opportunity to hear, among other things, several operas by Lotti, and he dedicated a violin concerto to the Dresden Konzertmeister J. G. Pisendel, a pupil of Vivaldi. From Frankfurt he sent his own operas to Leipzig and to Hamburg.

IV Hamburg

On 10 July 1721 Telemann was invited by the city of Hamburg to succeed Joachim Gerstenbüttel as Kantor of the Johanneum and as musical director of the five main churches in the city. In his letter to the Frankfurt authorities asking to be released from his contract Telemann explained that he had not applied for the post and so regarded the offer as an act of God. He must, however, have had some exploratory contacts with Hamburg, for in January 1721 his opera *Der geduldige Socrates* was given there; he had also contributed pieces to the performance of the opera pasticcio *Ulysses* on 7 July 1721. Presumably the prospect of finding an outlet for his compositions at the Hamburg Opera was a key factor in his decision to move. A definite disadvantage, however, was that the director of music was obliged to act as school Kantor. At his installation on 16 October 1721 Telemann delivered a Latin panegyric on the '*excellentia*' of church music.

Telemann's new post demanded unprecedented productivity. For each Sunday he was expected to write two cantatas and for each year a new Passion. Special cantatas were required for induction ceremonies, and oratorios for the consecration of churches. Still more cantatas had to be written and performed to mark civic celebrations, of which there were many; and, once a year, to entertain the guests of the commandant of the city's militia, Telemann had to provide the 'Kapitänsmusik', consisting of an oratorio and a serenata. The demands of his official duties did not, however, prevent Telemann from once again conducting a collegium musicum in public concerts, or from participating in operatic productions. At first he met strong opposition, and in July 1722 a group of city councillors tabled a motion forbidding the Kantor to take part in public performances of theatrical or operatic music. Telemann reacted by applying for the post of Kantor at the Thomaskirche, Leipzig, which conveniently had fallen vacant on Kuhnau's death. Having been chosen by the Leipzig authorities, he wrote asking to be released from his Hamburg contract, stating that he had decided to accept the Leipzig invitation because the post there was materially more advantageous and because the Hamburg public were not favourably disposed towards him. After lengthy deliberations the council refused to release Telemann. But his stipend was increased, and no further objections were raised to his public concerts or his involvement with the opera. In July 1723 Telemann told J. F. A. von Uffenbach that his public concerts were patronized by many high-ranking people, the most prominent citizens and the entire council. He was entirely responsible for the establishment of public concerts in the city. In 1722 he was appointed musical director of the Hamburg Opera, of which he had charge until it closed in 1738. As well as many operas by Telemann himself, works by Handel and Keiser were prominent in the programmes; for these he often provided additional material.

Telemann had published a number of his chamber works while he was still in Frankfurt; and in Hamburg, between 1725 and 1740, he brought out a further 44 publications, 43 of them under his own imprint. In this group an entire cycle of 72 sacred cantatas for the church year constitutes one published item, as does the three-part *Musique de table*, comprising 18 separate compositions. Telemann himself usually engraved the plates. Further, he was largely responsible for advertising the editions in the press and for soliciting subscriptions. In Berlin, Leipzig, Jena, Nuremberg, Frankfurt, Amsterdam and London, distribution was arranged through booksellers; elsewhere friends undertook this responsibility.

In autumn 1737 Telemann went to Paris, where he remained for eight months. The 1739 autobiography suggests that a group of musicians familiar with his music had invited him there and arranged performances of his works. One reason for the visit may have been Telemann's desire to forestall the printing of pirated editions of his music. Before 1734, Boivin had brought out six of his trio sonatas from a pirated manuscript, and in April 1736 Le Clerc was granted a royal warrant authorizing him to reprint five of Telemann's publications without the composer's consent. When Telemann arrived in Paris these had already appeared. He was given his own warrant, and during his stay he brought out two new editions, although he was powerless to prevent a further five pirated editions from Le Clerc after 1740. Performances of his works at court and in the Concert Spirituel seem to have won him great acclaim.

On 14 October 1740 Telemann offered for sale the plates of all his own editions of his works. In his biography this step is explained as a consequence of his decision to issue no more of his compositions but rather to devote the rest of his life to compiling books on musical theory. Certainly his musical output fell sharply between 1740 and 1755, though he continued to write Passions, Kapitänsmusiken and music for church consecrations, inductions, memorial services and civic occasions. Few church cantatas of this period have survived, apart from the two series published in 1744 and 1748–9 (which may have been written before 1740); and the Hamburg Opera was now closed.

1755 marks the beginning of a new phase in Telemann's creativity. Influenced perhaps by Handel, whom he had known since 1701 and with whom he still corresponded in old age, he turned once more, at the age of 74, to writing oratorios, choosing texts by the younger generation of poets, such as K. W. Ramler, F. G. Klopstock, J. A. Cramer, J. F. W. Zachariae and J. J. D. Zimmermann. Some of these late works were still frequently performed in Hamburg decades after Telemann's death in 1767.

CHAPTER TWO

Influence and reputation

Telemann not only lived through but helped to bring about a great change in German musical life. Until the 18th century a composer's output was largely dictated by the nature of the post he held, and the various spheres of musical activity were strictly defined. A Kantor did not write operas; public performances of music were generally connected with some institution. But Telemann refused to be fettered, as a composer, by the chains of his official duties; and he broke down the barriers between sacred and secular music. By organizing public concerts, he was trying to give music lovers the opportunity to hear all kinds of music, including some which had originally been composed to lend atmosphere to some special occasion and might otherwise have been heard by only a limited number of people. His concerts might thus include operatic excerpts as well as festive or funeral music. The Passion oratorios which he composed in addition to the liturgical Passions were sung not only in the city's smaller churches but even before paying audiences at public concerts. At a time when music publishing in Germany was still in its infancy, music lovers and self-taught musicians had great difficulty in obtaining printed scores. Telemann's eagerness to publish was prompted by a desire to ease this situation and provide a service to his fellow men. He not only increased sales, but also increased the possibilities for the performance of his music by reducing the scoring in the printed editions of his cantatas or by providing alternative instrumentation in his chamber music. Wherever possible, he avoided technical difficulties, for his constant aim was to achieve a wide dissemination and to foster the spread of music in the home as well as in the collegia musica. In some editions there are traces of Telemann the pedagogue, as when in the *Sonate metodiche* and the *Trietti methodici* he demonstrated the art of ornamentation in an instrumental line; or when, in the vocal *Singe-, Spiel- und Generalbass-Übungen*, the rules of continuo realization are given; or when, in the *Harmonischer Gottes-Dienst*, the principles of performing practice in recitative are elucidated. In the protracted struggle between Telemann and the Hamburg book publishers, the issue in dispute was not only the retailing of the textbooks but the composer's rights regarding the performance of his own works. Telemann's great achievement, through his public concerts, was to establish the composer's prerogative to do as he thought fit with his own compositions, even when they had originally been intended for some special occasion. He gave a new meaning to the post of civic director of music. By refusing to be confined to his contractual obligations, he reorganized the city's music to suit his own forward-looking ideas, and in so doing made his influence felt on musical life throughout Germany.

The decades between 1720 and 1760 were long regarded by musicologists as having been dominated by J. S. Bach. However, it was Telemann who played the leading role and ranked as one of the most famous and most important German composers. Even in his youth, as director of the Leipzig collegium musicum, he inspired and stimulated the most gifted music lovers among his fellow students to become professional musicians. Graupner, Pisendel and Heinichen all made music under his direction and were involved in his opera productions. Younger composers such as Fasch and Stölzel, who joined the collegium musicum under Telemann's successor and thus became familiar with some of his compositions, modelled their music on his. In Frankfurt and Hamburg the various civic musical events and the performances by the collegium musicum delighted visitors as well as the local audiences. But it was above all by publishing so many of his compositions that Telemann made them generally accessible; among the leading theorists of the day, not only Mattheson and Scheibe (who were in constant touch with him in Hamburg and directly influenced by him) but also Quantz and Marpurg could cite his works when setting out rules of composition and principles of style. Scheibe ranked Hasse, C. H. Graun, Telemann and Handel as the most advanced of the German composers, and praised their good taste and their achievement in bringing German music into high esteem. In his *Lexicon* J. G. Walther devoted four times as much space to Telemann as he did to his own kinsman J. S. Bach; while the poet Gottsched, in his *Ode zum Lobe Germaniens*, hailed Telemann and Handel as the most distinguished of German composers although as a resident of Leipzig he must have been familiar with the music of J. S. Bach. The subscription lists for some of Telemann's published works indicate that his compositions were also known and loved outside Germany. For the *Musique de table*

(1733), 52 of the 206 subscriptions came from abroad, 33 of them from France. Handel sent an order from London, and in several subsequent compositions (for example the Organ Concerto op.7 no.4) he borrowed and reworked many themes from the *Musique de table*. Another subscription list, that for the *Nouveaux quatuors* (Paris, 1738), attracted 237 orders, no fewer than 138 of them from France. Scheibe admired Telemann's compositions for their artlessness and unforced ease by comparison with the exaggerated contrapuntal artificiality and 'eye music' of the older composers. Telemann, in fact, was praised for not composing like Bach.

When the 19th century rediscovered Bach, it was Bach's concept of his official position and style of composing that were adopted as the criteria. A Kantor who had written operas came to be looked down on as merely a 'fashionable' composer, lacking in religious fervour. Although his works were hardly known, Telemann was criticized as superficial and excessively prolific. His output surpassed Bach's not only because he refused to be restricted by his official duties. Bach was content to produce no more than five cantata cycles during his 27 years in Leipzig; Telemann is known to have written at least 31. Clearly he believed that, because of his desire to reorganize the city's musical life in accordance with his own ideas, he must be seen to be all the more wholehearted in fulfilling his prescribed tasks. It was not until the 20th century, when he had long been written off as too facile, that research began to give a more general view of his creativity, to make the works more widely known, and to seek a more rational evaluation of him. This trend originated in the contentions of Max Schneider and Romain Rolland, which argue that Telemann's musical ideas were entirely different from those of Bach, that it is pointless to compare the two, and that Telemann should be seen as a forerunner of the Classical style. This clearly does not hold good for all the works or categories; but the present state of research allows for more detailed distinctions of judgment.

CHAPTER THREE

Works

I Sacred music

Although the publication of church cantatas was unusual in the 18th century, Telemann brought out four complete annual cycles and also published the arias from a further cycle, reducing the scoring so as to make them suitable for smaller church choirs and domestic worship. In 1752 Quantz could ironically claim that, until a few years before, there had still been some Kantors who could not bring themselves to perform one of Telemann's sacred works. The style of his cantatas met the demands of the music theorists. When he was still at Sorau, Telemann had met Erdmann Neumeister, originator of the modern 'madrigal' type of church cantata, designed to resemble nothing more or less than 'an operatic piece, combining recitatives and arias'. Mattheson thought that church music should aim to arouse the emotions in a specific way and to interpret the finer points of the text dramatically. To achieve this, the most appropriate style was that of the theatre, for even in church we are only human beings, susceptible to human representations. Both Mattheson and Scheibe expressed their admiration for the expression and harmony of Telemann's sacred music. Scheibe felt that sacred music demanded an elevated style the more vividly to bring out the imagery and 'affect' of the text. As an experienced opera composer Telemann was a master of the art of interpreting in musical terms the words and the sense of his text, and was quick to respond to any cue offered to him by his librettist. Although the edition of the cycle *Harmonischer Gottes-Dienst* requires one voice, one melody instrument and continuo, it demonstrates the broad range of possibilities inherent in theatrical church music. Words like 'hell', 'terror', 'revenge', 'torment', 'fear' etc are treated dramatically; but when the text speaks of 'grace', 'quiet enjoyment' or 'innocent trust', the music conveys the mood in smooth cantabile lines with the simplest of accompaniments. Here Telemann is a more straightforward composer than Bach. With its simple scoring and the restricted scope of each of its cantatas, *Harmonischer Gottes-Dienst* is exceptional in Telemann's cantata output. In each cycle his treatment of text, form and scoring are different. About 90 per cent of the surviving cantatas are for four or more voices, and about 60 per cent are accompanied by strings and woodwind; the cantatas for special feast days also have brass. When Scheibe called for harmony as well as expression, he meant that in the larger-scale works the middle parts should be unobtrusive, free from unvocal ornamentation and too frequent dissonances; over-elaborate counterpoint should be avoided, since it made the music obscure and unnatural. Telemann's cantatas fulfilled all these conditions.

In Hamburg one of Telemann's obligations was to write each year a new Passion, to be performed in turn in each of the five principal churches. In 46 years he only twice parodied Passions from previous years, and not until he was an old man did he occasionally re-use older recitatives and turba choruses. His Passions, too, clearly reflect the development of the genre in the 18th

Harmonischer

Gottes-Dienst/

oder

geistliche

CANTATEN

zum allgemeinen Gebrauche/

welche/

zu Beförderung so wol

der Privat-Haus-

als öffentlichen

Kirchen-Andacht/

auf die gewöhnlichen Sonn- und Fest-täglichen

Episteln durchs ganze Jahr

gerichtet sind,

und aus einer Singe-Stimme bestehen/ die entweder von einer Violine, oder Hautbois, oder Flûte traverse, oder Flûte à bec, nebst dem General-Basse, begleitet wird;

Auf eine leichte und bequeme Ahrt also verfasset/ daß nicht allein die, so zur Aufführung der Kirchen-Music gesetzet sind, und vor allen diejenigen/ so sich nur weniger Gehülfen darbey zu bedienen haben/ solche nützlich gebrauchen können/ sondern auch denen zur geistlichen Ergetzlichkeit/ die ihre Haus-Andacht musicalisch zu halten pflegen/ wie nicht weniger allen/ die sich im Singen/ oder im Spielen auf gedachten Instrumenten üben/ zur Erlangung mehrerer Fähigkeit;

In die Music gebracht, und zum Druck befördert

von

Georg Philipp Telemann/

Chori Musici Hamb. Direct.

In Verlegung des Autoris, und bey demselben, auch in den Leipziger-Messen im Kißnerischen Buch-Laden/ zu finden.

16. Title-page of Telemann's 'Harmonischer Gottes-Dienst' (Hamburg, 1725–6), with the opening of the cantata 'Unbegreiflich ist dein Wesen'

2
Am Feste der Heil. Drey-Einigkeit.
Violino ò Violetta.
Largo.
p.
Un be greiflich ist dein Wesen/ der du Eins in Dreyen
bist/ un be greiflich ist dein We sen/ :/:

century. The Gospel text is increasingly filled out with contemplative interpolations, and eventually even parts of the biblical text are altered and dramatized, with Christ's words rewritten as the texts for bravura arias. In the light of this development, Telemann's *St Luke Passion* of 1728 has a special historical significance: here he and his librettist (M. A. Wilkens) attempted to distinguish between the liturgical Passion and the Passion oratorio. The interruptions of the action of the Passion are neither random nor formless – a criticism levelled at Bach's Passions by M. Hauptmann in the 19th century. Before each of the five sections of the Passion story a parallel passage from the Old Testament is inserted, to act as a 'preparation', and is interpreted as an 'application of faith'. Da capo arias are found only in these sections. In the Gospel text only contemplative congregational chorales are interpolated. The turba choruses show concise rhythms and striking word interpretations. Unlike Bach, with his idiosyncratic recitative lines, Telemann, in his Passion recitatives and particularly in the 1728 *St Luke Passion*, kept to the rules laid down by the theorists and modelled on Italian opera recitative: rhythm and melodic line had to be subordinate to speech declamation. Only when the recitative was mostly in quavers could faster or slower declamation or the lengthening of individual notes give point to the text; and only when the melodic line consisted mainly of stepwise movement and repetition could melodic leaps and unusual intervals stand out and become significant. Where the recitative accompaniment generally pursued an uneventful course, the composer could at appropriate points in the text create additional rhetorical accents harmonically. By adopting these basic principles in his recitatives, Telemann was able to achieve the maximum effect with the simplest means.

Apart from the liturgical Passions with biblical texts, Telemann composed six Passion oratorios on freely written librettos. As early as 1716 he set the text by B. H. Brockes, which also served as a model for *Seliges Erwägen*, written in Frankfurt to his own text. This oratorio, whose dramatic accents, contrasting 'affects' and colourful language called for all the resources of theatrical composition, was given almost every year in Hamburg from 1728 in the concert hall and in the smaller churches. After Telemann's death there were 17 performances between 1786 and 1799 alone, although in 1755 he had also written settings of two more modern texts, Ramler's *Der Tod Jesu* and Zimmermann's *Betrachtung der neunten Stunde*.

II Secular vocal music

In Hamburg, Telemann's sacred and secular oratorios and his occasional compositions had many public performances, and their popularity helped to establish public concerts. In the oratorios and larger-scale cantatas Telemann followed traditional principles regarding the musical equivalents of rhetorical figures. He took as his starting-point a simple melodic line and easily accessible forms. His striking divergences from conventional musical language are always motivated by the text. Among the smaller secular cantatas of before 1740, some tend towards operatic virtuosity while others display a simple and folklike melodic line in their arias.

With his songs, Telemann revived a category which had fallen into oblivion during Germany's 'songless period'. He published his first songs earlier than

Sperontes (*Singende Muse an der Pleisse*, from 1736) and Gräfe (*Oden-Sammlungen*, from 1737). In the preface to 24 *Oden* (1741) he set out his theories about song composition in lighthearted terms: the melodic line of a song should be comfortable to sing, avoiding extremes of the vocal register and virtuoso ornamentation, it should accord with the sense of the text, and it should fit all the verses. To Telemann the greatest problem lay in determining the correct metric and periodic structure. His collection of *Oden* demonstrates this difficulty: the various texts (drinking-songs, comic songs, moral and pastoral songs) demand corresponding types of melody and accompaniment.

With his intermezzo *Pimpinone* Telemann was again a pioneer in a new development. In this work, written eight years before Pergolesi's *La serva padrona*, many elements of the *buffo* style are present, like the rapid 'babbling' on one note, the repetition of small motifs and the characterization by accompanying figuration in the orchestra. Telemann's surviving operas show that bourgeois German opera did not conform to any specific type. Apart from the intermezzos, he wrote both serious and light operas. The latter include the comic type (*Der geduldige Socrates*) and the satirical (*Der neumodische Liebhaber Damon*). His greatest success was undoubtedly achieved with *Socrates* (1721), whose text had been adapted by J. U. von König from the libretto by Nicolò Minato (1680); it shows many of the traits which later distinguished the *opera buffa* from the *opera seria*. There are 17 ensembles and choruses to 38 arias, the latter showing a diversity of formal structures. The four female singers appear more often in ensembles than in arias. A large-scale choral scene, comprising two choruses, an aria with chorus and a dance, opens the third act. The constant variety in the choice of accompanying instruments serves to highlight the characterization and the situations: a prince's opening aria is accompanied by a solo violin, and to mislead the audience as to which of the two leading ladies will finally prevail, the prima donna introduces herself in an aria with a modest accompaniment for continuo alone (her second aria has a flute obbligato). Changes in style are also used to convey character. When the rival ladies are pursuing the same end, they imitate one another in canon. Socrates' adversary is made ridiculous in a 'revenge' aria which is pushed to the extremes of parody. Such moments of comedy are repeatedly enhanced by effectively contrasting them with others of a more lyrical or seriously contemplative tone.

III Instrumental music

In evaluating Telemann's contribution to German music Romain Rolland described him as having introduced 'currents of fresh air'. This description applies above all to instrumental music. Fresh air had already been added to Germany's musical life through the spread of amateur music-making and music in the home; and Telemann accelerated this process by publishing a great quantity of instrumental music which, though technically not too demanding, offered scope for spirited and lively playing. When Telemann once said that he was no great lover of concertos, he had in mind only the purely virtuoso concerto. He many times exploited the inherent possibilities of concertante techniques in ensemble playing. In his concertos there is no rigid scheme dictating the number, disposition or relationship of the

17. Autograph MS from Telemann's opera 'Der geduldige Socrates', first performed in Hamburg, 1721 movements, nor the structure of the first movement. When he popularized the French overture in Germany, he turned what had been typical court music into a new form of light music, epitomized by the programme overture (for example *Don Quixote* and *Hamburger Ebb und Fluht*). But the fresh air was introduced above all in his rejection of the learned style and the formation of the *galant* style. Much (though not all) of Telemann's chamber and keyboard music shows *galant* characteristics: a simple melodic line with clear periodic divisions and transparent structure, in which the accompaniment occupies a purely subordinate role. He was particularly successful in developing a conversational style in his quartets, a form rarely used by other composers and employed by Telemann in each 'production' of the *Musique de table*. He was fond of using elements borrowed from folk music, such as rhythms and melodic phrases which he had first encountered in Polish and Hanakian (Moravian) music. His compositions of before 1740 are historically of particular interest. Side by side with their simple structure, the keyboard fantasias (published in 1732–3) show the beginnings of sonata form. Here Telemann used the technique of bringing together different motifs: some may be re-positioned in the reprise, while others fulfil a specific function as transitional, contrasting or epilogue motifs. The *Fugues légères*, also published before 1740 and described by Telemann as 'Galanterie-Fugen', show by their title the direction in which he was moving. A fugal but almost consistently two-part introduction is followed by a series of short fantasias and dance sections in which tremolando basses, bagpipe accompaniments, unison passages and Polish rhythms appear. This fugue cycle represents the antithesis of Bach's pattern of prelude and fugue. Following the sense of the requirements of Quantz and Marpurg, this is an illustration of the fact that residual traces of the contrapuntal style could still be used to enhance the *galant* style. Art should be combined with charm.

CHAPTER FOUR

Theory

Telemann's many and various official duties in no way hindered his activities as composer, editor and impresario. It was only in music theory that he was frustrated in his ambitious plans. Problems of theory preoccupied him all his life. In a letter to Mattheson, written in 1717, he announced his intention of writing a treatise on the most common instruments and the best ways of exploiting their individual characteristics. In 1728 a newspaper reported his plans for translating J. J. Fux's *Gradus ad Parnassum*, and a published catalogue of his printed works which appeared that year even listed its price. In the preface to *Der getreue Music-Meister* (1728–9) Telemann gave notice that, work permitting, he would publish theoretical analyses of some of his compositions in later fascicles. In 1731 another newspaper item stated that Telemann was writing a theoretical treatise on musical invention. The printed catalogue of 1733 lists a *Traité du récitatif* among his projected publications, while a treatise on composition was promised in 1735, combining the most important elements from the textbooks of Fux and Heinichen and including some discoveries of his own. After his visit to Paris, Telemann declared his intention of recording in print his impressions of French music and musical life. In the published biography a forthcoming treatise on composition (*Musicalischer Practicus*) is again referred to; and in the preface to the cantata cycle *Musicalisches Lob Gottes* Telemann said that he had intended to write at length on the application of the theatrical style to church music, on the composition of German recitative and on the use of dissonance. But he then limited himself to a few observations on the figuring of thoroughbass, and referred again to the *Musicalischer Practicus*, which was to throw further light on the subject.

Telemann, a self-taught musician, intended through his theoretical writings to make performance and composition more accessible to the amateur. The basic rules for setting out continuo parts are laid down in his detailed comments on the songs in the *Singe-, Spiel- und Generalbass-Übungen*. The appendix to the 1730 songbook not only gives directions for setting out continuo parts; it also gives a first introduction to writing inner parts where upper and lower ones are given. Telemann dealt with the basic principles of recitative composition in the foreword to the *Fortsetzung des Harmonischen Gottesdienstes*; and important rules on the proper performance of recitative had already been given in the preface to *Harmonischer Gottes-Dienst*. In many of his prefaces, and in particular in the autobiographies, Telemann gave indications of his attitude to music and to questions of musical aesthetics. Surprisingly, his contribution to the Sozietät der Musikalischen Wissenschaften, founded by Mizler, touched neither on composition nor on performing practice; instead he gave them his *Neues musicalisches System*. Here he attempted a theoretical demonstration of chromatic and enharmonic relationships, and tried to show the difference between B♯ and C, F𝄪 and G etc. But he did not mention all the possible alternatives and limited himself for each interval to four degrees – smallest, small, large and largest 2nd (C–D♭♭, D♭, D and D♯; thus the use of the double flat would preclude that of the double sharp, and vice versa). The criticism of this thesis by some members of the society was unjustified. Intervallic relationships cannot be demonstrated on an equal-tempered keyboard, and it was not Telemann's intention to develop a new temperament; nor had he contemplated using occasional enharmonic changes in harmonizing simple chorales unless called for by the text. Later Scheibe claimed priority for his own discovery of the interval system, though he conceded that he had discussed with Telemann in advance the details of his *Abhandlung von den musikalischen Intervallen* (1739). Only after Telemann's death did Scheibe also reveal that the *Critischer Musikus* stemmed from a joint project with Telemann, who had originally intended to compose every other piece himself, and had indeed seen and approved the first 14 pieces before printing. Even if this belated recognition contradicts other statements by Scheibe, it is certain that Scheibe's discussions with Telemann in Hamburg greatly stimulated his work on the *Critischer Musikus* and that his whole musical philosophy was influenced and strengthened by Telemann.

WORKS

Editions: *G. P. Telemann: Musikalische Werke* (Kassel and Basle, 1950–) [T]
G. P. Telemann: Orgelwerke, ed. T. Fedtke (Kassel, 1964) [F]

Catalogues: W. Menke: *Thematisches Verzeichnis der Vokalwerke von Georg Philipp Telemann* (Frankfurt am Main, 1982–3) [i, church cantatas; ii, other vocal] [M] M. Ruhnke: *Georg Philipp Telemann: Thematisch-Systematisches Verzeichnis seiner Werke: Instrumentalwerke* (Kassel, 1984–) [i, keyboard music, chamber music without bc or 1 inst with bc; ii, iii, other instrumental] [R]

M and R numbers are given at the end of each individual entry in the form, for example, of 1:2.
idem – denotes the same textual incipit, not necessarily the same text throughout.
Numbers in right-hand margins denote references in the text.

CHURCH CANTATAS
M – 1

284, 289, 291–2, 293, 294, 296, 298, 300–01

Principal sources: *B-Bc*, *D-B*/*Bds*, *Bdhm*, *Dlb*, *DS*, *F*, *LEm*, *LEt*, *SHk*, *SWl*, *DK-Kk*, *GB-Lbm*, formerly Königsberg.

* – version in printed cycle differs from original † – version in printed cycle differs from original; original lost. Where original and printed versions differ in title, the alternative is given.

Dates other than those of publication are of first performance.

Unless otherwise stated vocal forces are 4 or more voices. For unpublished works instrumental forces are 2 ob, str, bc, unless otherwise stated. For printed works scorings are as shown under cycles listed below unless otherwise stated. Scorings of original versions of some printed cantatas are unknown.

Harmonischer Gottes-Dienst, oder Geistliche Cantaten zum allgemeinen Gebrauche, 1v, 1 inst, bc (Hamburg, 1725–6); T ii–v [1725–6] — 296, 300, 301, *302–3*, 312

Auszug derjenigen musicalischen und auf die gewöhnlichen Evangelien gerichteten Arien (J. F. Helbig), 1v, bc (Hamburg, 1727) [cycle of 1726–7, arias only] [1727] — 300

Fortsetzung des Harmonischen Gottesdienstes (T. H. Schubart), 1v, 2 insts, bc (Hamburg, 1731–2) [undated cycle in reduced scoring] [1731–2] — 300, 312

Musicalisches Lob Gottes in der Gemeine des Herrn (E. Neumeister), 3vv, str, bc [with tpts, timp for festivals] (Nuremberg, 1744) [1744]; ed. K. Hofmann, 32 choruses in *Biblische Sprüche*, i (Stuttgart, 1973), ii (1978) — 294, 300, 312

Untitled cycle of cantatas (D. Stoppe), 1 solo v, 4vv, str, bc [with tpts, timp for festivals] (Hermsdorff, 1748–9) [1748–9] — 294, 300

*Abscheuliche Tiefe, 2 fl, str, bc (*1731–2), 1:1; Absteigende Gottheit (Dies is der Tag, 1:358); Ach bleib (1748–9), 1:3; Ach, dass der Herr, 1749, 1:4; idem, hn, str, bc, after 1740, 1:5; Ach, dass du den Himmel, 3 tpt, timp, str, bc, 1725, 1:7; †Ach ewiges Wort (1731–2), 1:9; Ach Gott, dein Zion klagt (1748–9), 1:10; Ach Gott, du bist gerecht (E. Neumeister), str, bc, 1719, 1:11; Ach Gott, es geht (1748–9), 1:12; Ach Gott vom Himmel (1748–9), 1:14; Ach Gott, wie beugt (Wenn langer Seuchen)

*Ach Gott, wie drückt (Betrübter Lohn, 1731–2), 1:16; Ach Gott, wie manches (Neumeister), 1721, 1:18; idem (Neumeister), 2 fl, 2 ob, str, bc, 1722, 1:19; idem (J. F. Helbig), 1724, 1:20; idem (1748–9), 1:21; Ach Herr, lehr uns, 2 ob, str, bc, 1:24; Ach Herr, wie ist meiner Feinde (Neumeister), 1717, 1:26; idem (von Lingen), 1723, 1:27; Ach, indem ich erblicke (Gott fähret auf, 1:644); Ach, Jesus geht (Neumeister), 2 fl, 2 ob, str, bc, 1732, 1:28; Ach, mein Herze (Neumeister), 1719, 1:29; † Ach Not (1731–2), 1:30; Ach reiner Geist (Ihr habt nicht, 1:904); Ach sagt mir nichts (1748–9), 1:31; Ach Seele (Der Herr Zabaoth)

Ach, sollte doch (Neumeister), 1722, 1:32; *Ach süsse Ruh (Sanftmutsvolle, zarte Triebe, 1731–2), 1:33; *Ach, welche Bitterkeit (Herzbrechend ist das Augenbrechen, 1731–2), 1:34; Ach, wer verkündigt mir ('Oratorium'), str, bc, 1:35; Ach, wie beisst mich (Neumeister), 2 fl, 2 ob, str, bc, 1722, 1:36; Ach, wie nichtig, 2 fl, 2 ob, str, bc, 1757, 1:37; idem, 3 rec, va da gamba, str, bc, 1:38; Ach, wie so lang, 1724, 1:40; Ach, wo bin ich, T, ob, str, bc, 1:41; idem, 2 fl, bn, va da gamba, str, bc, 1:42; Ach, wo flieh ich (Neumeister), 1719, 1:43; Ach, wo ist mein Jesus (Neumeister), 1732, 1:44; Ach, wo kömmt doch (1744), 1:45

Ach wundergrosser Siegesheld, ob, 2 tpt, timp, str, bc, 1750, 1:47; Ach

Zion (Wie lieget die Stadt so wüste); *Ach zu den tiefsten Jammerhöhlen (Die Bosheit dreht, 1731–2), 1:48; Alle, die gottselig (Neumeister), 1719, doubtful, 1:49; idem (G. Simonis), 1720, 1:50; idem, (Neumeister), fl, 2 ob, 2 hn, str, bc, 1722, 1:51; idem, 2 ob, 3 tpt, timp, str, bc, 1723, 1:52; idem (Neumeister), str, bc, 1:54; Alle eure Sorgen (Neumeister), 1719, 1:71; Alle gute Gaben, 1723, 1:55; Allein die Anfechtung (Helbig), 1725, 1:57; Allein Gott in der Höh sei Ehr, tpt, str, bc, 1:58, ed. K. Hofmann (Stuttgart, 1977); idem, S, B, B, str, bc, 1:59

Allein zu dir, 2 ob, 2 bn, 2 tpt, timp, str, bc, 1750, 1:60; idem, ob, ob d'amore, str, bc, 1:62; Alleluja, Herr Gott, 2 ob, 2 tpt, str, bc, 1731, 1:63; Allenthalben ist dies Leben, bc, str, 1:64; Aller Augen warten, 1719, 1:65; idem, 3 ob, bn, str, bc, 1720, 1:66; idem (B. Neukirch), 1725, 1:67; *Alles Fleisch ist Heu (Krache, sinke, morsche Hütte, 1727), 1:68; Alles, was ihr tut, fl, 2 tpt, timp, str, bc, 1756, 1:69; Alles, was von Gott, 2 ob, 3 tpt, timp, str, bc, 1723, 1:70; Allmächtiger, heiliger, starker Gott, 2 fl, 2 ob, str, bc, 1:72; All's Glück und Ungelücke, ob, str, bc, 1:73

Also hat Gott die Welt geliebet (Neumeister), fl, 2 ob, str, bc, 1719, 1:74; idem (Simonis), 2 ob, bn, str, bc, 1721, 1:75; idem (Neumeister), 1722, 1:76; idem (?von Lingen), 2 ob, 3 tpt, timp, str, bc, 1723, 1:77; † idem (Gott dem nichts verborgen, 1727), 1:79; idem (Simonis), 2 tpt, timp, str, bc, 1726, 1:80; idem (Daran ist erschienen, 1:168); idem (1744), 1:82; idem, 2 taille, str, bc, 1:85; idem, 3 tpt, timp, str, bc, 1:86; Also hoch (1748–9), 1:87; Also schweig mein Mund (Neumeister), 2 hn, str, bc, 1722, 1:89; Alter Adam (so leget nun ab von euch, 1:1369); Amen, amen, Lob und Ehre (1744), 1:91; idem (partly Neumeister), 1:92; Am göttlichen Segen, 2 fl, 2 ob, str, bc, 1:93; *Am guten Tag (Es klingelt oft kläglich, 1731–2), 1:94; Armselige Weisheit (Der natürliche Mensch); Auch der Mangel wird (Herr, wie sind deine Werke, 1:780)

Auf ehernen Mauern (M. Richey) (1725–6), 1:96, T iii; Auf ein gleicherhörtes Flehen, 1724, 1:97; Auf, erwachet meine Sinnen (1748–9), 1:98; Auf, fröhliches Zion, B, str, bc, 1:99; Auf Gott will ich S, B, vn, bc, 1:100; Auf grüner Auen (G. P. Telemann), 2 ob, 2 tpt, str, bc, 1722, 1:101; Auf ihr Christen insgemein (Neumeister), str, bc, 1:102; Auf ihr Priester, fl, str, bc, 1:103; Auf, lasset in Zions geheiligten Hallen, 2 fl, 2 ob, 2 tpt, timp, str, bc, 1756, 1:104; Auf mein Herze, ob, str, bc, 1721, 1:106; Auf, und lasset uns besingen, T, 2 vn, bc, 1:108; Auf Zion, 2 fl, ob, 2 tpt, timp, str, bc, 1760, 1:109; Auf zum Streiten (Seid stark, 1:1280)

Augenweide, Fleischeslust, str, bc, 1:110; Aus Gnaden seid ihr selig worden, 2 ob, 3 tpt, str, bc, 1724, 1:112; idem (1744), 1:113; Aussatz hat mich ganz gefressen, str, bc, 1:1740; *Aus Zion, ob, str, bc (Heult verruchte, 1727), 1:114; Barmherzigkeit kann uns (Seid allesamt gleichgesinnt); *Barmherzig und gnädig (Opfre Gott Preis, 1727), 1:116; Beglückte Mutter (Der Herr hat offenbaret, 1:261); Beglückte Zeit (1725–6), 1:118, T v; Begnadigte Seelen (to Büren) (1725–6), 1:119, T iv; Bei dem Herrn (Neukirch), 1725, 1:120; *Bekehret euch zu mir (Zerknische, du mein blödes Herz, 1731–2), 1:121; Belebende Lüfte, 2 ob, 2 tpt, timp, str, bc, 1763, 1:122; Bequemliches Leben, str, bc, 1:123; Beschämt und zittern (Suchet den Herrn, 1:1404); Bestelle dein Haus, S, B, 2 rec, str, bc, 1:124

Bete nur (Neumeister), str, bc, 1732, 1:125; Betrübter Lohn (Ach Gott, wie drückt); Beweget euch munter (Wohlan, ich will meinem Leben); Bist du denn so gar verlassen (Wenn jemand das Gesetz); Bittet, so wird euch gegeben (Neumeister), 1719, 1:127; Bittet, so wird euch gegeben (Simonis), 2 fl, 2 ob, str, bc, 1721, 1:128; †idem (Erleuchte uns, 1727), 1:129; Bleib, o Jesu, S, 2 vn, bc, 1:130; Blitz, der Herz und Geist (Verflucht sei jedermann); Blut, das Glut und Eifer (Christus ist kommen); Brannte nicht unser Herz (Helbig), 1725, 1:131; Brausende Stürme, 1724, 1:132; Brecht, heisse Seufzer, los (Wenn der Herr Friede gibt); Brich an und werde Licht, ob, 2 tpt, timp, bc, 1749, 1:133; Brich an den Hungrigen (Neumeister), 1722, 1:134

Christen heissen und nicht sein (Neumeister), 1:135; Christ ist erstanden (Neumeister), 1722, 1:136; Christum lieb haben ist besser (Neumeister), 2 fl, 2 ob, str, bc, 1719, 1:137; Christus der ist mein Leben, 1754, 1:138; Christus hat ausgezogen, 2 fl, 3 tpt, timp, str, bc, 1757, 1:139; Christus hat einmal . . . gelitten (Neumeister), 1722, 1:140; Christus hat gelitten (Neumeister), 1717, 1:141; Christus hat sich . . . gegeben (Neukirch), 1725, 1:142; Christus hat unsere Sünde selbst, 1723, 1:144; Christus ist aufgefahren (Helbig), 1723, 1:145; Christus ist der Glanz (Helbig), str, bc, 1725, 1:148; *Christus ist kommen (Blut, das Glut und Eifer, 1727), 1:149; Christus ist nicht eingegangen (Neumeister), 1717, 1:150

Christus ist um unserer Missetat willen (Simonis), 2 ob, bn, 2 tpt, str, bc, 1721, 1:151; Christus ist wahrlich der Prophet (Helbig), 1725, 1:152; Da die Zeit (G. Behrmann), 2 tpt, timp, str, bc, 1726, 1:154; *Dafür halte uns (Du bist mir schnödes Gut, 1731–2), 1:155; Da Jesu, deinen Ruhm (Es spielen die Strahlen), ed. K. Hofmann (Stuttgart, n.d.); Da Jesus nun merkte (Neumeister), 1717; 1:156; Danket dem Herrn, denn er ist freundlich (Neumeister), 2 ob, 2 hn, str, bc,

1716, 1:157; idem (von Lingen), str, bc, 1723, 1:158; idem (Neumeister), fl, 2 ob, 2 hn, str, bc, 1722, 1:159; Danket dem Herrn Zebaoth (Neumeister), fl, 2 ob, 2 chalumeau, str, bc, 1718, 1:163, ed. F. Schroeder (Munich and Leipzig, 1978)

Daran erkennen wir (Neumeister), bn, 2 tpt, timp, str, bc, 1:164; Daran ist erschienen (Neumeister), 1717, 1:165; idem (Simonis), 1720, 1:166; idem, str, bc, 1722, 1:167; *idem, 2 ob, 2 hn, str, bc (Also hat Gott, 1731–2), 1:168; Darum, wenn sie zu euch sagen, fl, ob, str, bc, 1:169; Das Blöken der Schafe, 2 tpt, str, bc, 1:170; Das Blut Christi des Sohnes Gottes (Simonis), 2 ob, hn, str, bc, 1721, 1:172; Das Blut Christi wird . . . reinigen (Helbig), 1723, 1:171; Das ew'ge Licht, str, bc, 1:173; Das heulende Winseln, 2 fl, 2 ob, 2 tpt, str, bc, 1755, 1:174; Da sie aber davon redeten (Neumeister), str, bc, 1717, 1:175

Das ist das ewige Leben (Helbig), 2 fl, 2 ob, str, bc, 1726, 1:176; idem (Neumeister), 1720, 1:177; Das ist die Freudigkeit (1744), 1:178; Das ist ein köstlich Ding (Neumeister), 2 ob, tpt, str, bc, 1719, 1:180; Das ist je gewisslich wahr (Simonis), 1717, 1:181; Das ist je gewisslich wahr (Neumeister), 1723, 1:182; idem (Helbig), 1727, 1:183 [= BWV141], ed. in *J. S. Bach: Werke*, xxx (1884); *idem, 2 fl, 2 ob, str, bc (Mein Glaube ringt, 1731–2), 1:184; Das ist meine Freude, str, bc, 1:186; idem, B, 1:187; idem (Helbig), fl, 2 ob, bn, str, bc, 1723, 1:188

Das ist sein Gebot (1744), 1:189; Das Kreuz ist eine Liebesprobe (Neumeister), 1719, 1:190; Das macht Gottes Vaterherz (1748–9), 1:191; Das Manna deiner Speise (G, C, Lehms), B, 2 fl, str, bc, 1:192; Das Reich Gottes (Helbig), 1723, 1:193; Dass Herz und Sinn (1725–6), 1:194, T v; Das sollst du wissen (1744), 1:195; Das weiss ich fürwahr (Neumeister), 1717, 1:196; idem (Simonis), 1717, 1:197; idem (1744), 1:198; Das Wetter rührt (Richey), (1725–6), 1:199, T iv

Das Wort Jesus Christus (Neumeister), 1718, 1:200; Das Wort ward Fleisch, str, bc, 1722, 1:202; *idem, 2 ob, 2 tpt, timp, str, bc (Wie hoch bist du, 1727), 1:203; Davids Sohn, str, bc, 1:204; Dazu ist erschienen (Simonis), 1721, 1:205; idem, 2 ob, 3 tpt, timp, str, bc, 1722, 1:206; idem (Helbig), 1723, 1:207; idem, str, bc, 1727, 1:208; Deine Liebe, str, bc, 1:209; Deiner Sanftmut Schild (1748–9), 1:211; Deines neuen Bundes Gnade (Wilkens) (1725–6), 1:212, T iv; Deine Toten werden leben (Wilkens) (1725–6), 1:213, T iii; Dein gnädig Ohr (1748–9), 1:214; Dein Schade ist verzweifelt böse (Neumeister), ob, str, bc, 1720, 1:215

Dein Wort ist meinem Munde (Neumeister), fl, 2 ob, str, bc, 1721, 1:216; Dein Wort ist meines Herzens Trost (Neumeister), 1732, 1:217; Dem Herren musst du trauen, 1:218; Dem höchsten Gott zu loben, B, hn, 2 vn, str, bc, 1:219; Demut ist der Tugend Krone (Gott widerstehet den Hoffärtigen); Den Christen mischt Christus (Herr, warum trittest du); Denen, die Gott lieben (Neumeister), 1720, 1:220; Den Frommen gehet das Licht auf, 1723, 1:221; Dennoch bleibe ich stets an dir (Neumeister), 1720, 1:223; idem (Helbig), 1725, 1:224; idem (1748–9), 1:225; Dennoch bleib ich immer stille (Telemann), 2 ob, 3 tpt, timp, str, bc, 1721, 1:222; Denn, so jemand das ganze Gesetz (Neukirch), 1725, 1:226

Den Reichen von dieser Welt (Neukirch), 1725, 1:227; idem, str, bc, 1:228; *Der Allergrösste von Weisheit (Erwäg, o Mensch, 1731–2), 1:229; Der am Oelberg zagende Jesus [= Die stille Nacht]; Der arme Mensch, 1757, 1:231; Der Engel des Herrn (Neumeister), 2 hn, str, bc, 1719, 1:232; idem (Helbig), 1728, 1:235; Der Feind ist viel, 2 fl, ob, hn, str, bc, 1758, 1:237; Der feste Grund Gottes (Simonis), 2 ob, 2 chalumeau, str, bc, 1721, 1:238; Der Friede Gottes, 2 ob, 2 hn, tpt, str, bc, 1723, 1:239; Der Geist des Herrn (Neukirch), str, bc, 1725, 1:242

Der Geist gibt Zeugnis, 2 tpt, timp, str, bc (1748–9), 1:243; Der geliebte und verlorene Jesus [= Meine Rede bleibt vertrübt]; Der Gerechte kommt (Neumeister), 1717, 1:245; Der Gerechte muss viel leiden (Neumeister), 1720, 1:246; idem, 2 ob, 3 tpt, timp, str, bc, 1723, 1:247; †Der Gerechten Seelen (Verliere nur gebrochen dich, 1727), 1:248; Der Glaube muss dauern (Wer zu Gott kommen will); Der Gott des Friedens, 1721, 1:249; idem (Helbig), 2 fl, ob, str, bc, 1723, 1:250; Der Gottlose ist wie ein Wetter (Simonis), 2 ob, bn, str, bc, 1717, 1:251; Der Gottlose lasse, 1723, 1:252; Der Gottlose lauret, 1:253

Der Gott unsers Herrn (Helbig), 1723, 1:254; Der Herr bewahret (Simonis), 1721, 1:257; Der Herr hat gesagt (Neukirch), 1725, 1:260; †Der Herr hat offenbaret (Beglückte Mutter, 1727), 1:261 [= Ermuntre dich]; idem, 2 ob, 2 tpt, timp, str, bc, 1762, 1:262; Der Herr ist mein getreuer Hirte (Neumeister), 2 ob, tpt, str, bc, 1722, 1:263; Der Herr ist mein Hirte (Neumeister), 1722, 1:264; idem, 2 ob, 3 tpt, timp, str, bc, 1723, 1:265; idem (Helbig), 1725, 1:266; idem (1744), 1:268; *Der Herr ist nahe allen (Ich seufze, 1727), 1:273; idem (Helbig), 1728, 1:274

Der Herr ist nahe bei denen, 1722, 1:275; *idem (Freud und Hoffnung, 1727), 1:276; Der Herr ist Sonn und Schild, 3 tpt, timp, str, bc, 1725, 1:277; Der Herr ist unser Gott, str, bc, 1:278; Der Herr Jesus wird offenbart werden (Neukirch), 1725, 1:279; Der Herr kennet (Simonis), 2 fl, 2 ob, str, bc, 1721, 1:280; *idem (Ich weine, 1731–2), 1:281;

idem (von Lingen), 1723, 1:282; idem (1744), 1:283; Der Herr lebet, 2 tpt, timp, str, bc (1748–9), 1:284; Der Herr regieret (1748–9), 1:285; Der Herr sprach, str, bc, 1:287; Der Herr verstösset nicht (Neumeister), 1722, 1:288

Der Herr weiss die Gottseligen, 1724, 1:289; idem, 1724, 1:290; Der Herr wird dich schlagen, 1:292; Der Herr wird die Elenden, 1723, 1:291; Der Herr wird ein Neues im Lande schaffen, 1719, 1:293; †Der Herr Zebaoth (Ach Seele, 1731–2), 1:294; Der Himmel ist offen, der Himmel ist mein (Neumeister), 2 ob, bn, str, bc, 1722, 1:295; Der Himmel ist offen, mein Jesus (Neumeister), 1732, 1:296; Der Himmelskönig (Es ist keine Obrigkeit); Der Himmel wird heiter, ob, 2 tpt, timp, str, bc, 1757, 1:297; †Der himmlischen Geister, S, vn, bc (1731–2), 1:298

Der höchste Gott (Neumeister), 1722, 1:300; Der jüngste Tag (Neumeister), 2 ob, bn, str, bc, 1718, 1:301; idem (Neumeister), str, bc, 1:302; *Der Kern verdammter Sünder (Was zeigen freche Höllenkinder, 1731–2), 1:303; *Der mit Sünden beleidigte Heiland, 2 ob, 3 tpt, timp, str, bc (1731–2), 1:306; *Der natürliche Mensch (Armselige Weisheit, 1727), 1:308; †Der Regen Gottes (1731–2), 1:312; Der Reichtum macht (Wilkens) (1725–6), 1:313, T ii; Der Segen des Herrn, 1725, 1:309; idem (Neumeister), str, bc, 1719, 1:310; idem (Neumeister), 1725, 1:311; idem (1744), 1:316

Der Sohn Gottes (Neumeister), 1717, 1:317; *Der Stein, den die Bauleute (Jesu bleibt, 1727), 1:319; Der sterbende Jesus [= Jesus liegt in letzten Zügen]; Der Tod ist verschlungen (Simonis), 2 ob, 2 cl/2 tpt, timp, str, bc, 1721, 1:320; idem (Helbig), 1723, 1:322; Der treue Freund, 2 fl, 2 ob, str, bc, 1757, 1:324; *Des Königs Tochter, ob, str, bc (Mein Freund ist mein, 1727), 1:327; Des wütenden Meeres (Ertönet bald herrlich); †Dich, den meine Seele (1731–3), 1:328; Dich rühmen die Welten (Eschenburg), 2 ob, 3 tpt, timp, str, bc, 1762, 1:329; Die auf den Herrn hoffen, 1721, 1:330 [= 7:8]; Die Bosheit dreht (Ach zu den tiefsten Jammerhölen); Die Bosheit siehet oft (Die Glut des Zorns, 1731–2), 1:331; Die Ehe soll ehrlich (Neumeister), 1717, 1:332; idem (Helbig), 1725, 1:333

Die Ehre des herrlichen Schöpfers (Wilkens) (1725–6), 1:334, T v; *Die Engel sind . . . Geister, 2 vn, bc (Tronen der Gottheit, 1727), 1:335; Die Furcht des Herrn ist der rechte Gottesdienst (Helbig), 1725, 1:336; Die Furcht des Herrn ist Ehre, 3vv, bc, 1737, 1:337; Die G'bot uns all gegeben sind, 2 fl, ob, str, bc, 1755, 1:338; Die Glut des Zorns (Die Bosheit siehet oft); Die Gnadentüre stehet offen (1748–9), 1:339; Die Gottes Gnad alleine, ob, 3 tpt, timp, str, bc, 1750, 1:341; *Die Gottlosen ziehen das Schwerdt, ob, str, bc (Feinde, berst vor Grimm, 1727), 1:343; Die Gott vertrauen (Helbig), 2 fl, 2 ob, str, bc, 1723, 1:344; *Die Grube ist von gestern (Kommt, ihr aufgeblasenen, 1727), 1:345

Die Güte des Herrn, S, B, str, bc, 1:346; Die Hauptsumme der Gebote (1744), 1:347; Die Hirten bei der Krippe [= Hier schläft es]; Die ihm vertrauen (1744), 1:348; Die Kinder des Höchsten (Wilkens) (1725–6), 1:349, T v; Die Liebe gegen meinen Gott, fl, str, bc, 1:350; Die mit Tränen säen, S, B, str, bc, 1:352; Die Opfer, die Gott gefallen, 2 ob, 3 tpt, timp, str, bc, 1724, 1:353; Dieser Jesus, 1:355; Dies ist der Gotteskinder Last (Wilkens) (1725–6), 1:356, T iii; †Dies ist der Tag (Absteigende Gottheit, 1727), 1:358; idem (1744), 1:359; idem, T, vn, bn, bc, 1:1741; Die, so ihr den Herrn fürchtet (Neumeister), 1717, 1:362; Die stärkende Wirkung (Kenzler) (1725–6), 1:363, T iv

Die stille Nacht (Der am Oelberg zagende Jesus), B, 2 fl, 2 ob, str, bc, 1:364; Die Sünd hat uns verderbet, 2 fl, 2 ob, str, bc, 1:365; Die Sünd macht leid, fl, ob, 3 tpt, timp, str, bc, 1749, 1:366; Die Wahrheit fällt auf die Gasse (Neumeister), 2 ob, tpt, str, bc, 1717, 1:367; Die Wahrheit ist ein edles Kind, 1757, 1:368; Die Weisheit ruft, 2 fl, 2 ob, str, bc, 1757, 1:370; Die Welt bekümmert sich, 1:371; Die Welt kann ihre Lust, ob, str, bc, 1:372; Die Welt vergehet, 2 fl, ob, str, bc, 1758, 1:373; Donnerode [= Wie ist dein Name so gross and 6:3]; Drei sind, die da zeugen im Himmel (Neumeister), 2 tpt, 2 trbn, str, bc, 1719, 1:374; idem, 2 ob, 2 tpt, 2 hn, str, bc, 1725, 1:375; idem (Neukirch), str, bc, 1:376

Drückt euch (Neumeister), bn, str, bc, 1:378; Du aber, was richtest du (von Lingen), 1723, 1:379; Du bist erschrecklich (Neumeister), 1720, 1:382; Du bist ja (Du teure Liebe); Du bist mein Vater (1748–9), 1:384; Du bist mir schnödes Gut (Dafür halte uns); Du bist verflucht (1725–6), 1:385, T iii; Dünke dich nicht weise sein (Neumeister), ob, str, bc, 1720, 1:386; Du fährest mit Jauchzen (Richey) (1725–6), 1:387, T iii; Du Gott, dem nichts verborgen, 2 fl, 2 ob, str, bc, 1757, 1:389; *Du Hirte Israel, ob, str, bc (Rufst du, süsse Hirtenstimme, 1727), 1:391; Du machst mir (Was hilft der Erden Lust), 1:392

Du o schönes Weltgebäude, 2 fl, 2 ob, str, bc, 1754, 1:394; Durch Adams Fall, ob, str, bc, 1726, 1:395; idem, str, bc, 1:396; Durch Christi Auferstehungskraft (1748–9), 1:397; Durch Christum habt ihr gehöret (Neumeister), 1717, 1:398; Durchsuche dich (Wilkens) (1725–6), 1:399, T iv; *Durch tausend . . . Ränke (Euch wackelhafte

Hoffartsberge, 1731–2), 1:400; Durch Trauren, 2 fl, 2 tpt, timp, str, bc, 1750, 1:401; Du riefest einst, 2 fl, ob, 2 tpt, timp, str, bc, 1761, 1:402; Du sollst lieben Gott (Simonis), 1717, 1:403; *Du teure Liebe (Du bist ja, 1731–2), 1:404

Du Tochter Zion, 1724, 1:405; *idem, fl, 2 ob, 2 hn, str, bc (Jesus kommt, 1731–2), 1:406; idem, 3 tpt, str, bc (1748–9), 1:407; idem, 2 hn, str, bc, 1:408; Du unbegreiflich höchstes Gut, 2 ob, 3 tpt, timp, str, bc, 1747, 1:409; Edler Geist (J. G. Hermann), str, bc, 1:410; Ehre sey Gott in der Höhe, 2 fl, 2 ob, 2 tpt, timp, str, bc, 1756, 1:411; idem, 3 tpt, timp, str, bc, 1:412, ed. G. Fock (Kassel, 1969); Ehr und Dank, 2 tpt, str, timp, bc (1748–9), 1:413; Eilt hinweg, 1742, 1:414; Eilt zu, 2 fl, 2 ob, str, bc, 1:415; Ein Arzt ist uns gegeben (1748–9), 1:416; Ein Aussatz ist die Sünde (Neumeister), str, bc, 1722, 1:417

Einen solchen Hohenpriester (Helbig), 1728, 1:418; Ein feste Burg, B, vn, bc, 1:419; idem, 3 tpt, timp, str, bc, 1:420; Ein gläubiges Flehn, 1:421; Ein gütiges Herz, str, bc, 1:422; Ein heisser Durst (So töricht ist die Welt); Ein Herz, mit seinem (Lasset uns beweisen); *Ein Jammerton (1731–2), 1:424; Ein jeder läuft (Wilkens), (1725–6), 1:425, T ii; Ein Kindelein (Neumeister), str, bc, 1721, 1:426, ed. K. Schultz-Hauser (Berlin, 1963); *Ein lispelnd . . . Gedränge (1731–2), 1:427; Ein Richter muss im Urteil sprechen, 1:429; Ein sanftes Erfreuen, 2 ob d'amore, str, bc, 1:430; Eins bitte ich vom Herren (Neumeister), str, bc, 1717, 1:431; Ein seliges Kind Gottes (So ziehet nun an, 1:1391); Ei nun, mein lieber Jesu (1748–9), 1:432

Eins ist not, str, bc, 1:433; Ein ungefärbt Gemüte (Neumeister), 2 ob, 2 hn, str, bc, 1722, 1:434; Ein Weiser rühme sich nicht (Simonis), 1720, 1:435; Ein zartes Kind (1725–6), 1:436, T v; Ei, warum sollt ich dich lassen (1748–9), 1:437; Endlich wird die Stunde schlagen (Wilkens) (1725–6), 1:440, T v; Engel, Menschen, Himmel (Gelobet sei der Herr, 1:601), Entzückende Lust, A, va da gamba, str, bc, frag., 1:442; Erbarme dich mein, 1721, 1:443; Er, der Herr des Friedens (Neumeister), ob, str, bc, 1719, 1:444; Er, der Messias (Helbig), 1727, 1:445; Erfreue dich (Nun aber, die ihr); Ergeuss dich zur Salbung (Wilkens) (1725–6), 1:447, T iv

Ergötzt euch nur (Neumeister), fl, ob, bn, str, bc, 1722, 1:448; Erhalte mich (Wilkens) (1725–6), 1:449, T v; Erhalt mein Herz (1748–9), 1:450; Erhalt uns Herr (Neumeister), 1722, 1:451; Er hat alles wohlgemacht (1744), 1:452; Er hat ein Gedächtnis gestiftet, ob, str, bc, 1758, 1:453; Erhebet den Herrn, 2 ob, 3 tpt, str, bc, 1:455; Erhebet euch, 1724, 1:457; Erhöhet die Täler (Wilkens), str, bc, 1725, 1:458; Erhöre mich (Neumeister), 2 ob, cornettino, 3 tpt, str, bc, 1717, 1:459; Er ist auferstanden, T, B, 2 tpt, str, bc, 1:460; Er ist mein (1748–9), 1:461

Er kam, lobsinget ihm, fl, ob, 3 tpt, timp, str, bc, 1759, 1:462; † Erkennet, dass der Herr Gott ist (Kehret wieder, 1727), 1:463; Er kennt die rechten Freudenstunden (1748–9), 1:464; Erkenntliche Christen, 1724, 1:465; Erleuchte uns (Bittet, so wird euch gegeben); Ermuntert euch, 1724, 1:466; Ermuntre dich [= Der Herr hat offenbaret, 1:261]; Er neigte den Himmel, ob, 3 tpt, timp, str, bc, 1762, 1:467; Eröffne dich, ob, 3 tpt, timp, str, bc, 1764, 1:468; Erquickendes Wunder (Wilkens) (1725–6), 1:469, T v; Erquicktes Herz, B, vn, vc, bn, bc, 1:470, ed. F. Schroeder (Stuttgart, 1960)

Erscheine, Gott (Wilkens) (1725–6), 1:471, T ii; Erschrick, im Geiz versenkter Sünder (von Lingen), 1723, 1:472; Erschrick mein Herz vor dir, str, bc, 1:473; Erstanden ist der heilge Christ, ob, bn, 2 tpt, timp, str, bc, 1737, 1:474; Erster Anfang, letztes Ende, str, bc, 1723, 1:475; *Ertönet bald herrlich, ob, hn, tpt, str, bc (Des wütenden Meeres, 1731–2), 1:476; Ertönt, ihr Hütten, 2 ob, 2 hn, str, bc, 1723, 1:478; *Ertrage nur das Joch (1731–2), 1:479; Erwachet, entreisst euch (Wilkens) (1725–6), 1:480, T iii; Erwachet zum Kriegen (Wilkens) (1725–6), 1:481, T v

Erwacht aus eurem Sorgen (Herr Gott Zebaoth, 1:751); Erwäg o Mensch (Der Allergrösste von Weisheit); Erwecke dich Herr (Simonis), 1721, 1:482; Erwecke dich, mein (Wir danken dir, Gott); Er wird trinken vom Bach, 2 fl, 2 ob, 3 tpt, timp, str, bc, 1:484; Erzittre Welt vor dem Gericht, str, bc, 1723, 1:485; *Es danken dir Gott die Völker (Lebendiger Gott, 1727), 1:486; Es erhub sich ein Streit, 2 tpt, str, bc, 1:487; idem, ob, hn, str, bc, 1:488; *Es fähret Jesus auf, 2 ob, 3 tpt, timp, str, bc (1731–2), 1:489; Es füllen der Allmacht (Grosse Städte); Es gleitet mein Fuss, str, bc, 1:491

Es hat mich umgeben (von Lingen), 1723, 1:492; Es ist das Heil uns kommen her (Neumeister), 1719, 1:494; Es ist das Herz (Simonis), 1717, 1:495; Es ist dir gesagt (Neukirch), 1725, 1:496; idem (1744), 1:497; Es ist ein elend jämmerlich Ding (Neukirch), 1725, 1:498; Es ist eine Stimme, 2 fl, str, bc, 1758, 1:499; Es ist ein Gott (Neukirch), 1725, 1:500; Es ist ein grosser Gewinn (Neumeister), str, bc, 1717, 1:501; idem (Neumeister), 1720, 1:502; idem (Helbig), 1723, 1:503; Es ist ein köstlich Ding (Helbig), 1723, 1:504; Es ist ein schlechter Ruhm (Wilkens) (1725–6), 1:506, T v

Es ist erschienen (Helbig), 2 fl/ob, str, bc, 1722, 1:507; idem, ob, str, bc,

1754, 1:508; Es ist gewisslich an der Zeit (1748–9), 1:509; Es ist gewisslich an der Zeit, S, A, str, bc, 1:511; Es ist gut, 1724, 1:512; *Es ist keine Obrigkeit (Der Himmelskönig, 1727), 1:513; *Es ist um aller Menschen Leben (1731–2), 1:514; Es ist umsonst (Simonis), 2 ob, bn, str, bc, 1717, 1:515; *idem (Wer mit Gott den Anfang, 1727), 1:516; Es jauchzen die Engel, 2 fl, ob, 2 tpt, timp, str, bc, 1758, 1:517; Es kann nicht anders sein (Neumeister), 1729, 1:518; Es klingelt oft kläglich (Am guten Tag)

Es komm mein End, 2 fl, rec, ob, str, bc, 1758, 1:519; Es kommt die Stunde (Helbig), 1728, 1:520; idem, str, bc, frag., 1:522; Es kommt die Zeit heran, fl, 3 tpt, timp, str, bc, 1:521; Es sei denn, dass jemand (Helbig), 1724, 1:524; Es sei ferne von mir rühmen (Simonis), fl, 2 ob, str, bc, 1721, 1:526; Es sind mancherlerlei Gaben (Neukirch), 1725, 1:527; Es sind schon die letzten Zeiten (1748–9), 1:528; idem, B, ob, str, bc, 1:529; Es soll auf diesen Tag (Helbig), str, bc, 1725, 1:530; *Es spielen die Strahlen (Da Jesu, deinen Ruhm, 1731–2), 1:531; Es spricht, der solches Zeuget (1744), 1:532; Es spricht der unweisen Mund (Neumeister), 1719, 1:533

Es werden nicht alle, die zu mir sagen (Neukirch), 1725, 1:535; Es werden sich viele falsche Propheten erheben (von Lingen), 1723, 1:536; Es wird des Herrn Tag kommen (Simonis), 1717, 1:537; idem, rec, 2 ob, str, bc, 1759, 1:538; Es wird ein Durchbrecher, 2 ob, 3 tpt, timp, str, bc, 1757, 1:539; Es wird ein Tag sein, fl, ob, str, bc, 1:541; Es wird ein unbarmherzig Gericht (Simonis), 1717, 1:542; Es wird geschehen (Neukirch), 1725, 1:543; Es woll uns Gott genädig sein, 1761, 1:544 (Stuttgart, n.d.); Euch wackelhafte Hoffartsberge (Durch tausend . . . Ränke); Euch zuvörderst (Neukirch), str, bc, 1725, 1:545

Ew'ge Quelle (Wilkens) (1725–6), 1:546, T iii; Feiert, Kinder des Höchsten, 2 fl, 2 ob, 2 tpt, timp, str, bc, 1760, 1:547; Feinde, berst vor Grimm (Die Gottlosen ziehen das Schwerdt); Findet nun im Grabe Ruh, str, bc, 1:548, ed. H. Burckhardt in *Kantorei* (n.p., n.d.); Fleuch der Lüste Zauberauen (1725–6), 1:549, T ii; Flüchtige Schätze (Seid allezeit bereit); Flügel her, fl, ob, hn, str, bc, 1750, 1:552; Fraget nicht (Neumeister), ob, str, bc, 1722, 1:553; Fragst du, Jesu, B, 2 vn, vc, bc, 1:554; Freu dich sehr, fl, 2 ob, str, bc, 1721, 1:555; Freud und Hoffnung (Der Herr ist nahe bei denen, 1:276); Freuet euch, die ihr mit Christo (1744), 1:557; Freuet euch, ihr Himmel, fl, 2 tpt, timp, str, bc, 1758, 1:558

Freuet euch mit Jerusalem, 2 fl, 2 ob, 2 tpt, timp, str, bc, 1762, 1:559; Friede, Friede!, 1724, 1:561; Frohlocket, ihr Engel, 2 fl, 2 ob, 2 tpt, timp, str, bc, 1757, 1:564; Frohlocket, ihr Völker, 2 ob, 2 hn, 2 tpt, timp, str, bc, 1760, 1:565; Frohlocket und jauchzet (Neumeister), tpt, timp, str, bc, 1:567; Frohlocket vor Freuden (M. Brandenburg), 2 ob, 2 tpt, timp, str, bc, 1723, 1:568; Fürchtet den Herrn (Helbig), 1723, 1:570; idem, fl, 2 ob, str, bc, 1:571; Fürchtet euch nicht (Helbig), 1725, 1:572; Fürchtet Gott (von Lingen), 1723, 1:573; Furcht und Zittern, fl, 3 ob, str, bc, 1:577

Für Schmeicheln, List und Heuchelei, 1:574; Fürwahr, er trug unsre Krankheit (Helbig), 1728, 1:575; idem, 2 fl, str, bc, 1724, 1:576; Gebenedeiter Weibersamen (Gelobet sei der Herr, 1:605); Gebet dem Kaiser (Neumeister), 1719, 1:578; idem, 1724; 1:579; Gebet, so wird euch gegeben (Neumeister), va da gamba, str, bc, 1726, 1:580; *Gedenke an den (Helbig) (Wolken, ward ihr leer, 1727), 1:585; Gedenke des Sabbattages, 1723, 1:581; Gedenke doch, wie herrlich hoch, fl, hn, str, bc, 1750, 1:582; Gedenke doch, wie ich so elend (1744), 1:583; Gedenket an den (Helbig), 1725, 1:584

Gedenket an Jesum (Neumeister), 1717, 1:586; Geduldig sein, 1:587; Geduld, wenn Menschen sich zu Teufel machen (Neumeister), 1719, 1:589; Gefährten zum Ewgen, ob, 3 tpt, str, bc, 1758, 1:590; Gehet ein (Helbig), 1724, 1:592; Gehet hin, 2 fl, 2 ob, str, bc, 1757, 1:593; Geht heraus, str, bc, ?1726, 1:594; Gelobet sei der Herr (Neumeister), 3 tpt, timp, str, bc, 1719, 1:596; idem (?von Lingen), 2 ob, 3 tpt, timp, str, bc, 1722, 1:597; idem (Helbig), 1723, 1:598; idem (Neukirch), str, bc, 1725, 1:599; idem, 1723, 1:600; †idem (Engel, Menschen, Himmel, 1727), 1:601; idem, 3 fl, 2 ob, 4 hn, timp, str, bc, 1733, 1:602; *idem (Gebenedeiter Weibersamen, 1727), 1:605; Gelobet sei des Herrn Nam', ob, hn, str, bc, 1750, 1:603

Gelobet sei Gott, 2 fl, 2 ob, 2 ob d'amore, 3 tpt, str, bc, 1725, 1:604; idem, 1724, 1:606; idem (Simonis), 1717, 1:607; idem, 1723, 1:608; idem (1744), 1:610; Gelobet seist du, Jesus Christ (Neumeister), 1718, 1:611; idem (Neumeister), str, bc, 1:612; Gen Himmel, zu dem Vater mein (Neumeister), 1719, 1:613; Gerechter Gott, str, bc, 1:614; Gesegnet ist die Zuversicht (Neumeister), 1719, 1:616; Gesegnet sei die Zuversicht (Neumeister), 2 fl, str, bc, 1:617, ed. A. Dürr [as Gesegnet ist die Zuversicht, T, B, 2 fl, 2 vn, bc] (Kassel, 1954); Gib auch den Göttern (Jedermann sei untertan der Obrigkeit, 1:962), 1:620; Gib, dass ich mich nicht (1748–9), 1:621; Gib deiner Gnade Sonnenschein (Gott Zebaoth, wende dich); Gib mildiglich dein' Segen (1748–9), 1:622; Gib mir ein fröhliches Herz (Neumeister), ob, str, bc, 1722, 1:623; Gib uns Segen (1748–9), 1:624

Glaubet, hoffet (1725–6), 1:626, T v; Gläubst du?, str, bc, 1:625; Glaubt nicht einem jeglichen Geist (Neukirch), 1725, 1:627; Gleichwie der Blitz aufgehet, 2 ob, 2 vn, bc; 1:628; Gleichwie der Regen (Neumeister), 1719, 1:630; Gott, bei dir ist die lebendige Quelle, 2 fl, 2 ob, str, bc, 1:633; Gott, dem nichts verborgen (Also hat Gott die Welt geliebet, 1:79); Gott der Hoffnung erfülle euch (Neumeister), 2 ob, 2 tpt, 2 hn, timp, str, bc, 1717, 1:634 [=BWV218], ed. in *J. S. Bach: Werke*, xli (1894); Gott der liebt nicht nur die Frommen, str, bc, 1:635; Gott der Vater wohnt uns bei, ob, hn, str, bc, 1739, 1:636; Gott, du erhörest Gebet (Helbig), 1725; 1:637

Gott, du lässest mich erfahren, 1757, 1:638; Gottesfurcht, der Weisheit Quelle, fl, 2 ob, str, bc, 1:639; Gottes Liebe (1748–9), 1:640; Gottes Wort (Neumeister), 1722, 1:641; Gott fähret auf (Neumeister), 2 ob, 2 tpt, str, bc, 1717, 1:642; *idem (Ach, indem ich erblicke, 1727), 1:644; idem, 3 tpt, timp, str, bc (1744, 1:645; idem, 2 fl, 2 ob, 2 tpt, timp, str, bc, 1756, 1:646; idem, 3 tpt, timp, str, bc, 1:647; idem, 3 tpt, timp, str, bc, 1:648; Gott hält gewiss, 2 fl, 2 ob, str, bc, 1754, 1:650 Gott hat berufen (Helbig), 1723, 1:656; Gott hat Geduld (Helbig), 2 fl, 2 ob, str, bc, 1723, 1:653

Gott hat Jesum erhöhet, ob, str, bc, 1727, 1:654; Gott hat uns Güter ausgetan (Neumeister), str, bc, 1:658; Gott, ich weihe dir, S/T, ob, str, bc, 1:659; Gott ist das ewige Leben (Helbig), 1724, 1:657; Gott ist die Liebe (von Lingen), 1723, 1:660; †idem (Welt, verlange nicht, 1727), 1:661 [2 other pieces extant: 1st chorus as part of J. C. Seibert's Welt, verlange nicht mein Herze; Hilf, dass durch dieser Mahlzeit]; Gott ist ein rechter Richter (Helbig), 1728, 1:663; Gott ist unsre Zuversicht, str, bc, 1725, 1:665; idem, 2 fl, hn, tpt, str, bc, 1:668; Göttlichs Kind, lass mit (Kündlich gross, 1:1020), ed. K. Hofmann (Stuttgart, 1977)

Gottlob, die Frucht hat sich gezeiget (Wilkens) (1725–6), 1:670, T v; Gottlob! Nun geht das Jahr (Neumeister), fl, 2 ob, str, bc, 1721, 1:671; Gott Lob und Dank (Neumeister), 1732, 1:672; idem (Neumeister), T, vn, bc, 1:673; idem, 2 tpt, timp, str, bc, 1:674; Gott mein Ruhm, 2 vn, bn, bc, 1:676; Gott nimmt die Farren (Kommet herzu); Gott schützt uns (Neumeister), tpt, 2 hn, str, bc, frag., 1:677; Gott, schweige doch nicht (Neumeister), 2 ob, 3 bn, str, bc, 1722, 1:678; *Gott segne den Frommen (Sei still, 1731–2), 1:679

Gott sei mir gnädig, 1720, 1:681, ed. T. Fedtke (Stuttgart, 1963) [= 2 frags.; see *Mf*, xviii, 1965, p.412]; Gott steh mir bei (Neumeister), str, bc, 1:682; Gott, strafe nicht, str, bc, 1:683; Gott straft den kranken Menschen (Neumeister), 1720, 1:684; Gott unser Heiland will (1744), 1:686; Gott Vater Sohn und heilger Geist (Neumeister), ob, str, bc, 1:688; Gott verlässt die Seinen nicht (Neumeister), 2 fl, 2 ob, 3 hn, str, bc, 1722, 1:689; Gott weiss, ich bin von Seufzen müde (Ich winselte); *Gott widerstehet den Hoffärtigen (Demut ist der Tugend Krone, 1727), 1:691; Gott, wie ist dein Name, str, bc, 1:692

Gott will, dass allen Menschen (Helbig), 1722, 1:693; Gott will Mensch und sterblich werden (Wilkens) (1725–6), 1:694, T iii; Gott wird geben (1744), 1:696; Gott Zebaoth, in deinem Namen, 1:698; *Gott Zebaoth, wende dich, ob, str, bc, (Gib deiner Gnade Sonnenschein, 1727), 1:699; Greulich sind die letzten Zeiten (Neumeister), 1722, 1:700; Grosse Gottheit, T, bc, frag., 1:701; *Grosse Städte (Es füllen der Allmacht, 1731–2), 1:702; Gross sind die Werke, 2 fl, ob, 2 bn, str, bc, 1757, 1:703; Gute Nacht, 2 ob, 2 tpt, timp, str, bc, 1:704 Habe deine Lust an dem Herrn (Helbig), 1723, 1:705

Habt ihr nicht gesehen, str, bc, 1725, 1:708; Habt nicht lieb die Welt (Neumeister), 2 fl, 2 ob, str, bc, 1719, 1:709; idem (Helbig), 1724, 1:710; idem (Neukirch), 1725, 1:711; Halleluja! Lobet den Namen, 3 tpt, timp, str, bc (1744), 1:713; Halleluja! Singet dem Herrn (von Lingen), 2 ob, 3 tpt, timp, str, bc, 1723, 1:714; Halt ein (Wilkens) (1725–6), 1:715, T ii; Haltet fest (Simonis), 1717, 1:716; Halt im Gedächtnis (1744), 1:717; Hast du denn, Jesu, 2 fl, 2 ob, str, bc, 1:718; Hast du, Herr (Siehe da, eine Hütte); Hat Gott nicht zu seiner (Wisset ihr nicht, 1:1688)

Heile mich (Warum währet doch); Heilige Flamme, 2 fl, 2 ob, 2 hn, timp, str, bc, 1762, 1:719; Heiliger Samen, 1:720; Heiliges Fest, 2 tpt, timp, str, bc, 1:721; Heilige sie Vater (Helbig), 1725, 1:722; Heilig, heilig ist der Herr Zebaoth, 2 ob, 3 tpt, timp, str, bc, 1:724; Heilig ist der Herr Zebaoth (Neumeister), 2 ob, 2 tpt, timp, str, bc, 1720, 1:726; idem, 3 tpt, str, bc, frag., 1:727; Heilig ist Gott der Herr, SATB only, 1:728; Hemmet den Eifer (Wilkens) (1725–6), 1:730, T ii; Herr Christ, den rechten (1748–9), 1:731; Herr Christ, der einge Gottessohn (Neumeister), str, bc, 1722, 1:732; idem, 2 rec, 2 ob, str, bc, 1758, 1:733

Herr, deine Augen sehen, 1:734; Herr, dein Wort erhält, 1:735; Herr, die Demut fällt, 1724, 1:736; *Herr, die Wasserströme erheben sich (Hörst du garnicht, 1727), 1:737; Herr, erhöre meine Stimme, B, 2 vn, vc, bc, 1:738; Herr, gehe nicht ins Gericht, str, bc, 1:740; Herr Gott, barmherzig (Neukirch), 1725, 1:741; Herr Gott, der du uns hast (?von Lingen), str, bc, 1729, 1:742; Herr Gott, dich loben wir, fl, bn, str, 1:745; Herr Gott, du König (?von Lingen), 3 tpt, timp, str, bc, 1728, 1:746

Herr Gott, wer wird es dir genug verdanken (?von Lingen), str, bc, 1729, 1:749; Herr Gott Zebaoth (Helbig), 1728, 1:750; *idem (Erwacht aus evrem Sorgen, 1731–2), 1:751; Herr, ich bin beide (Simonis), fl, 2 ob, str, 1:753; Herr Jesu Christ, dich zu uns wend (Neumeister), str, bc, 1719, 1:755; Herr Jesu Christ, gross ist die Not (Neumeister), 1722, 1:757; Herr Jesu Christ, ich schrei zu dir (Neumeister), 1718, 1:758; Herr Jesu, der du wunderbar, 1:761, doubtful; Herr, lehre uns bedenken (Neumeister), 2 ob, va d'amore, va da gamba, str, bc, 1720, 1:763; *Herr, meiner Sehnsucht (1731–2), 1:764

Herr, nun lass in Frieden (1748–9), 1:766; Herr, schau, die Seele (Lieblicher Saiten); *Herr, schau doch (1731–2), 1:767; Herr, segne meinen Tritt (1748–9), 1:768; Herr, sei mir gnädig, denn mir ist angst (Simonis), fl, 2 ob, str, bc, 1717, 1:769; Herr sei mir gnädig, heile meine Seele, 1724, 1:770; Herr, strafe mich nicht (von Lingen), 2 ob, 3 tpt, timp, str, bc, 1723, 1:771; Herr, streu in mich (So leget ab alle Unsauberkeit); *Herr, warum trittest du, 2 fl, 2 ob, str, bc (Den Christen mischt Christus, 1731–2), 1:774; Herr, was muss ich tun, 2 tpt, timp, str, bc, 1:775; Herr, wie lange willt du (Neumeister), 1722, 1:777; idem (partly Neumeister), str, bc, 1:778

Herr, wie sind deine Werke (von Lingen), fl 2 ob, str, bc, 1723, 1:779; *idem (Auch der Mangel wird, 1727), 1:780; Herr, wir liegen vor dir (Neumeister), fl, str, bc, 1720, 1:781; idem (Helbig), 1723, 1:782; Herzbrechend ist das Augenbrechen (Ach, welche Bitterkeit); Herzlich lieb hab ich dich, 2 fl, 2 ob, str, bc, 1757, 1:783; Herzlich tut mich verlangen (Neumeister), str, bc, 1719, 1:784, ed. K. Hofmann (Stuttgart, 1980); Herzliebster Jesu (Helbig), 1725, 1:785; Herzog meiner Seligkeit, 2 fl, str, bc, 1740–50, 1:786; Heult, verruchte (Aus Zion); Heute geht aus seiner Kammer, 1756, 1:787

Heut ist die werte Christenheit, 2 tpt, str, bc, 1:789; Heut lebst du, ob, str, bc, 1750, 1:790; Heut schleusst er wieder auf, ob, 2 tpt, timp, str, bc, 1758, 1:791; Heut triumphieret Gottes Sohn, 2 fl, 3 tpt, str, bc, 1:793; Hier hab ich meine Plage (1748–9), 1:794; Hier ist mein Herz, S, T, str, bc, 1:795; Hier schläft es (Die Hirten bei der Krippe) (sacred oratorio, K. W. Ramler), 2 rec, 2 ob, 2 hn, 3 tpt, timp, str, bc, 1759, 1:797; Hilf, dass durch dieser Mahlzeit [= part of Gott ist die Liebe, 1:661]; Hilf, dass ich sei von Herzen froh, ob, str, bc, 1:798

Hilf Herr, die Heiligen haben abgenommen (Simonis), 1717, 1:799; idem (Helbig), 1723, 1:800; †idem (Seel und Leib sind fest, 1727), 1:801; Hilf Herr, o hilf, 1:803; Hilf Jesu, hilf, S, str, bc, 1:802; Hirt' und Bischof uns'rer Seelen (Wilkens) (1725–6), 1:805, T iii; Hochselige Blicke (Selig sind die Augen); Horche nur, 2 fl, 2 ob, str, bc, 1:806; Hörst du garnicht unser Flehen (Herr, die Wasserströme); Hosianna dem Sohne David, 2 fl, 2 ob, str, bc, 1720, 1:808; idem (1744), 1:809; Hosianna, dieses soll die Losung sein, 2 fl, 2 ob, 3 tpt, timp, str, bc, 1737, 1:810; Hütet euch (Neumeister), 1716, 1:811

Ich armer Mensch, fl, 2 ob, str, bc, 1:812; Ich bin arm und elend (1744), 1:814; Ich bin das A und das O, 2 fl, 2 ob, 2 tpt, timp, str, bc, 1763, collab. Cron, 1:815; Ich bin der erste (Neumeister), 2 ob, 2 tpt, timp, str, bc, 1717, 1:816; Ich bin der Herr, 1717, 1:817; Ich bin der Weg (Helbig), str, bc, 1724, 1:818; Ich bin dir, Herr, (von Lingen), str, bc, 1729, 1:819; Ich bin getauft (to Büren) (1725), 1:820, T iv; Ich bin getrost (Neumeister), 2 ob, 2 tpt, str, bc, 1:821

Ich bin ja Herr, 2 ob, ob d'amore, bn, tpt, str, bc, 1754, 1:822; Ich bin vergnügt (Neumeister), 1719, 1:823; Ich danke dir (1748–9), 1:824; Ich fahre auf (Simonis), 2 fl, 2 ob, 2 hn, str, bc, 1721, 1:825; Ich freue mich, T, B, ob d'amore, str, bc, 1:826; Ich fürchte keinen Tod (Neumeister), fl, 2 ob, str, bc, 1719, 1:827; Ich gehe voll Freuden (Neumeister), 2 ob, 2 tpt, str, bc, 1:829; Ich girre, seufze, klage (von Lingen), 1723, 1:830; Ich glaube aber, 2 ob, hn, bc, 1:831; Ich habe Lust (Neumeister), 1719, 1:833; idem (Neumeister), 2 fl, 2 ob, str, bc, 1722, 1:834; idem (1744), 1:836

*Ich habe Wächter über euch gesetzt, 2 ob, bn, str, bc (Über der Propheten Blut, 1731–2), 1:839; Ich halte aber dafür (Simonis), fl, 2 ob, str, bc, 1721, 1:840; Ich halte es dafür (1744), 1:841; Ich hatte mich verirret (1748–9), 1:842; Ich hatte viel Bekümmernis (Neumeister), 1717, 1:843; Ich hebe meine Augen auf (Helbig), 1725, 1:845; Ich hoffe darauf (Simonis), 2 fl, 2 ob, timp, str, bc, 1721, 1:847; idem (1744), 1:848; Ich komme, str, bc, 1724, 1:849; Ich lebe, 2 fl, ob, 2 tpt, timp, str, bc, 1754, 1:850

Ich muss auf den Bergen weinen, fl, ob, str, bc, 1723, 1:851; Ich muss im Leben immer wandeln, str, 1:852; Ich muss seufzen (Neumeister), 1732, 1:853; Ich recke meine Hand aus (Neumeister), 1722, 1:854; Ich rufe dich . . . an (von Lingen), str, bc, 1729; 1:855; Ich rufe zu dir, str, bc, 1:857; Ich sage euch (Neukirch), 1725, 1:858; Ich schaue bloss (Wilkens) (1725–6), 1:859, T v; Ich schicke mich (Neumeister), S, 2 vn, bc, 1732, 1:860; Ich schlafe, S, B, 2 vn, bc, 1:861; Ich sehe mich mit Finsternissen (?von Lingen), str, bc, 1729, 1:863; Ich seh euch fast (Seht, wir gehn hinauf)

Ich seufze, winsle, klage (Der Herr ist nahe allen, 1:273); Ich suchte des Nachtes im meinem Bette, 2 ob, 3 tpt, timp, str, bc, 1723, 1:866; †Ich taumle vor Freuden (1727), 1:867; Ich traue auf Gott (1748–9), 1:868; Ich traue Gott, 1732, 1:869; Ich war tot und siehe (Helbig), 1725, 1:872; Ich weine, durch der Armut (Der Herr kennet, 1:281); Ich weiss, dass mein Erlöser lebt (Neumeister), 1717, 1:873; idem (Neumeister), str, bc, 1722, 1:874; idem (Neumeister), str, bc, 1732, 1:875; idem (Neumeister), str, bc, 1:876; idem, S/T, vn, bc, 1:877 [= BWV160], ed. in *J. S. Bach: Werke*, xxxii (1886); idem, 2 fl, 2 ob, str, bc, 1:879

Ich weiss wohl (Helbig), str, bc, 1725, 1:880; †Ich werde fast entzückt (1731–2), 1:881; Ich werfe mich zu deinen Füssen (Neumeister), 2 ob, theorbo, str, bc, 1722, 1:882; Ich will den Kreuzweg gerne gehen (Neumeister), B, vn, bc, 1732, 1:884; Ich will dich erhöhen (Simonis), 2 ob, 2 tpt, str, bc, 1717, 1:885; idem; ob, str, bc, 1:886; Ich will mich mit dir verloben (Neumeister), 2 fl, 2 ob, str, bc, 1720, 1:888; idem, (Helbig), 1723, 1:889; Ich will mit Danken kommen, 1759, 1:890 [= O wie selig ist die Stunde]

Ich will rein Wasser über euch sprengen, ob, str, bc, 1:891; Ich will singen von der Gnade, ob, bn, str, bc, 1:893; *Ich winselte, 2 fl, 2 ob, str, bc (Gott weiss, ich bin . . . müde, 1731–2), 1:894; Ihr Christen, wollt ihr selig sein, 1:896; Ihr, deren Leben mit banger Finsternis (Wilkens) (1725–6), 1:897, T iv; Ihr, die ihr Christi (1748–9), 1:898; Ihr frommen Christen (?von Lingen), 1729, 1:900 [= Treuer Gott]; Ihr Gerechten, freuet euch (1744), 1:901; *Ihr habt nicht einen knechtischen Geist, 2 ob, 3 tpt, timp, str, bc (Ach reiner Geist, 1731–2), 1:904; idem, 2 fl, rec, 2 ob, 2 tpt, timp, str, bc, 1757, 1:905

Ihr in Sünden tote Glieder (So einer für alle gestorben); Ihr Lieben, gläubet nicht (Kantate wider die falschen Propheten) (1744), 1:908, ed. W. Bergmann (London, 1967); Ihr Lieben, lasset uns untereinander Lieb haben (Helbig), 1723, 1:909; †Ihr schüchternen Blicke (1731–2), 1:912; Ihr seid alle Gottes Kinder (Neumeister), 1716, 1:914; idem (Neumeister), 2 ob, bn, str, bc, 1718, 1:915; idem (Neumeister), str, bc, 1:916; Ihr seligen Stunden (Richey) (1725–6), 1:917, T iv; Ihr sollt geschickt sein (Neumeister), 1720, 1:918

Ihr Völker, bringet her (Neumeister), str, bc, 1719, 1:919; idem (Neumeister), 1721, 1:920; Ihr Völker hört, (Wilkens) (1725–6), 1:921, T ii; Ihr werdet, aus Gottes Macht (von Lingen), 1723, 1:922; Ihr werdet weinen und heulen, 1:923; †Ihr Wölfe droht (1731–2), 1:924; Im sechsten Monat, 2 ob, tpt, str, bc, 1:927; In allen meinen Taten, 1:928; In Christo gilt weder Beschneidung (Neumeister), 1720, 1:929; idem (Helbig), 1722, 1:930; In Demut gehen (Neumeister), str, bc, frag., 1:933; In den Seilen deiner Liebe, T, B, 2 ob, 2 vn, bc, 1:934

In der Welt habet ihr Angst, 1728, 1:936; In dich wollst du mich (1748–9), 1:938; In dulci jubilo (Neumeister), 2 ob, hn, str, bc, 1719, 1:939, ed. F. Stein (Berlin, 1957) [= later version of Lobt Gott ihr Christen, 1719]; In Ephraim sind allenthalben Lügen, ob, str, bc, 1757, 1:940; In gering- und rauhen Schalen (Wilkens) (1725–6), 1:941, T ii; In Gott vergnügt zu leben, B, rec/ob, hn, str, bc, 1:942; Ist auch ein Kreuz bitter, 1:944; Ist eine Stunde bald zugegen (Neumeister), 1732, 1:945; Ist Gott versöhnt (1748–9), 1:946; Ist Widerwärtigkeit den Frommen eigen (Wilkens) (1725–6), 1:948, T ii; Ja, selig sind, die Gottes Wort (1744), 1:949; Jauchze du Tochter Zion, 2 fl, 2 ob, 2 tpt, timp, str, bc, 1762, 1:950

Jauchzet dem Herrn alle Welt, ob, 3 tpt, timp, str, bc, 1733, 1:951; Jauchzet, frohlocket (Wilkens) (1725–6), 1:953, T v; Jauchzet, ihr Christen (Wilkens) (1725–6), 1:955, T iii; Jauchzet, ihr getreuen Herzen, S, T, vn, vc, bc, 1:956; Jauchzet ihr Himmel (1744), 1:957, ed. K. Hofmann (Stuttgart, 1974); idem, fl, ob, bn, str, bc, 1756, 1:958; Jauchzt dem Höchsten alle Welt, 2 ob, 2 tpt, timp, str, bc, 1:959; Jedermann sei untertan der Obrigkeit (Helbig), 1723, 1:960; †idem (Gib auch den Göttern, 1731–2), 1:962; Jesu bleibt (Der Stein, den die Bauleute); Jesu durch dein Auferstehen (Neumeister), 2 ob, bn, str, bc, 1:963

Jesu, in deinem Namen, vn, vc, bc, 1:964; Jesu meine Freude (Neumeister), 1719, 1:965; idem (Neumeister), 4 fl, 2 ob, str, bc, 1722, 1:966; idem (Neumeister), str, bc, 1722, 1:967; idem, 2 ob, 2 bn, str, bc, 1754, 1:970; idem, 1:973; Jesu meiner Seele Weide, S, 2 fl, 2 ob, bc, 1:974; Jesu meine Zuversicht, fl, rec, ob, ob d'amore, hn, str, bc, 1754, 1:984; Jesus Christus ist kommen (1744), 1:975; Jesus Christus unser Heiland (Neumeister), 1722, 1:976; Jesus Christus wohnt uns bei (Neumeister), 2 ob, 2 hn, str, bc, 1732, 1:977; Jesus ist der Heilands-Name, fl, ob, str, bc, 1:980; idem, 2 fl, 2 ob, str, bc, 1:981; Jesus kommt (Du Tochter Zion, 1:406)

Jesus Liebe mich vergnüget, S, vn, bc, 1:982; Jesus liegt in letzten Zügen (Der sterbende Jesus), B, 1:983; Jesus nimmt die Sünder an, fl, 2 ob, str, bc, 1:985; Jesus sei mein erstes Wort, 1722, 1:986; Jesus tilget alle Sünden, str, bc, 1:987; Jesu, wirst du bald erscheinen (Neumeister), ob, cl, cornett, 3 tpt, str, bc, 1719, 1:988; *Jetzt geht der Lebensfürst (1731–2), 1:989; Kann euch dieser Donnerschlag (Wir müssen alle offenbaret werden, 1:1671); *Kann man auch Trauben lesen (Schmücke dich nur, 1727), 1:990; Kantate wider die falschen

Propheten [= Ihr Lieben, gläubet nicht]; Kaum ist der Heiland, B, str, bc, 1:991; Kaum wag ich es, str, bc, 1762, 1:992; Kehret wieder (Erkennet, dass der Herr Gott ist)

Kehre wieder (Wo ist denn dein Freund hingegangen); Kein Hirt kann, 2 vn, bc, 1:993; Kein Vogel kann (Wilkens) (1725–6), 1:994, T iv; Kinder, es ist die letzte Stunde (von Lingen), 1723, 1:997; *Kommet herzu (Gott nimmt die Farren, 1727), 1:998; Komm, Geist des Herrn, 2 ob, 3 tpt, str, bc, 1759, 1:999; Komm Gnadentau, 2 fl, 3 tpt, timp, str, bc, 1:1000; Kommt alle, die ihr traurig seid (Neumeister), 1722, 1:1003; Kommt alle, die von so manchem Sündenfalle (Neumeister), 1722, 1:1004; Kommt, Christen, str, bc, 1723, 1:1005; Kommt, die Tafel, str, bc, 1:1006

Kommt her zu mir (Neumeister), 1:1007; idem (Neumeister), 1720, 1:1008; Kommt herzu, lasset uns, str, bc, frag., 1:1009; Kommt, ihr aufgeblasenen (Die Grube ist von gestern); Kommt, ihr, Schäflein, fl, 2 ob, str, bc, 1757, 1:1010; Kommt, lasset uns anbeten, str, bc, 1723, 1:1012; Kommt und lasst, str, bc, 1:1013; Kommt, verruchte Sodoms-Knechte, S, B, 2 ob, 4 bn, str, bc, 1:1014; Krache, sinke, morsche Hütte (Alles Fleisch ist Heu); Kräftiges Wort (O wie ist die Barmherzigkeit des Herrn); Kündlich gross (Neumeister), 1716, 1:1017; idem (Helbig), 1724, 1:1019; *idem, 2 ob, 3 tpt, timp, str, bc (Göttlichs Kind, lass mit, 1731–2), 1:1020

Lasset uns aufsehen, 1724, 1:1023; *Lasset uns beweisen (Ein Herz, mit seinem, 1727), 1:1024; Lasset uns den Herrn preisen, 2 trbn, timp, str, bc, 1750, 1:1025; Lasset uns Gott lieben (Helbig), 1723, 1:1026; idem, 1724, 1:1027; Lasset uns nicht eitler Ehre geizig sein (Neumeister), 2 ob, 2 bn, str, bc, 1720, 1:1028; Lasset uns rechtschaffen sein (Helbig), 2 ob, 2 bn, str, bc, 1723, 1:1029; Lasset uns von Gott singen, S, B, vn, vc, bc, 1:1031; Lass mich an andern üben, B, 1:1033; Lass mich beizeit mein Haus bestellen (1748–9), 1:1034; Lass mich, Jesu, dich begleiten (Wer ist der, so von Edom kommt); Lass mich o mein Gott (1748–9), 1:1037; Lasst uns eins ums ander singen, 3 vv, 1:1036 [= chorus in 2:13]

Lass vom Bösen und tue Gutes, 1724, 1:1038; Lauter Wonne, lauter Freude (Wilkens) (1725–6), 1:1040, ed. G. Braun (Stuttgart, 1963), T v; Lebendiger Gott (Es danken dir Gott die Völker); Lebensfürst, auf dein Erblassen (Man singet mit Freuden, 1:1085); Leben wir, so leben wir dem Herrn, 2 fl, 2 ob, bn, str, bc, 1756, 1:1041; Lehre uns bedenken (1744), 1:1042; idem (1744), 1:1043; Liebe, die vom Himmel stammet (Wilkens) (1725–6), 1:1044, T ii; Liebe, Liebe, muss der Christen Merkmal sein, 1724, 1:1045; Liebet eure Feinde (Simonis), 1717, 1:1046; idem (Helbig), 1723, 1:1047; *Lieblicher Saiten, 2 fl, 2 ob, str, bc (Herr, schau, die Seele, 1731–2), 1:1048

Liebster Jesu, kehre wieder (aria), S, 2 fl, str, bc, 1:1051, ed. F. Schroeder (Berlin, 1963); Liebster Jesu meine Lust, 2 ob/2 taille, bn, str, 1:1052; Lobe den Herrn meine Seele (Neumeister), 2 tpt, str, bc, 1723, 1:1053; idem, 1723, 1:1054; idem, S, B, str, bc, 1:1056; Lob Ehr und Preis, fl, ob, 3 tpt, timp, str, bc, 1760, 1:1057; Lobet den Herrn alle Heiden (Neumeister), 1717, 1:1058; idem, 3 tpt, timp, str, bc (1744), 1:1059, ed. K. Hofmann (Stuttgart, n.d.); idem (1744), 1:1060

Lobet den Herrn, alle seine Heerscharen (Simonis), 2 ob, 3 tpt, timp, str, bc, 1721, 1:1061; Lobet den Herrn, ihr seine Engel, 2 fl, 2 ob, 2 tpt, timp, str, bc, 1756, 1:1063; Lobsinget dem Herrn, 1724, 1:1064; †idem (Mein Herze muss an Freuden, 1727), 1:1065; Lobt Gott ihr Christen, 1719 [= earlier version of In dulci jubilo]; idem (Neumeister), 2 ob, 2 tpt, 3 trbn, timp, str, bc, 1723, 1:1066, ed. A. Adrio (Berlin, 1947); Lobt ihn mit Herz und Munde (1748–9), 1:1067; Locke nur, Erde (Richey) (1725), 1:1069, T v

Mache dich auf (von Holten), 3 tpt, str, bc, 1725, 1:1070; Machet Bahn dem, 2 fl, 2 ob, 2 tpt, str, bc, 1:1073; Machet die Tore weit (Helbig), 1722, 1:1074, ed. T. Fedtke and K. Hofmann (Stuttgart, 1975); idem, 2 fl, 2 ob, hn, tpt, str, bc, 1:1075; Machet euch Freunde (Neumeister), str, bc, 1722, 1:1076; idem, 1077; Machet keusch eure Seelen (Helbig), 1723, 1:1078; Mach mir stets zuckersüss (1748–9), 1:1080; Mächtiger Heiland (Werfet Panier auf im Lande); Man muss nicht zu sehr trauern (Neumeister), 1720, 1:1082

Man singet mit Freuden (Neumeister), 2 ob, 2 tpt, timp, str, bc, 1719, 1:1084; †idem (Lebensfürst, auf dein Erblassen, 1727), 1:1085; idem, 3 tpt, timp, str, bc, 1734, 1:1086; idem, 3 tpt, timp, str, bc (1744), 1:1087; idem, 2 ob, bn, 3 tpt, timp, str, bc, 1:1088; Maria stand auf in den Tagen (Neumeister), 1719, 1:1089; Mein Alter kömmt, str, bc, 1:1090; Meine Augen sehen (1744), 1:1093; Meine Brüder, seid stark, 1723, 1:1094; Meine Liebe lebt in Gott (Lehms), Bar, ob, vn, vc, bc, 1725, 1:1095; Meinem Jesum will ich singen, fl, str, bc, frag., 1:1096; Meinen Jesum lass ich nicht (Neumeister), 2 ob, bn, str, bc, 1724, 1:1097

Meinen Jesum will ich lieben, 2 ob, hn, str, bc, 1:1098; Meine Rede bleibt betrübt (Der geliebte und verlorene Jesus), S, ob, str, bc, 1:1099; Meines Bleibens ist nicht hier (Neumeister), 1732, 1:1100; idem (Neumeister), A, 2 vn, bc, 1:1101; Meine Schafe hören meine

Stimme (Neumeister), ob, str, bc, 1719, 1:1102; idem (1744), 1:1103; Meine Seele erhebt den Herrn, 1717, 1:1104; idem, 1722, 1:1105; idem, str, bc, 1723, 1:1106; †idem (So schön, so zärtlich, 1731–2), 1:1107; idem (1744), 1:1108

Meine Seele harret nur auf Gott, 2 fl, 2 ob, 2 bn, str, bc, 1:1109; Meine Seele trägt Verlangen (Neumeister), vn, vc, bc, 1:1110; Mein Freund ist mein (Des Königs Tochter); Mein Glaube ringt in letzten Zügen (Das ist je gewisslich wahr, 1:184); Mein Gott, ich bin gesinnt (Telemann), S, B, ob d'amore, vn, vc, bc, 1725, 1:1112; Mein Gott, ich schäme mich (Simonis), 1717, 1:1114; Mein Herz ängstet sich (Neumeister), 2 fl, 2 ob, str, bc, 1722, 1:1115; Mein Herze, heg Barmherzigkeit, ob, str, bc, 1:1116; Mein Herze muss an Freuden (Lobsinget dem Herrn, 1:1065)

Mein Herz, warum betrübst du dich, hn, vn, va, bc, 1:1117; Mein Herz, was kränkest du dich (Neumeister), str, bc, 1:1118; Mein Jesu, ist dirs denn verborgen (Neumeister), 1719, 1:1119; Mein Jesu, meines Herzens Freude (Neumeister), T, vn, bc, 1732, 1:1120; Mein Jesus ist mein Leben, S/T, vn, bc, 1:1122; Mein Jesus ist mein treuer Hirt (Neumeister), 2 fl, 2 ob, str, bc, 1:1123; Mein Jesus nimmt die Kranken an, str, bc, 1:1124; Mein Jesus starb, bn, 2 hn, str, bc, 1:1125; Mein Kind, verwirf die Zucht (Simonis), 1717, 1:1128

Mein Kind, willtu Gottes Diener sein (Neumeister), 1716, 1:1129; idem (Neumeister), 1718, 1:1130; Mein lieber Gott allein (Neumeister), str, bc, 1:1131; Mein Schade ist verzweifelt böse, S, B, fl, ob, str, bc, 1:1133; Mein Schutz und Hülfe (1748–9), 1:1132; Mein Sünd mich werden kränken (Neumeister), 1719, 1:1134; Me miserum, miseriarum conflictu, A, str, bc, 1:1135; Michael, wer ist wie Gott, 2 ob, 2 tpt, timp, str, bc, 1764, 1:1136; Mir hat die Welt (Neumeister), 2 ob, 2 hn, str, bc, 1719, 1:1137; idem (1748–9), 1:1138; Mir stehet keine Weltlust an (Neumeister), str, bc, 1:1139; Mit Fried und Freud, str, bc, 1:1140

Mit Gott im Gnadenbunde stehen, S, B, 2 ob, 2 chalumeau, 2 hn, glock, str, bc, 1722, 1:1141; Möcht ich, Jesu, fl, ob, 2 tpt, timp, str, bc, 1763, 1:1144; †Muntre G'danken, fliehet (1731–2), 1:1145; Muss nicht ein Mensch (Neumeister), 1722, 1:1146; †Nach ausgelöschtem Feindschafts-Feuer (1731–2), 1:1147; Nach dem Weinen, 1724, 1:1148; Nach dir will ich mich sehnen (1748–9), 1:1149; †Nach Finsternis und Todesschatten (1731–2), 1:1150; Nehmt hin, ihr Reichen (Neumeister), STB, str, bc, 1:1152; Nichts, nichts kann mich (1748–9), 1:1153; Nicht viel Weise (Helbig), 1725, 1:1156

Niedrigkeit ist ein Spott, 2 fl, 2 ob, str, bc, 1:1157; Nimm von uns, Herr, du Treuer Gott (Neumeister), 1722, 1:1159; Nimm nicht zu Herzen, 2 fl, 2 ob, str, bc, 1:1160; *Nun aber, die ihr (Erfreue dich, 1727), 1:1161; Nun aber gehe ich hin (Neumeister), str, bc, 1719, 1:1162; idem (Simonis), fl, 2 ob, str, bc, 1721, 1:1163; Nun freut euch (1748–9), 1:1167; Nun ist das Heil (Mayer), 2 tpt, timp, str, bc, 1726, 1:1170; idem, 2 fl, ob, 2 tpt, timp, str, bc, 1754, 1:1171; idem, 2 fl, 2 ob, 2 hn, 2 tpt, timp, str, bc, 1761, 1:1172

Nun ist das Reich, fl, ob, 2 tpt, str, bc, 1762, 1:1173; Nun komm der Heiden Heiland (Neumeister), 2 ob, 2 tpt, timp, str, bc, 1718, 1:1174; idem (Neumeister), 2 ob, 2 tpt, str, bc, 1721, 1:1175; idem (Neumeister), str, bc, 1:1177; idem (Neumeister), str, bc, 1:1178, ed. R. Fricke (Hameln, 1930); Nun kommt die grosse Marterwoche (Neumeister), 1719, 1:1179; Nun lernet dich mein Herze kennen, 2 fl, 2 ob, str, bc, 1757, 1:1181; Nunmehr hab ich ausgeruft, 1:1182; Nun weicht ihr trüben Trauerstunden (Neumeister), 2 vn, bc, 1:1183; Nun wir denn sind gerecht worden (Helbig), 1:1184

Nur getrost und unverzagt, 2 solo vv, 2 ob, str, bc, 1:1185; Ob bei uns ist der Sünden viel, 1:1186; O ein gütiges Befehlen, 1724, 1:1188; O Ewigkeit, (Neumeister), 1723, 1:1189; O fröhliche Stunden (aria), T, 2 ob, 2 hn, str, bc, 1:1190; O Gotteslamm (1748–9), 1:1191; O Gottes Sohn, str, bc, 1:1192; O Gott, gedencke mein, T, B, str, bc, 1:1193; O Gott, wie gross ist deine Güte, 1724, 1:1194; O grosse Lieb (1748–9), 1:1195; O grosser Gott von Macht (Neumeister), 1719, 1:1196; Ohne Glaube ist's unmöglich (1744), 1:1199; O Jesu Christ, dein Kripplein (1748–9), 1:1200, ed. G. Braun (Stuttgart, 1966)

O Jesu, meine Wonne, str, bc, 1:1201; O Jesu, treuer Hirte (1748–9), 1:1202; O Land, höre (Neumeister), fl, 2 ob, str, bc, 1720, 1:1203; O liebster Gott (Neumeister), frag., 1:1204, O mein Gott, vor den ich trete (Neumeister), fl, 2 ob, str, bn, bc, 1719, 1:1205; idem (Neumeister), 1:1206; O Mensch, bedenke stets dein Ende, str, bc, 1:1207; O Mensch, wer du auch immer bist, str, bc, 1:1209; Opfre Gott Dank, ob, str, bc, 1747, 1:1210; Opfre Gott Preis (Barmherzig und Gnädig); O schnöde Wollust dieser Erden (Wie fehlet doch ein Herz); O selig Vergnügen, A, B, 2 fl, 2 vn, bc, 1:1212; O setze alles Leid hintan, 2 fl, ob, 2 tpt, timp, str, bc, 1756, 1:1214

O tausendmal gewünschter Tag (Neumeister), str, bc, 1719, 1:1215; O weh, schaut der Egypter Heer, ob, 3 hn, str, bc, 1732, 1:1216; O wie herrlich wirds im Himmel (Neumeister), str, bc, 1722, 1:1217; *O wie ist die Barmherzigkeit des Herrn (Kräftiges Wort, 1727), 1:1219; O wie selig ist die Stunde, 2 fl, ob, str, bc, 1759, 1:1220 [= Ich will mit Danken kommen]; O wie wird der Mensch beliebt, 2 hn, str, bc, 1:1221; Packe dich, gelähmter Drache (Wilkens) (1725–6), 1:1222, T v; Posaunen wird man hören gehn, 2 ob, trbn, str, bc, 1:1223; Prangende Lilien (Wohl dem, der den Herrn fürchtet); Prang im Golde, stolze Welt (Wir sind allesamt wie die Unreinen, 1:1675)

Redet untereinander mit Psalmen (Neumeister), 2 ob, 3 tpt, timp, str, bc, 1719, 1:1225; idem (Neukirch), str, bc, 1725, 1:1226; Reiner Geist, lass doch mein Herze (Lehms), B, fl, vn, vc, bc, 1725, 1:1228; †Reiss los, reiss vom Leibe (1727), 1:1229; Rufst du, süsse Hirtenstimme (Du Hirte Israel); Ruft es aus, 3 tpt, timp, str, bc, 1:1230; Sage mir an (Simonis), fl, 2 ob, str, bc, 1721, 1:1231; Sage nicht, ich bin ein Christ, 1757, 1:1232; *Saget dem verzagten Herzen (So kommt denn auch, 1727), 1:1233; Saget der Tochter Zion, 2 ob, 3 tpt, timp, str, bc, 1723, 1:1235; Sanftmutsvolle, zarte Triebe (Ach süsse Ruh); Sanftmut und Geduld (H. F. von Uffenbach), 1728, 1:1236

Schaffe in mir Gott, 2 ob, 3 tpt, timp, str, bc, 1723, 1:1238; idem, ob, str, bc, 1:1240; idem, str, bc, 1:1241; Schaue Zion die Stadt, 2 fl, ob, 2 tpt, timp, str, bc, 1761; 1:1242; Schau nach Sodom nicht zurücke (C. Steetz) (1725–6), 1:1243, T iv; *Schau Seele, Jesus geht (1731–2), 1:1244; Schaut die Demut Palmen tragen (1725–6), 1:1245, T iii; Schicket euch in die Zeit (Neumeister), 2 ob, 2 chalumeau, str, bc, 1720, 1:1247; Schmecket und sehet, wie freundlich (1744), 1:1250; idem, 2 pic, 2 fl, 2 ob, 3 tpt, timp, str, bc, 1:1251

Schmeckt und sehet unsers Gottes (Wilkens) (1725–6), 1:1252, T ii; Schmücke dich nur (Kann man auch Trauben lesen); Schmücke dich, o liebe Seele (Neumeister), 1721, 1:1253; idem, ob, str, bc, 1:1254; Schmücket das Fest, fl, 2 ob, 3 tpt, timp, str, bc, 1:1255; Schmückt das frohe Fest (Richey) (1725–6), 1:1256, T iv; Schwing dich auf zu deinem Gott (Neumeister), 1722, 1:1257; Seele, lerne dich erkennen (Wilkens) (1725–6), 1:1258, T ii; Seel und Leib sind fest (Hilf Herr, 1:801); Segensreicher Gang (Wir haben die Hoffnung); Sehet an die Exempel (Simonis), 2 ob, hn, 3 tpt, str, bc, 1721, 1:1259

Sehet auf und hebet (Neumeister), 1721, 1:1260; Sehet nun zu (Simonis), 1717, 1:1262; Sehet, wir gehn hinauf (Neumeister), ob, str, bc, 1719, 1:1263; *idem (Ich seh euch fast, 1731–2), 1:1264; idem, str, bc, 1:1265; *Seid allesamt gleichgesinnt (Barmherzigkeit kann uns, 1727), 1:1266; *Seid allezeit bereit (Flüchtige Schätze, 1727), 1:1267; Seid barmherzig (Neumeister), 4 ob, str, bc, 1719, 1:1268; Seid dankbar (1744), 1:1269; Seid getrost, fürchtet euch nicht, 1757, 1:1270; Seid getrost und hoch erfreut (1748–9), 1271; Seid ihr mit Christo auferstanden, 2 ob, 3 tpt, timp, str, bc, 1723, 1:1272

Seid nüchtern (Neumeister), 1717, 1:1273; idem (Simonis), 2 fl, 2 ob, 2 hn, str, bc, 1721, 1:1274; idem (Neumeister), 1719, 1:1275; idem (Helbig), 1723, 1:1276; idem (1744), 1:1278; Seid stark (Neumeister), 1717, 1:1279; *idem (Auf zum Streiten, 1727), 1:1280; Sei du mein Anfang, fl, ob, str, bc, 1755, 1:1282; Seid wacker alle Zeit (Simonis), 1720, 1:1281; Sei getreu bis in den Tod (Simonis), 2 ob, 2 bn, str, bc, 1720, 1:1283; idem, str, bc, 1:1284; Sei Jesu treu bis in den Tod (Neumeister), 1731, 1:1287; Sein starker Arm, 3 tpt, str, bc, 1:1288; Sei still, zerreiss der Nahrung Netze (Gott segne den Frommen)

Sei zufrieden meine Seele, B, vn, bc, 1:1290; Selig ist der Mann, 1723, 1:1291; idem, str, bc, 1:1292; Selig ist der und heilig (Helbig), 1723, 1:1293; *Selig sind die Augen (Hochselige Blicke, 1731–2), 1:1294; Selig sind, die Gottes Wort hören (Neumeister), 1717, 1:1295; idem (von Lingen), 1722, 1:1296; Selig sind die Toten (Simonis), 1717, 1:1298; idem (Simonis), 2 ob, 2 bn, str, bc, 1721, 1:1299; idem (von (Lingen), 1721, 1:1300; idem, ob d'amore, bn, str, bc, 1757, 1:1302; idem, str, bc, 1:1303; Selig sind, die zum Abendmahl (Neumeister), 2 fl, str, bc, 1719, 1:1304; idem, 3 tpt, timp, str, bc, 1723, 1:1305; idem (Neumeister), 2 fl, 2 ob, str, bc, 1725, 1:1306; idem (Simonis), 2 tpt, timp, str, bc, 1726, 1:1307; idem (1744), 1:1308; idem, S, T, str, bc, 1:1743

Sie gehen in die Welt, ob, hn, str, bc, 1750, 1:1312; Siehe an meinen Jammer und Elend (Neukirch), 1724, 1:1313; *Siehe da, eine Hütte (Hast du, Herr, 1727), 1:1314; Siehe da, ich lege in Zion (Helbig), 1724, 1:1315; Siehe, das ist Gottes Lamm (Neumeister), 1717, 1:1316, ed. R. Kubik (Stuttgart, 1982); idem (Helbig), 1725, 1:1317; idem (1744), 1:1318, ed. K. Hofmann (Stuttgart, 1972); idem, 3 ob, str, bc, 1:1320; Siehe, der Herr kommt, 1724, 1:1322; Siehe, des Herrn Auge, 1723, 1:1324

Siehe, eine Jungfrau ist schwanger (Neumeister), 2 ob, 2 bn, str, bc, 1717, 1:1326; idem (Helbig), 1725, 1:1327; Siehe, es hat überwunden (Helbig), 2 ob, 3 tpt, timp, str, bc, 1723, 1:1328 [= BWV219], ed. in *J. S. Bach: Werke*, xli (1894); idem, 2 ob, 3 tpt, timp, str, bc, 1:1329; Siehe, es kömmt ein Tag, str, bc, 1723, 1:1330; Siehe, ich komme, str, bc, 1:1331; idem, str, bc, 1:1332; Siehe, ich verkündige euch, fl, ob, 2 tpt, timp, str, bc, 1755, 1:1333; idem, 2 fl, 2 ob, 2 tpt, timp, str, bc, 1761, 1:1334

Sie ist gefallen, Babylon (Helbig), 1725, 1:1336; *idem (Sünder, die sich selbst, 1727), 1:1337; Sie verachten das Gesetz (1744), 1:1339; Sind sie nicht allzumal dienstbare Geister, str, bc, 1724, 1:1340; Sing Dank und Ehr, 2 fl, ob, 2 tpt, timp, str, bc, 1758, 1:1341; Singet dem

Herrn ein neues Lied (Ps xcviii, ?Telemann), 2 tpt, timp, 2 harp, str, bc, 1708, 1:1748; idem (Helbig), 2 ob, 2 tpt, timp, str, bc, 1725, 1:1343; idem (Peticius), 2 fl, 2 ob, 2 tpt, timp, str, bc, 1761, 1:1344; idem, 2 hn, tpt, harp, str, bc; 1:1345; Singet Gott, lobsinget seinem Namen, 2 ob, 3 tpt, timp, str, bc, 1728, 1:1346

Singet um einander, 2 fl, 2 ob, 2 tpt, timp, str, bc, 1764, 1:1347; *So der Geist des, der Jesum von den Toten (Weil ich glaube, 1727), 1:1348; So du freiest, str, bc, 1723, 1:1349; So du mit deinem Munde bekennest, 2 ob, 3 colascione, timp, glock, str, bc, 1723, 1:1350; So du mit deinem Munde bekennest (Helbig), 1725, 1:1351; idem, str, bc, 1:1352; idem (chorus) [= BWV145], ed. in *J. S. Bach: Werke*, xxx (1884) and in *J. S. Bach: Neue Ausgabe sämtlicher Werke*, I/x (1955); †So einer für alle gestorben (Ihr in Sünden tote Glieder, 1727), 1:1353

So fahre hin (1748–9), 1:1354; So Feuer als Flamme, 1:1356; So gehest du, mein Jesu, hin, 2 ob d'amore, str, bc, 1:1744; So grausam mächtig ist der Teufel (Wir haben nicht mit Fleisch und Blut); So hoch hat Gott geliebet (von Lingen), str, bc, 1729, 1:1358; So ihr den Menschen (von Lingen), 2 ob, 3 tpt, timp, str, bc, 1723, 1:1359; So ist der Mensch gesinnt, 1737, 1:1361; So jemand Christi Wort wird halten, 1724, 1:1362; So kommt denn auch (Saget dem verzagten Herzen); So lasset uns nicht schlafen (1744), 1:1364; So lasset uns nun hinzutreten, str, bc, 1:1363; *So leget ab alle Unsauberkeit, 2 fl, 2 ob, str, bc (Herr, streu in mich, 1731–2), 1:1366

So leget nun ab von euch (Simonis), 1717, 1:1367; idem, 2 ob, 3 tpt, timp, str, bc, 1723, 1:1368; *idem (Alter Adam, 1731–2), 1:1369; So leget nun von euch ab (Helbig), 1724, 1:1370; Soll ich nicht von Jammer sagen, B, fl, str, bc, 1724, 1:1371; Sollt ein christliches Gemüte, B, fl, ob, str, bc, 1721, 1:1373; Sollt uns Gott nun können lassen, 1:1374; So macht's die Welt (Neumeister), S, vn, bc, 1:1375; So nun das alles soll zergehen (Helbig), 1723, 1:1376; So richtet denn des Herrn Wege, doubtful, 1:1379; So schön, so zärtlich (Meine Seele erhebt den Herrn, 1:1107); So spricht der Herr: Man höret, 1:1383

So spricht der Herr Zebaoth (Simonis), 1717, 1:1384; *So töricht ist die Welt, 2 fl, 2 ob, str, bc (Ein heisser Durst, 1731–2), 1:1385; So wie das alte Jahr (?von Lingen), 3 tpt, timp, str, bc, 1729, 1:1386; So wir denn nun haben (1744), 1:1387; So wir sagen, dass wir Gemeinschaft (Helbig), 1723, 1:1388; So wir sagen, wir haben keine Sünde (Helbig), 1723, 1:1389; So ziehet nun an (Helbig), 2 fl, 2 ob, str, bc, 1723, 1:1390; *idem (Ein seliges Kind Gottes, 1731–2), 1:1391; Sprich nicht im Mangel (1748–9), 1:1392; Sprich nur ein Wort (1748–9), 1:1393; Spricht der Herr aber also (Neumeister), 1717, 1:1394; Stehe auf, Nordwind, 2 rec, str, bc, 1:1397; Stern aus Jacob (Neumeister), 2 tpt, str, bc, 1722, 1:1398

Sticht dich so mancher falscher Feind (Neumeister), str, bc, frag., 1:1400; Stille die Tränen (Richey) (1725), 1:1401, T iv; Stimmt an mit vollen Chören (Neumeister), frag., 1:1402; Suchet den Herrn (Neukirch), 1725, 1:1403; *idem, ob, str, bc (Beschämt und zittern, 1727), 1:1404; idem (1744), 1:405; idem (Neumeister), 1758, 1:1406 [= 1:1405 with different text]; Sünder, die sich selbst (Sie ist gefallen, 1:1337); Süsser Trost für meine kranke Seele (von Lingen), 2 ob, str, bc, 1723, 1:1407; †Süsse Ruh in herben Leiden (1727), 1:1408; Tag und Stunden (Wohl dem, des Hülfe der Gott Jacobs)

Tausend Segens Proben (Neumeister), 1:1409; Todesangst und Höllenschr. (Wohl dem, dem die Übertretungen, 1:1700); Tönet die Freude, 2 fl, 2 tpt, timp, str, bc, 1757, 1:1410; Trachtet am ersten (Helbig), 1724, 1:1411; idem (1744), 1:1412; Trag mit Geduld, T, B, vn, bc, 1:1413; Trauret ihr Himmel, 2 fl, 2 ob, 2 tpt, timp, str, bc, 1760, 1:1414; Traurigs Herz, verzage nicht, 1:1415; Treuer Gott, ich muss dir klagen, 1:1416 [= Ihr frommen Christen]; Trifft menschlich und voll Fehler sein (1725), 1:1417, T iv; Tritt Arbeit und Beruf (Neumeister), 2 ob, 2 hn, str, bc, 1722, 1:1418; idem (Neumeister), 2 hn, str, bc, 1:1419

Tritt auf die Glaubensbahn (S. Franck), str, bc, 1:1420; Triumph! Denn mein Erlöser lebt, 2 ob, 2 hn, 2 tpt, timp, str, bc, 1724, 1:1421; Triumphierender Versöhner (Wilkens) (1725–6), 1:1422, T iii; Triumph, lobsinget dem, 2 tpt, timp, str, bc, 1758, 1:1423; Triumph, Triumph, ihr Frommen (?von Lingen), 3 tpt timp, str, bc, 1729, 1:1424; Tronen der Gottheit (Die Engel sind . . . Geister); Tröstet mein Volk, 2 tpt, timp, str, bc, 1:1425; Über der Propheten Blut (Ich habe Wächter . . . gesetzt); Umschlinget uns (Wilkens) (1725–6), 1:1426, T v; Unbegreiflich ist dein Wesen (to Büren) (1725–6), 1:1745, T iv; Und alle Engel stunden, str, 3 tpt, bc (1744), 1:1427; Und als der Tag der Pfingsten, fl, ob, 2 tpt, timp, str, bc, 1758, 1:1428

302–3

Und als er nahe hinzu kam (Simonis), 1717, 1:1429; Und da die Engel gen Himmel (Neumeister), 1716, 1:1430; Und das Wort ward Fleisch (1744), 1:1431; Und der Herr Zebaoth, 2 fl, 2 ob, str, bc, 1:1432; Und die Apostel sprachen (Neumeister), 1717, 1:1433; Und es erhub sich ein Streit (Neumeister), 2 tpt, timp, str, bc, 1717, 1:1434; Und Gott ruhte, str, bc, 1:1436; Und siehe, eine Stimme (1744), 1:1437; Und sie redeten miteinander (Neumeister), ob, str, bc, 1722, 1:1438; Unempfindlich muss man sein (aria), S, str, bc, 1:1439; Unerschaffnes Licht (Wer Arges tut); Unschuld und ein gut Gewissen (Neumeister), 1722, 1:1440; idem (Neumeister), str, bc, 1:1441

Unser keiner lebet ihm selber (Helbig), 1723, 1:1442; idem (1744), 1:1443; Unser Leben währet, 2 fl, rec, 2 ob, str, bc, 1759, 1:1444, ed. W. Menke (Hamburg, n.d.); Unser Trost ist der, dass wir (1744), 1:1445; Unser Wandel ist im Himmel, 2 ob, 2 hn, str, bc, 1:1449; Uns ist ein Kind geboren (Neumeister), 2 ob, 2 tpt, timp, str, bc, 1716 1:1450; idem (Neumeister), 2 fl, 2 ob, 3 tpt, timp, str, bc, 1718, 1:1451; idem (Simonis), 2 ob, 2 hn, str, bc, 1720, 1:1452; idem, 3 tpt, timp, str, bc (1744), 1:1453; idem, 2 tpt, timp, str, bc (1748–9), 1:1454; idem (Neumeister), inc., 1:1455; Unverzagt in allem Leide (Wilkens) (1725–6), 1:1456, T v; Unverzagt und ohne Grauen (1748–9), 1:1457

Valet will ich dir geben (Neumeister), 1722, 1:1458; Vater, Gott von Ewigkeit, 2 fl, ob, str, bc, 1758, 1:1459; Vater unser im Himmelreich (Neumeister), 1722, 1:1460; idem, 1:1461; Vater unser in dem Himmel, 2 ob, bn, str, bc, 1:1462; †Verdammet, fluchet ihr Gesetze (1731–2), 1:1463; Verdammte Brut (Neumeister), 2 ob, 2 bn, str, bc, 1732, 1:1464; Verdammter Wankelmut (Neumeister), frag., 1732, 1:1465; *Verflucht sei jedermann (Blitz, der Herz und Geist, 1727), 1:1466; Verfolgter Geist, wohin (Wilkens) (1725), 1:1467, T v; Verirrte Sünder, kehret (?von Lingen), str, bc, 1728, 1:1469; Verlass doch einst o Menschenkind (?von Lingen), str, bc, 1729, 1:1470

Verliere nur gebrochen dich (Der Gerechten Seelen); Verlöschet, ihr Funken (Wilkens) (1725–6), 1:1471, T iv; Versuchet euch selbst (Helbig), 1723, 1:1473; Victoria, mein Jesus ist erstanden, B, tpt, str, bc, 1:1746; Victoria, Triumph, 3 tpt, timp, str, bc, 1:1475; Viel Kreuze liegt (Neumeister), str, bc, 1:1476; Viel sind berufen, 2 ob, 3 tpt, timp, str, bc, 1723, 1:1478; *Viel tausend sind (1731–2), 1:1479; Vor allen Dingen ergreifet, str, bc, 1:1481; *Vor allen Dingen habt unter einander (Wolken, regnen, 1727), 1:1482

Vor des lichten Tages Schein (Wilken) (1725–6), 1:1483, T v; Vor Gott gilt keine Heuchelei, S, B, 2 fl, vn, bc, 1:1484; *Vor Wölfen in der Schafe Kleider (1727), 1:1485; Wache auf, der du schläfest, 2 fl, 2 ob, 2 bn, str, bc, 1:1486; Wachet auf, ruft uns die Stimme (1748–9), 1:1487; idem (contrapunctischer Choral) (chorus), SBB, str, bc, ?1754, 1:1488; idem, 2 fl, 2 ob, 3 tpt, timp, str, bc, 1721, 1:1492; Wachet und betet (Helbig), 1725, 1:1489; idem, S, B, vn, bc, 1:1490; Wachset in der Gnade (1744), 1:1491

Wahrlich ich sage euch: ein Reicher (Neumeister), fl, 2 ob, str, bc, 1722, 1:1494; idem (Neumeister), str, bc, 1:1495; Wahrlich ich sage euch, so ihr den Vater (Simonis), 1721, 1:1493; idem, 2 rec, 2 ob, str, bc, 1:1497; Wandelt in der Liebe (Wilkens) (1725–6), 1:1498, T iii; Warum betrübst du dich, meine Seele (Neumeister), 2 ob, va da gamba, str, bc, 1722, 1:1499; Warum betrübst du dich mein Herz (Neumeister), ob, str, bc, 1:1500; Warum sollt ich traurig sein (Neumeister), 1731, 1:1501; Warum verstellst du (Wilkens) (1725–6), 1:1502, T ii; *Warum währet doch unser Schmerz (Heile mich, 1727), 1:1503

Was betrübst du dich, meine Seele (Simonis), 1717, 1:1505; idem, 2 ob, 3 tpt, timp, str, bc, 1723, 1:1506; Was fehlt dir doch (Neumeister), fl, 2 ob, str, bc, 1722, 1:1507; Was frag ich nach der Welt, 2 fl, ob, 3 tpt, timp, str, bc, 1754, 1:1508; Was für ein jauchzendes Gedränge, 1:1509; Was gibst du denn, fl, 2 ob, str, bc, 1:1510; Was gleicht dem Adel (1725–6), 1:1511, T v; Was Gott im Himmel will (1744), 1:1512; Was Gott tut, S, B, vn, bc, 1:1747; Was hast du, Mensch! (Simonis), 1717, 1:1513; Was hat das Licht (Simonis), 1717, 1:1514; †Was hilft der Erden Lust (Du machst mir, 1731–2), 1:1515; Was ist das Herz (1725–6), 1:1516, T ii

Was ist dein Freund, 1:1517; *Was ist ein Mensch, str, bc (Zerbrechliche Gefässe, 1727), 1:1520; Was ist mir doch das Rühmen nütze (Wilkens) (1725–6), 1:1521, T ii; Was ist schöner als Gott dienen (1748–9), 1:1522; Was Jesus nur mit mir (Neumeister), 1719, 1:1523; idem (Neumeister), 1:1525; Was Jesus tut (Neumeister), S, vn, bc, ?1732, 1:1526; Was mein Gott will (Neumeister), 1719, 1:1529; Was meinst du, ob, str, bc, 1:1530; Was suchet ihr (von Lingen), str, bc, 1723, 1:1531; Was vom Fleisch geboren wird, 2 fl, 2 ob, str, bc, 1762, 1:1533; Was zeigen freche Höllenkinder (Der Kern verdammter Sünder); Weg mit Sodoms gift'gen Früchten (Wilkens) (1725–6), 1:1534, T iii

Weg nichtige Freude, str, bc, 1:1535; Weiche, Lust und Fröhlichkeit, S/T, vn, va, bc, 1:1536, ed. F. Schroeder (Stuttgart, 1966); Weichet fort aus meiner Seele (Neumeister), 2 ob, 2 hn, str, bc, 1722, 1:1537; Weicht, ihr Sünden (Mayer) (1725–6), 1:1538, T iv; Weich, verborgner Pharisäer, 1:1539; Weide mich auf grüner Auen (Wir gingen alle in der Irre); Weil ich den rufe (Simonis), 1717, 1:1540; Weil ich glaube (So der Geist des); *Weine, nicht, siehe, 2 ob, 3 tpt, timp, str, bc (Zürne nur, 1731–2), 1:1541; Weint, weint, betrübte Augen (L. Mizler), 2 fl, ob, str, bc, 1754, 1:1542; Welcher unter euch (Simonis), 2 fl, 2 ob, str, bc, 1721, 1:1545

Welch Getrümmel erschüttert, 2 fl, 2 ob, 2 tpt, timp, str, bc, 1757, 1:1546; Welt, hinaus, T, vn, vc, bc, 1:1547; Welt, verlange nicht (Gott ist die Liebe, 1:661); Wende dich zu mir (1744), 1:1550; Wende meine Augen ab (Neumeister), 1721, 1:1551; Wenn aber der Tröster kommen wird (Helbig), 1725, 1:1552; Wenn aber des Menschen Sohn, fl, ob, str, bc, after 1755, 1:1553; Wenn Angst und Not (Neumeister), str, bc, 1:1554; Wenn böse Zungen stechen (1748–9), 1:1555; *Wenn der Herr Friede gibt (Brecht, heisse Seufzer, los; 1727), 1:1556; Wenn du deine Gabe (1744), 1:1557; Wenn ich ein gut Gewissen habe (Neumeister), str, bc, 1719, 1:1561

Wenn Israel am Nilusstrande (Wilkens) (1725–6), 1:1562, T iv; *Wenn jemand das Gesetz (Bist du denn so gar verlassen, 1727), 1:1563; *Wenn langer Seuchen, fl, 2 ob, str, bc (Ach Gott, wie beugt, 1731–2), 1:1564; Wenn meine Sünd mich kränken, str, bc, 1754, 1:1565; Wenn mich die böse Rott anfällt, 2 ob, hn, str, bc, 1757, 1:1566; Wenn mir angst ist (1744), 1:1567; Wenn wir in höchsten Nöten (Neumeister), 3 ob, 3 tpt, timp, str, bc, 1719, 1:1568; idem (1748–9), 1:1569; Wenn wir nicht Kreuz (1748–9), 1:1570; *Wer Arges tut (Unerschaffenes Licht, 1727), 1:1572; Wer bei Gott in Gnaden ist (Neumeister), 1722, 1:1573

Wer bringt dir nicht Ehre, B, fl, 2 ob, 2 hn, 2 tpt, str, bc, 1:1574; Wer da saget (Simonis), 1717, 1:1575; Werde munter mein Gemüte (Neumeister), str, bc, 1:1576; Wer denkt, die Krone des Lebens zu erben (Neumeister), 1731, 1:1577; Wer der Barmherzigkeit (1744), 1:1578; Werd ich dann zu deiner Rechten (Neumeister), 1722, 1:1579; *Werfet Panier auf im Lande, 2 ob, 2 tpt, timp, str, bc (Mächtiger Heiland, 1727), 1:1580; Wer Gott meint was abzuringen (Neumeister), str, bc, 1:1581; Wer hofft in Gott, str, bc, 1:1582; Wer ist aber, der die Welt überwindet, str, bc, 1722, 1:1583

Wer ist, der dort von Edom (1725–6), 1:1584, T iii; *Wer ist der, so von Edom kommt (Lass mich, Jesu, dich begleiten, 1727), 1:1586; Wer ist der, so von Sodom kommt, 1722, 1:1585; Wer ist wohl wie du Jesu (1748–9), 1:1587; Wer Jesum kennt (Neumeister), fl, 2 ob, str, bc, 1722, 1:1588; Wer mich liebet (Simonis), fl, 2 ob, 2 cl, hn, str, bc, 1721, 1:1589; idem (Neumeister), 1722, 1:1590; idem, 2 fl, ob, 2 tpt, str, bc, 1754, 1:1591; Wer mit Gott den Anfang (Es ist umsonst, 1:516); Wer nur den lieben Gott (Neumeister), 1722, 1:1593; Wer sehnet sich (Wilkens) (1725–6), 1:1594, T iv

Wer seinen Bruder hasset, 1723, 1:1595; idem, fl, 2 ob, str, bc, 1:1596; Wer sich auf seinen Reichtum, str, bc, 1:1597; Wer sich des Armen erbarmet, 1723, 1:1598; Wer sich des Armen erbarmt, 2 fl, 2 ob, bn, 2 tpt, timp, str, bc, 1757, 1:1599; Wer sich erhöhet (Helbig), 2 fl, 2 ob, str, bc, 1724, 1:1603; Wer sich rächet (Neumeister), 1719, 1:1600; idem (1744), 1:1601; idem, 1:1602; Wer sich vor dir, 2 fl, 2 ob, str, bc, 1:1605; Wertes Zion, sei getrost (Neumeister), 2 ob, 3 tpt, timp, str, bc, 1722, 1:1606; idem (partly Neumeister), SS or TT, B, str, bc, 1:1607

Wer weiss, wie nahe, 1749, 1:1609; idem (1748–9), 1:1610; idem (1748–9), 1611; Wer will uns scheiden (Neumeister), 1717, 1:1613; *Wer zu Gott kommen will (Der Glaube muss dauern, 1727), 1:1614; *Wer zweifelt (1731–2), 1:1615; Wie der Hirsch schreiet (Simonis), 1717, 1:1616; idem, 3 tpt, timp, str, bc, 1722, 1:1617; idem, 1:1618; Wie fehlet doch ein Herz (O schnöde Wollust dieser Erden, 1731–2), 1:1620; Wie freudig seh ich, 2 fl, 2 hn, 3 trbn, timp, str, bc, 1:1621; Wie hoch bist du (Das Wort ward Fleisch, 1:203); Wie ist dein Name so gross (Donnerode), 1:1624 [= 6:3]; †Wie, kehren sich (1731–2), 1:1625

Wie lieblich sind deine Wohnungen (Neumeister), 1719, 1:1627; idem (1744), 1:1628; *Wie lieget die Stadt so wüste (Ach Zion; 1727), 1:1629; Wie ofte hört man nicht, 2 fl, str, bc, 1:1633; *Wie schmerzlich drückt (1731–2), 1:1634; Wie schön wirds nicht im Himmel sein (Neumeister), fl, 2 ob, bn, va da gamba, str, bc, 1719, 1:1640; Wie sich ein Vater, 1724, 1:1641; Wie Spinnen Gift aus Blumen saugen (Brandenburg), 1725, 1:1644; Wie teuer ist deine Güte (Simonis), 2 fl, 2 ob, 2 cl, str, bc, 1720, 1:1646; Wiewohl er Gottes Sohn war, 2 fl, 2 ob, 3 tpt, timp, str, bc, 1757, 1:1648; Wie würd es uns ergehen, str, bc, 1729, 1:1649

Willkommen, du Licht, 2 ob, 2 hn, str, bc, 1:1651; Willkommen, segenvolles Fest (Neumeister), 2 ob, 3 hn, 3 tpt, timp, str, bc, 1719, 1:1652; idem (Neumeister), str, bc, 1:1653; Wir aber, die wir des Tages Kinder sind, 1:1654; *Wir danken dir, Gott, ob, str, bc (Erwecke dich, mein, 1727), 1:1657; Wird mein Jesus nicht bald kommen (Neumeister), 1731, 1:1658; *Wir gingen alle in der Irre (Weide mich auf grüner Aven, 1731–2), 1:1659; Wir glauben an den heiligen Geist (Neumeister), fl, 2 ob, str, org obbl, bc, 1719, 1:1660; Wir haben den funden, str, bc, 1724, 1:1661; †Wir haben die Hoffnung (Segenreicher Gang, 1727), 1:1662

Wir haben ein festes prophetisches Wort (Helbig), str, bc, 1725, 1:1663; Wir haben hier keine bleibende Statt, 2 fl, bn, str, bc, 1734, 1:1665 [? = 4:9]; Wir haben hier keine bleibende Stätte, str, bc, 1:1666; *Wir haben nicht mit Fleisch und Blut (So grausam mächtig, 1731–2), 1:1667; Wir liegen, grosser Gott (Neumeister), 2 ob, 2 bn, str, bc, 1722, 1:1668; Wir müssen alle offenbaret werden

(Helbig), 1723, 1:1670; †idem (Kann euch dieser Donnerschlag, 1727), 1:1671; Wir müssen alle offenbar werden (Neumeister), 1720, 1:1669; idem (1744), 1:1672

Wir sind allesamt wie die Unreinen (Neukirch), str, bc, 1725, 1:1674; *idem (Prang im Golde, stolze Welt, 1727), 1:1675; Wir sind allzumal Sünder (Simonis), 1721, 1:1676; Wir sind Gottes Mitarbeiter, 1757, 1:1677; Wir sollen selig werden (Neumeister), 1722, 1:1678; Wir wissen, dass denen (Helbig), 1723, 1:1681; idem (1744), 1:1682; Wir wissen, dass Gott, 1723, 1:1683; Wisset ihr nicht, dass alle (Helbig), 2 ob, 2 bn, str, bc, 1723, 1:1685; Wisset ihr nicht, 3 tpt, timp, str, bc (1744), 1:1686; idem (Helbig), 2 fl, 2 ob, str, bc, 1724, 1:1687; †idem (Hat Gott nicht, 1727), 1:1688

Wo bleibt die brüderliche Lieb (1748–9), 1:1689; Wo find ich meinen Jesum wieder, 1724, 1:1690; Wohin ich nur die Augen wende (Neumeister), 1719, 1:1692; idem, str, bc, 1:1693; Wohlan alle, die ihr durstig seid (Neukirch), 1724, 1:1694; *Wohlan, ich will meinem Lieben (Beweget euch munter, 1731–2), 1:1696; Wohlauf Herz, sing und spring (1748–9), 1:1697; Wohl dem, dem die Übertretung, 3 trbn, str, bc, 1:1701; Wohl dem, dem die Übertretungen (Helbig), 1723, 1:1698; idem, 1724, 1:1699; *idem (Todesangt und Höllenschr., 1727), 1:1700; *Wohl dem, der den Herrn fürchtet (Prangende Lilien, 1727), 1:1703; idem, str, bc, 1:1704

Wohl dem, der in Gelassenheit, 1732, 1:1705; Wohl dem, der mit Geduld (Neumeister), 1:1707; Wohl dem, des Hilfe der Gott Jacobs (Simonis), 1721, 1:1708; *Wohl dem, des Hülfe der Gott Jacobs (Tag und Studen, 1727), 1:1709; Wohl dem Volk, das jauchzen kann, 2 fl, 2 ob, 3 hn, str, bc, 1761, 1:1710; Wohl her nun, 2 ob, 3 tpt, timp, str, bc, 1739, frag., 1:1713; *Wo ist denn dein Freund hingegangen (Kehre wieder, 1727), 1:1716; Wo ist denn mein Jesus hin, B, vn, bc, 1:1717; Wo ist ein solcher Gott, str, bc, 1:1719

Wo ist solch ein Gott (Helbig), 1723, 1:1720; idem (1744), 1:1721; Wo Jesus Hunger merket, 1:1722; Wolken regnen (Vor allen Dingen habt unter einander); Wolken, ward ihr leer (Gedenke an den); Wollüstiger Sinnen, 2 fl, 2 ob, str, bc, 1:1723; Wo soll ich fliehen hin, fl, 2 ob, bn, str, bc, 1:1724; idem, S, 2 vn, bc, 1:1725; Wünschet Jerusalem Glück (Neumeister), 2 ob, 2 hn, str, bc, 1717; 1:1726; idem, STB, 2 ob, bn, 2 tpt, timp, str, bc, 1734, 1:1727; Zerbrechliche Gefässe (Was ist ein Mensch); Zerknische, du mein blödes Herz (Berkehret euch zu mir)

Zerreiss das Herz, ed. F. Schroeder (Stuttgart, 1963) [not in M]; *Zeuch ohn verzug (1731–2), 1:1728; Zions Hülf und Abrams Lohn, ob, hn, str, bc, 1750, 1:1730; Zion spricht: Der Herr hat mich verlassen (Helbig), 1725, 1:1731; Zischet nur, stechet (Wilkens) (1725–6), 1:1732, T iii; Zorn und Wüten (Neumeister), fl, 2 ob, str, bc, 1722, 1:1733; idem (Neumeister), str, bc, 1:1734; Zu dir flieh ich (1748–9), 1:1735; Zu Mitternacht ward ein Geschrei (Helbig), 1723, 1:1736; Zürne nur, du alte Schlange (Weine nicht, siehe); Zween Jünger gehn nach Emmaus (Neumeister), 2 fl, 2 ob, str, bc, 1719, 1:1738

Other frags.: Da hebt die lichte Siberwolke, 2 ob, 3 tpt, str, bc; Gott hat alles wohlgemacht (aria)

105 cantatas (texts only); 252 cantatas, lost

CANTATAS FOR CHURCH CONSECRATIONS, ETC — 292, 294
M – 2

Principal sources: *D-B/Bds*, *Dlb*, *DS*, *MÜG*; for 4 or more vv, orch, bc, unless otherwise stated.

Wohlan alle, die ihr durstig sind, 1726, 2:1; Herr, ich habe lieb die Stätte deines Hauses, str, bc, 1729, 2:2; Siehe da, eine Hütte Gottes (M. Richey), 1739, 2:3, ed. in Rhea; Ich halte mich zu deinem Altar, 1742, 2:4; Kommt, lasset uns anbeten (H. G. Schellhaffer), 1745, 2:5; Heilig, heilig ist Gott, 1747, 2:6; Zerschmettert die Götzen, 1751, 2:7; Singet Gott, lobsinget seinem Namen (Ballhorn), 1756, 2:9; Wie lieblich sind deine Wohnungen (E. Neumeister), 1757, 2:11; Komm wieder, Herr (J. J. D. Zimmermann), 1762, 2:12; Wie lieblich sind doch deine Wohnungen, frag., 2:13 [incl. Lasst uns eins, 3vv, 1:1036]

1 cantata (text only); 1 cantata lost

MUSIC FOR THE INSTITUTIONS OF PRIESTS — 292, 294
M – 3

Principal sources: *D-B/Bds*, *Dlb*, *F*, *RUh*; for 4 or more vv, orch, bc, unless otherwise stated.

So sind wir nun Botschafter, str, bc, 1749, inc., 3:56; Alles was Odem hat (chorus), SATB, 1749, 3:57; Wie lieblich sind auf den Bergen, 1754, 3:61; Nimm Dank und Weisheit, SAB, 1758, 3:67; Veni sancte spiritus, 1739, 3:82; idem, str, bc, 1756, 3:83; idem, 1760, 3:84; idem, S, B, bc, 1760, 3:85; idem, inc., 3:86; idem, bc, 3:87; idem, SSB, bc, 3:88; idem, SSS, bc, 3:89; idem, 3:90; idem, 2 choirs a cappella, 3:91; Komm heiliger Geist, SSB, str, bc, ?1742, 3:92; idem, frag., 1727, 3:93; idem, str, bc, ?1730, 3:94; Wahrheit und das Licht, frag., 3:95

?55 other works (texts only); ?23 works lost

FUNERAL CANTATAS 294

M – 4

Principal sources: *A-Wn*, *B-Bc*, *D-B/Bds*, *Dlb*, *NL-DHgm*, *US-Wc*; for 4 or more vv, orch, bc, unless otherwise stated.

Selig sind die Toten, str, bc, 1722, 4:1; Ach wie nichtig, ach wie flüchtig (C. F. Weichmann), 1723, frag., 4:2; Das Leben ist ein Rauch (M. Richey), 1728, 4:3; Ich hab, gottlob, das mein vollbracht (Richey), 1729, 4:5; Ach wie nichtig, ach wie flüchtig (H. Sillem), 1735, 4:6; In dunkler Nacht (J. J. D. Zimmermann), 1733, 4:7; Dränge dich an diese Bahre (Richey), 1739, 4:8; Wir haben hier keine bleibende Statt, 1739, 4:9 [? = 1:1665]; Ich hoffete aufs Licht (Zimmermann), 1745, 4:13; Lieber König, du bist tot, B, 1760, 4:15; Du aber Daniel, gehe hin, 4:17, ed. G. Fock (Kassel, 1968); Ein Mensch ist in seinem Leben, str, bc, 1729, 4:18; Nun geh ich aus der Sorgen Banden, B, bc, 4:20; Wie so kurz ist unser Leben, S, 4:21; Gottlob, es ist vollbracht (chorus), 4:22; Herr, nimmst du mir was Liebes ab, A, B, frag., 4:23

?8 other cantatas (texts only); 1 cantata lost

PASSION ORATORIOS, PASSIONS 291, 294, 295, 301, 304–5

M – 5

Principal sources: *B-Bc*, *D-B/Bds*, *Bdhm*, *LEm*, *SWl*, *DK-Kk*; for solo vv, chorus, orch, bc.

Der für die Sünden der Welt gemarterte und sterbende Jesus (Brockes-Passion) (B. H. Brockes), 1716, rev. 1722, 5:1; ed. H. Winschermann and F. Buck (Hamburg, 1964) 290, 305

Seliges Erwägen (Telemann), 1722, 2 versions extant, 5:2 305

Die Bekehrung des Römischen Hauptmanns Cornelius, 5:3 [see 'Sacred oratorios']

Die gekreuzigte Liebe, oder Tränen über das Leiden und Sterben unseres Heilandes (J. U. von König), 1731, 5:4

Betrachtung der neunten Stunde an dem Todestag Jesu (J. J. D. Zimmermann), 1755; 5:5 305

Der Tod Jesu (K. W. Ramler), 1755; 5:6 305

1 Passion for each year, 1722–67; extant works: St Mark (B. H. Brockes), 1723, 5:8; St Luke (M. A. Wilkens), 1728, 5:13, T xv; St Matthew, 1730, 5:15, ed. K. Redel (Vaduz, n.d.); St John, 1733, 5:18; St John, 1737, 5:22; St John, 1741, 5:26; St Luke (J. Gesenius), 1744, 5:29, ed. F. Schroeder (Stuttgart, 1966); St John (Zimmermann), 1745, 5:30, version pubd as Music von Leiden und Sterben des Welt Erlösers (Nuremberg, 1745–9); St Matthew (J. Rist), 1746, 5:31; St Luke, 1748, 5:33; St John, 1749 [= parody of 5:26], 5:34; St Matthew, 1750, 5:35; St Mark, 1755, 5:40; St John, 1757, 5:42; St Matthew, 1758, 5:43; St Mark, 1759, 5:44, ed. K. Redel (Vaduz, 1963); St Luke, 1760, 5:45; St John, 1761, 5:46; St Matthew, 1762, 5:47; St Luke, 1764, 5:49; St John, 1765, 5:50; St Matthew, 1766, 5:51; St Mark, 1767, 5:52 304

1, text only; 23 lost

SACRED ORATORIOS 294, 305

M – 6

Principal sources: *A-Wn*, *B-Bc*, *D-B/Bds*, *SWl*; for solo vv, chorus, orch, bc.

Davidische Gesänge (J. U. von König), 1718 (text only), 6:1

Freundschaft gehet über Liebe, 1720 (text only), 6:2

Die Bekehrung des Römischen Hauptmanns Cornelius (A. J. Zell), 1731 (text only) [= 5:3]

Donnerode (J. A. Cramer), pt 1, 1756, pt 2, 1762, 6:3 [= Wie ist dein Name so gross, 1:1624]; T xxii

Sing, unsterbliche Seele and Mirjam und deine Wehmut, from *Der Messias* (F. G. Klopstock), 1759, 6:4; ed. G. Godehart (Celle, n.d.)

Das befreite Israel (F. W. Zachariae), 1759, 6:5; T xxii

Die Hirten bei der Krippe zu Bethlehem (K. W. Ramler), 1759 [= Hier schläft es, 1:797]

Die Auferstehung und Himmelfahrt Jesu (Ramler), 1760, 6:6

Die Auferstehung (Zachariae), 1761, 6:7; ed. W. Menke (Hamburg, 1967)

Der Tag des Gerichts (C. W. Alers), 1762, 6:8; ed. in DDT, xxviii (1907/*R*)

PSALMS

M – 7

Principal sources: *D-B*, *Dlb*, *F*; some settings use only part of psalm text.

Ach Herr, strafe mich nicht (Ps vi); A, str, bc, 7:1, ed. W. Steude (Leipzig, 1966); idem (Ps vi), S, 9b, bn, vn, bc, 7:2, ed. K.

Hofmann (Stuttgart, 1978); idem (Ps vi), A, SATB, 2 vn, bn, bc, 7:3; Alleluja, singet dem Herrn (Ps cxlix, cxlvii), 2 ob, str, bc, 7:4; Danket dem Herrn (Ps cxviii), SATB, 2 ob, bn, 2 hn, str, bc, 7:5; Das ist ein köstlich Ding (Ps xcii), SATB, str, bc, 7:6; Deus judicium tuum (Ps lxxii), SATB, 2 fl, 2 ob, str, bc, 7:7; Die auf den Herren hoffen (Ps cxxv), S, A, 2 ob, str, bc, 7:8 [= 1:330]; idem (Ps cxxv), T, B, 2 ob, str, bc, 7:9

Dies ist der Tag (Ps cxviii), SATB, 2 ob, 2 hn, str, bc, 7:10; Dominus ad adjuvandum me (Ps lxx), SATB, str, bc, 7:11; Domine, dominus noster (Ps viii), SATB, str, bc, 7:12; Exaltabo te deus (Ps cxlv), S, B, bc, 7:13; Ich danke dem Herrn (Ps cxi), SATB, 2 ob, tpt, str, bc, 7:14, ed. K. Hofmann (Stuttgart, 1977); Ich hebe meine Augen (Ps cxxi), T, vn, bc, 7:15, ed. K. Hofmann (Stuttgart, 1978); idem (Ps cxxi), SATB, 2 ob, str, bc, 7:16; idem (Ps cxxi), SBB, ob, str, bc, 7:17

Ich will den Herrn loben (Ps xxxiv), S, A, vc, bc, 7:18, ed. K. Hofmann (Stuttgart, 1978); Ihr seid die Gesegneten des Herrn (Ps cxv), SATB, ob, str, bc, 7:19; Jauchzet dem Herrn alle Welt (Ps c), B, ob, tpt, str, bc, 7:20, ed. K. Hofmann (Stuttgart, 1974); idem (Ps c), STB, 2 fl, str, bc, 7:22; Jehova pascit me (Ps xxiii), S, B, 2 fl, 2 ob, str, bc, 7:23; Jubilate deo, omnia terra (Ps c), SATB, str, bc, 7:24; Laudate Jehovam omnes gentes (Ps cxvii), SATB, str, bc, 7:25, ed. E. Valentin (Kassel, 1936)

Laudate pueri dominum (Ps cxiii), S, str, bc, 7:26, ed. F. Schroeder and K. Hofmann (Stuttgart, n.d.); Lobe den Herrn meine Seele (Ps civ), SATB, 2 ob, 2 tpt, str, bc, 7:27; Lobet den Herrn alle Heiden (Ps cxvii), SATB, str, bc, 7:28 [= 9:20]; idem (Ps cxvii), TTB, vc, bc, 7:29; Singet dem Herrn ein neues Lied (Ps xcvi), SATB, str, bc, 7:30, ed. K. Hofmann (Stuttgart, 1978); Der Herr ist König (Ps xciii), B, ob, hn, str, bc, 7:31; Schmecket und sehet (Ps xxxiv), S, vn, bc, 7:32

1 psalm, lost; ?7 others, lost

MOTETS

M – 8

Principal sources: *A-Wgm*, *D-B*/*Bds*, *Dlb*, formerly Königsberg; for SATB, bc, unless otherwise stated.

Amans disciplinam, S, T, bc, 8:1; Amen, Lob und Ehre, SAB, bc, 8:2, ed, in Cw, civ (1967), ed. W. Menke (Stuttgart, 1967); Danket dem Herrn, 2 choirs, 8:3, ed. W. Hobohm (Wolfenbüttel, 1967); Der Gott unsers Herrn, 8:4, ed. in Cw, civ (1967), ed. W. Menke (Stuttgart, 1967); Der Herr gibt Weisheit, SAB, vc, bc, 1756, 8:5; Der Herr ist König, SATB, 8:6, ed. W. Menke (Stuttgart, 1967); Ein feste Burg, 1730, 8:7, ed. W. Menke (Stuttgart, 1967), ed. G. Graulich (Stuttgart, n.d.); Es segne uns Gott, 8:8, ed. in Cw, civ (1967), ed. W. Menke (Stuttgart, 1967); Halt, was du hast, 2 choirs, bc, 8:9, ed. W. Menke (Stuttgart, 1967), ed. P. Horn (Stuttgart, 1980); Jauchzet dem Herrn alle Welt, 2 choirs, 8:10 [= BWV Anh.160], ed. in Cw, civ (1967), ed. K. Hofmann (Stuttgart, n.d.); Laudate dominum, 8:11; Non aemulare cum viris, S, B, bc, 8:12; Selig sind die Toten, 8:13, ed. F. Jöde (Wolfenbüttel, 1931); Und das Wort ward Fleisch, SATB, 8:14, ed. W. Menke (Stuttgart, 1967); Werfet Panier auf, 8:15, ed. in DDT, xlix–l (1915/*R*); Wohl dem, der den Herr fürchtet, S, A, bc, 8:16, ed. K. Hofmann (Stuttgart, n.d.)

MASSES, ETC

M – 9

Principal sources: *B-Bc*, *D-B*/*Bds*, *F*; for SATB unacc. unless otherwise stated.

Masses on chorale melodies: Ach Gott vom Himmel sieh darein, 9:1; Allein Gott in der Höh sei Ehr, 9:2, ed. W. Menke (Stuttgart, 1967); Christ lag in Todesbanden, 9:3; Durch Adams Fall, 9:4; Ein Kindelein so löbelich, SATB, bc, 9:5, ed. K. Schultz-Hauser (Heidelberg, 1964); Erbarm dich mein, o Herre Gott, SATB, bc, 9:6; Es wird schier der letzte Tag kommen, SATTB, str, bc, 1751, 9:7; Es woll' uns Gott genädig sein, 9:8; Gott der Vater wohn uns bei, 9:9; Komm heiliger Geist, 9:10; Komm heiliger Geist, 9:11

Missa alla siciliana, SATB, 2 vn, bc, 9:12; Missa, SATB, str, bc, 9:13; Missa, A, bn, 2 vn, vc, bc, 9:14; Missa brevis, S, SATB, 9:15; 2 short masses (Ky-Gl)

Sanctus, Pleni sunt coeli, SATB, 3 tpt, timp, str, bc, 9:16

Heilig ist der Herr, unacc. male voice chorus 2vv

Magnificat (Lat.), SATBB, 3 tpt, timp, str, bc, 9:17; Magnificat (Ger.), SATB, 2 rec, 2 ob, 2 hn, str, bc, 9:18, ed. K. Hofmann (Stuttgart, n.d.); Kleines Magnificat (Ger.), S, vn, va, fl, vc [= BWV Anh.21; see H.-J. Schulze (1968)]

Amen (chorus), SATB, 2 tpt, str, bc, 9:19; Lobe den Herrn alle Heiden, 9:20 [see 7:28]

MISCELLANEOUS SACRED VOCAL
M – 10

(*principal sources: A-Wn, B-Bc, D-F, Mbs, GB-Lbm*)

Fast allgemeines Evangelisch-Musicalisches Lieder-Buch, 4vv, bc (Hamburg, 1730, 2/1751/*R*), 10:1 312

[12] Canones, 2–4 vv (Hamburg, 1735–6), 10:2–13; ed. F. Stein (Berlin and Darmstadt, 1954)

XI dicta biblica, 2vv, 2 vn, bc, 10:21–31

Die Begnadung (Kaum wag' ich es, dir Richter), 1v, bc, in *Unterhaltungen*, ii (Hamburg, 1766), 328

7 arias, lost

WEDDING CANTATAS AND SERENADES
M – 11

Principal sources: *B-Bc, D-B, F, MÜG, SWl*; for 4 or more vv, orch, bc, unless otherwise stated.

Ihr lieblichen Täler (serenade), 1727, 11:1; Herr Gott dich loben wir, Nun hilf uns Gott (cantatas, M. Richey), O erhabnes Glück der Ehe (serenade, Richey), all 1732, 11:15

10 cantatas: Drei schöne Dinge sind, 11:22; Ein wohl gezogen Weib, 11:23; Es woll uns Gott genädig sein, SSB, str, bc, 11:24; Herr, hebe an zu segnen, S, B, str, bc, 11:25; Liebe, was ist schöner, S, T, 2 ob, str, bc, 11:26; Lieblich und schön sein, ATB, orch, bc, 11:27; Lustig bei dem Hochzeits-Schmause, 1v, bc, lost, part ed. H. Leichtentritt (Berlin, 1905), 11:28; Sprich treuer Himmel (Cantate Nuptialis), B, orch, bc, 11:30; Wem ein tugendsam Weib, S, B, orch, bc, 11:31; Was Gott einmal gesegnet hat, frag., 11:32

15 works (texts only); 5 works, lost

MUSIC FOR BIRTHDAYS
M – 12

Principal sources: *B-Bc, D-DS, F, DK-Kk*; for 4 or more vv, orch, bc, unless otherwise stated.

De danske, norske og tydske undersaatters glaede (cantata), S, B, SATB, str, bc (Hamburg, 1757), ed. W. Hobohm (Leipzig, n.d.), 12:10

Auf Christenheit, begeh ein Freudenfest and Auf ihr treuen Untertanen (cantatas, J. G. Pritius), 1716, 12:1; Germania mit ihrem Chor (serenade, Pritius), 1716, 12:1c; Willkommen, schöner Freudentag (serenade), 1718, 12:3; Unsre Freude wohnt in dir (serenade), 1723, 12:4; Erklingt durch gedoppelt annehmliche Töne (serenade), inc., 1724–30, 12:6; Kommt mit mir, ihr süssen Freuden (Die Plaisir) (serenade), 1725, 12:7; Grossmächtiger Monarch der Britten (cantata), *c*1760, S, B, orch, bc, 12:11

41 other works (texts only)

MUSIC FOR POLITICAL CEREMONIES
M – 13 292, 294

Principal sources: *B-Bc, D-B, F, Ha*; for SATB, orch, bc, unless otherwise stated.

Zwo geistlicht Cantaten (J. G. Hamann); Sei tausendmal willkommen, S, str, bc; Du bleibest dennoch unser Gott, S, B, str, bc (Hamburg, 1731), 13:9

Zeuch, teures Haupt, 1722, 13:5b; Gebeut, du Vater der Gnade, 1744, 13:13; Geschlagene Pauken, auf! auf!, 1744, 13:14; Holder Friede, heilger Glaube (oratorio, J. J. D. Zimmermann), 1755, 13:18; Halleluja, amen, Lob und Ehr, 1757, 13:19; Hannover siegt, der Franzmann liegt (oratorio), S, orch, bc, 1758–61, 13:20; Bleibe, lieber König (cantata), B, orch, *c*1760, 13:21; Herr, wir danken deiner Gnade (aria), B, str, bc, ?1762, 13:22; Von Gnade und Recht (Ps ci), frag., 13:25

14 works (texts only); 7 works, lost

MUSIC FOR HAMBURG AND ALTONA SCHOOLS
M – 14

(*principal sources: B-Bc, D-B/Bds, F*)

Cantatas: Heilig, heilig, 4vv, orch, bc, 14:3c; Laetare juvenis in juventute tua, 4vv, 2 vn, bc, 1758, 14:11; Gott, man lobet dich in der Stille, 5vv, orch, bc, 1763, 14:12; In omni tempore dedit confessionem, 4vv, orch, bc, 14:20; Studiosa salve te corona, frag., 14:21

Motet: Gehet hin zur Ameise, 2vv, bc, 1756, 14:17

Duets with insts: O quam lata vis, 14:7; Deus aperi coelorum (C. F. Hunold), 14:13; O terra felicissima (Hunold), 14:14; Friede, dich preisen, 1755, 14:16

Arias with insts: Wo sich Ruh und Friede küssen, frag., 14:4; In der ersten Unschuld, 1752, 14:9; Friede, dich grüssen, 14:15; Fein säuberlich müsst ihr, 14:18; Holder Frühling, frag., 14:19

3 cantatas (texts only); 3 works, lost

KAPITÄNSMUSIKEN 292, 294

M – 15

Principal sources: *D-B/Bds*, *SWl*. Sacred oratorio and secular serenata for the yearly celebrations of the Hamburg militia commandant; for solo vv, chorus, orch, bc.

Complete oratorio and serenade pairs: Freuet euch des Herrn, ihr Gerechten and Geliebter Aufenthalt (J. P. Praetorius), 1724, 15:2; Jauchze, jubiliere und singe and Zu Walle, zu Walle! (G. P. Telemann, J. G. Hamann), 1730, 15:5; Preise Jerusalem den Herrn and So kömmt die kühne Tapferkeit (J. J. D. Zimmermann), 1736, 15:9; Wohl dem Volke and Es locket die Trommel (J. F. Lamprecht), 1738, 15:11; Der Herr ist meine Stärke and Schlagt die Trommel (F. W. Roloffs), 1742, 15:13; Vereint euch, ihr Bürger and Freiheit! Göttin, die Segen und Friede begleiten (N. D. Giesecke, W. A. Paulli), 1744, 15:15; Danket dem Herrn and Ihr rüstigen Wächter (H. Schellhaffer, M. Richey), 1755, 15:20; Herr, du bist gerecht and Wir nähren, wir zieren (Paulli), 1760, 15:23; Der Herr Zebaoth ist mit uns and Trompeten und Hörner erschallet (Paulli), 1764, 15:25

Mit innigstem Ergetzen (serenade), 1728, 15:4; Wohl dem Volk (oratorio, Paulli), frag., 1756, 15:21; Freuet euch des Herrn (oratorio), frag., 1761, 15:24; Schliesst die Kette der Einigkeit fest (chorus), 15:27

9 oratorio and serenade pairs (texts only); 6 oratorios, 7 serenades, lost

SECULAR CANTATAS, ETC 305

M – 20

Principal sources: *A-Wgm*; *B-Bc*; *D-B/Bds*, *DS*, *LEm*, *SHk*, *SWl*; *DK-Kk*

For SATB, orch, bc: Die Tageszeiten (F. W. Zachariae), 1757, 20:39, ed. A. Heilmann (Wolfenbüttel, 1934); Der Schulmeister, 20:57, ed. F. Stein (Kassel, 1956)

For S, B, orch, bc; Alles redet itzt und singet (B. H. Brockes), 1720, 20:10, ed. W. Menke (Kassel, 1955); Der May (K. W. Ramler), 20:40

For S, B, ob, str, bc: Geht, ihr unvergnügtem Sorgen, 20:50; Gute Nacht, du Ungetreuer, 20:52

Solo cantatas:

Ich kann lachen, weinen, scherzen (M. von Ziegler), S, bc, in Der getreue Music-Meister (Hamburg, 1728–9/*R*1981), 20:15

Sechs Cantaten, 1v, fl, rec, ob, str, bc (Hamburg, 1731): Dich wird stets mein Herz erlesen; Mein Vergnügen wird sich fügen; Mein Schicksal zeigt mir; Dein Auge tränt; Lieben will ich; In einem Tal, ed. K. Hofmann as Tirsis am Scheidewege (Kassel, 1981); 20:17–22

VI moralische Cantaten (D. Stoppe), S, bc (Hamburg, 1735–6): Die Zeit; Die Hoffnung; Das Glück; Der Geiz; Die Falschheit; Grossmut; 20:23–8; ed. K. Janetzky (Leipzig, 1978)

6 moralische Kantaten (J. J. D. Zimmermann), S, vn/fl, bc (Hamburg, 1736–7): Die Zufriedenheit; Tonkunst; Das mässige Glück; Die Liebe; Die Landlust, ed. R. Ermeler as Kleine Kantate von Wald und Au (Kassel, 1943); Die Freundschaft; 20:29–34

Trauer-Music eines kunsterfahrenen Canarienvogels (Canarien-Cantate), S, str, bc, 1737, ed. W. Menke (Kassel, 1977), 20:37; Ino (Ramler), S, orch, bc, 1765, ed. in DDT, xxviii (1907/*R*), 20:41; La Tempesta (P. Metastasio), S, hn, str, bc, 20:42

Amor heisst mich freudig lassen, S, bc, 20:43; Bin ich denn so gar verlassen, S, vn, bc, 20:44; Bist du denn gar von Stahl und Eisen, 1v, ob/vn, bc, 20:45; Bleicher Sorgen Kummer-Nächte, B, 2 vn, bc, 20:46; Der Mond zog nach und nach, T, bc, 20:47; Die Hoffnung ist mein Leben, B, str, bc, ed. W. Menke (Kassel, n.d.), 20:48; Du angenehmer Weiberorden, S, str, bc, ed. W. Hobohm (Leipzig, 1966), 20:49; Gönne doch dem freien Munde, S, vn, vc, bc, 20:51; Ha, ha, wo will wi hüt noch danzen, S, vn, vc, bc, ed. W. Hobohm (Leipzig, 1971), 20:53

Haltet ein, ihr schönsten Blicke, S, bc, 20:54; Ich hass und fliehe, S, bc, 20:55; Ich liebe dich wie meine Seele, S, bc, 20:56; In den Strahlen jener Sonne, S, vn, bc, 20:75; Mein Herze lachet vor Vergnügen, B, bc, 20:58; Parti mi lasci, S, fl, vn, vc, bc, 20:61; Pastorella venga bella, S, vn, vc, bc, doubtful, 20:62; Per che vezzoso, S, vn, vc, bc, frag., 20:63; Reicher Herbst, ihr kühlen Lüfte, T, vn, bc, 20:64; Ruht itzt sanft, ihr zarten Glieder, S, bc, 20:65

Sagt, ihr allerschönsten Lippen, B, bc, 20:66; Seufzen, Kummer, Angst und Tränen, A, bc, 20:67; So bald wird man das nicht vergessen, S, bc, 20:68; Soll die Marter meiner Seelen, S, bc, 20:69; Süsse Hoffnung, wenn ich frage (Die Hoffnung des Wiedersehens), S, 2 bn, str, bc, ed. W. Menke (Kassel, 1954), 20:70; Unbestand ist das Gift, S, bc, 20:71; Unbestand ist das Gift, S, vn, bc, 20:72; Voglio amarti, o caro nume, S, vn, bc, 20:73; Von geliebten Augen brennen, S, vn, bc, 20:74

14 cantatas, 2 serenades (texts only); 4 works, lost

OPERAS
M – 21 — 285, 292, 306
(*principal sources: D-B/Bds, F, Hs, Mbs, SHs*)

Adonis, 1708, 1 aria extant, 21:4
Narcissus, 1709 [?also 1701], 3 arias extant, 21:5
Mario, 1709, 2 arias extant, 21:6
Die Satyren in Arcadien, Leipzig, 1719, rev. 1724, as Der neumodische Liebhaber Damon (?G. P. Telemann); T xxi, 21:8 — 306
Der geduldige Socrates (J. U. von König, after Minato), Hamburg, 1721; T xx, 21:9 — 291, 306–7, *308*
Sieg der Schönheit (C. H. Postel, C. F. Weichmann, Telemann), 1722, rev. 1725 and 1732 as Genserich; 21:10
Belsazar (J. Beccau), 1723, 3 arias extant; 1 in Der getreue Music-Meister (Hamburg, 1728–9/*R*1981); 21:11
Pimpinone (intermezzo, J. P. Praetorius), 1725 (Hamburg, 1728); ed. in EDM, 1st ser., vi (1936); 21:15 — 306
La Caprizziosa e il Credulo (Praetorius), 1725, 4 arias extant; 21:16
Adelheid, 1725–6, part pubd as Lustige Arien aus der Opera Adelheid (Hamburg, 1727–8); 21:17
Die wunderbare Beständigkeit der Liebe oder Orpheus, 1726, inc., 21:18; see M. Ruhnke's report in *Hamburger Jb für Musikwissenschaft*, v (Hamburg, 1981), 20
Calypso, 1727, 1 chorus extant in Der getreue Music-Meister (Hamburg, 1728–9/*R*1981); 21:19
Sancio, 1727, 4 arias extant in Der getreue Music-Meister (Hamburg, 1728–9/*R* 1981), 1 ed. in HM, xii (1949); 21:20
Die verkehrte Welt (Praetorius), 1727, 1 aria, 1 scene extant in Der getreue Music-Meister (Hamburg, 1728–9/*R*1981); 21:23
Miriways (J. S. Müller), 1728; 21:24
Emma und Eginhard, oder Die Last-tragende Liebe (C. G. Wend), 1728, destroyed, 1907 copy in *US-Wc*, 5 arias and 1 duet in Der getreue Music-Meister (Hamburg, 1728–9/*R*1981); 3 arias, 1 duet ed. in HM, xii (1949); 21:25
Aesopus bei Hofe (G. Mattheson), 1729, 3 arias extant in Der getreue Music-Meister (Hamburg, 1728–9/*R*1981), 1 ed. in HM, xii (1949); 21:26
Flavius Bertaridus (Wend) 1729; 21:27
Don Quichotte, der Löwenritter (serenade, D. Schiebeler), 1761, ed. B. Baselt (Kassel, n.d.); 21:32
Adam und Eva (Richter), 7 pieces extant; 21:33
Herkules und Alceste, 3 pieces and frags. extant; 21:34
Herodes und Mariamne, frags.; 21:35
Serenade (untitled), 8 pieces incl. 1 aria in 20:17, 1 in 20:21; 21:36
11 other operas, 3 other intermezzos (texts only); 1 opera, lost — 283
Other pieces: 50 arias, 4 duets from operas and secular cantatas; 21:101–54

MUSIC FOR OPERAS BY OTHER COMPOSERS
M – 22 — 291, 293
(*principal sources: A-Wn, D-B/Bds, SWl, S-Skma*)

Duet for the opera pasticciò Ulysses, 1721, 22:1; Intermezzo and 6 arias for G. F. Handel's Tamerlano, 1725, 22:2; 14 arias ?and recitatives for Handel's Ottone, 1726, 22:3; 3 pieces for N. Porpora's Siface, 1727, 22:4; 4 arias for R. Keiser's Masagniello furioso, 1727, 22:5; Overture and 2 arias for Keiser's Nebucadnezar, 1728, 22:6; Aria for Keiser's Janus, 1729, 22:7; Recitatives and arias for Handel's Riccardo Primo, 1729, 22:8; Recitatives for Orlandini's Ernelinda, 1730, 22:9; Recitative and aria for Porpora's Aeneas, 1731, 22:10; Music for Handel's Cleofida [Poro], 1732, doubtful, 22:11; 3 arias for F. Chelleri's Judith, 1732, 22:12; Aria ?and overture for Handel's Almira, 1732, 22:13

OPERATIC PROLOGUES
M – 23
(*principal source: D-SWl*)

Rondinella a cui rapita, from prol to Das Lob der Musen, 1737, 23:12; 11 other prols (texts only)

OTHER SECULAR ORATORIOS
M – 24
(*principal sources: D-B, F, PL-GD, USSR-Ml*)

Admiralitätsmusik: Unschätzbarer Vorwurf erkenntlicher Sinnen (M. Richey), 1723, 24:1; Serenade, 1765 (sinfonia only), 24:4; Durch des Krieges Trutz und Macht (aria), 24:5
2 works (texts only)

SONGS, ETC
M – 25 — 305–6
(*principal sources: B-Bc, D-B/Bds, Hs, US-Wc*)

2 songs in Der getreue Music-Meister (Hamburg, 1728–9/*R*1981); incl. Das Frauenzimmer verstimmt sich immer, ed. in HM, xii (1949); 25:37–8
48 songs in Singe-, Spiel- und Generalbass-Übungen (Hamburg, 1733–4/*R*1981); ed, M. Seiffert (Kassel, n.d.); 25:38–85 — 296, 312
24 Theils ernsthafte, theils scherzende Oden (Hamburg, 1741), ed. in DDT, lvii (1917/*R*); 25:86–109 — 306

3 songs in F. W. Birnstiel: Oden mit Melodien (Berlin, 1753); 25:110–12
Other songs etc pubd in 18th-century anthologies

Batholomaeus (quodlibet), 3vv, bc, *D-B*; 25:113
36 songs in J. C. Losius: Singende Geographie (Hildesheim, 1708) [handwritten anon. adds.]; ed. in Hoffmann (1962); 25:1–36 — 284
Canon, 6vv, 1735; 25:114

OVERTURES, ETC — 309

Principal sources: *B-Bc*, *D-B/Bds*, *Dlb*, *DS*, *KA*, *MÜu*, *ROu*, *SWl*, *DK-Kk*; thematic catalogue and editions in Hoffmann (1969).

3 ovs. (e, D, B♭) in Musique de table (Hamburg, 1733); T xii–xiv; ed. in DDT, lxi–lxii (1927/*R*) — 293, 297–8
Six ouvertures à 4 ou 6 (Hamburg, 1736), destroyed; incl. 2 (g, a) ed. in Perlen alter Kammermusik (Leipzig, n.d.)
122 ovs., incl: 6 in T x; 1, a, ed. H. Büttner (Leipzig, 1936); 1, D, ed. in HM, cvii (1953); 1, E♭, ed. in NM, clxxvii (1959); 1, G, ed. W. Hobohm (Leipzig, 1968); 5 ed. H. Winschermann (Hamburg, n.d.); 3 frags. of ovs.; 1 ov. destroyed
4 syms.; 2 divertimentos, ed. in Musikschätze der Vergangenheit (Berlin, 1936–7)

CONCERTOS — 291, 307

Principal sources: *A-Wgm*, *B-Bc*, *D-B/Bds*, *Dlb*, *Ds*, *MÜu*, *PA*, *ROu*, *SWl*, *S-Skma*, *US-Wc*; thematic catalogue in Kross (1969).

Conc., A, fl, vn, str; conc., F, 3 vn, str; conc., E♭, 2 hn, str: all in Musique de table (Hamburg, 1733), T xii–xiv, ed. in DDT, lxi–lxii (1927/*R*) — 293, 297–8
47 concs. for 1 solo inst, str: 21 for vn, 12 in T xxiii, 1 (G) ed. F. Schroeder and F. Rübart (Zurich, 1965), 1 (a) ed. K. Grebe (Hamburg, 1967); 11 for fl, incl. 1 (G) ed. in HM, cxxxi (1955); 8 for ob; 2 for ob d'amore; 2 for rec, incl. 1 (F) ed. in HM, cxxx (1955); 2 for va, incl. 1 (G) ed. in HM, xxii (1949); 1 for hn; 1 for tpt
25 concs. for 2 solo insts, str: 8 for 2 vn, 1 (G) ed. W. Lebermann (Mainz, 1970); 4 for 2 hn; 3 for 2 fl, incl. 1 (a) ed. in NM, clxvii (1953); 2 for 2 chalumeaux; 2 for rec, va da gamba; 1 for 2 va, ed. K. Flattschacher (Kassel, n.d.); 1 for 2 ob d'amore; 1 for rec, fl; 1 for rec, bn; 1 for ob, vn; 1 for vn, vc
9 concs. for 3 solo insts, incl. conc., D, 3 tpt, ed. G. Fleischhauer (Leipzig, 1968); conc., D, 2 fl, vn, ed. F. Schroeder (Hamburg, 1973); conc., A, 2 fl, bn, ed. G. Fleischhauer (Leipzig, 1977); conc., D, vn, tpt, vc, ed. H. Töttcher and K. Grebe (Hamburg, 1965)
6 concs. for 4 solo insts, incl. conc., B♭, 2 fl, ob, vn; ed. G. Fleischhauer (Leipzig, 1974); conc., D, 2 fl, vn, vc, ed. G. Fleischhauer (Leipzig, 1978)
8 concerti grossi, incl. conc., F, a 7, ed. F. Brüggen and W. Bergmann (London, 1967), conc., F, rec, ob, 2 hn, bn, ed. F. Schroeder (Adliswil, 1972)

CHAMBER MUSIC — 293, 296, 309

Principal sources: *A-Wgm*, *Wn*; *B-Bc*; *D-B/Bds*, *BMs*, *Dlb*, *DS*, *F*, *Gs*, *LEm*, *PA*, *ROu*, *Sl*, *SWl*; *DK-Kk*; *F-Pn*; *GB-Lbm*; *S-Skma*.

(*without bc*)
R – 40

Sonata, D, va da gamba, in Der getreue Music-Meister (Hamburg, 1728–9/*R*1981); 40:1
12 fantaisies, fl (Hamburg, 1732–3); T vi, 40:2–13
[12] Fantaisie, vn (Hamburg, 1735); T vi, 40:14–25
12 fantasies, b viol (Hamburg, 1735–6), lost; 40:26–37
Sonates sans basse, 2 fl/vn/rec (Hamburg, 1727); T viii, 40:101–6
Sonata, B♭, 2 fl/va da gamba, ed. in HM, xi (1949), 40:107; Suite, D, 2 vn, ed. in HM, xi (1949), 40:108; Carillon, F, chalumeau, chalumeau/rec/fl, 40:109; Menuett, 2 hn, 40:110; Sonata, 2 insts, ed. in HM, xi (1949), 40:111; all in Der getreue Music-Meister (Hamburg, 1728–9/*R*1981)
XIIX Canons mélodieux, ou VI Sonates en duo (Paris, 1738); T viii, 40:118–23
Duo, 2 vn/fl/ob, livre II (Paris, 1752); T vii, 40:124–9
Sei duetti, 2 fl; T vii, 40:130–35
Quartet, A, 2 vn, va, vle, ed. in HM, cviii (1969); 40:200
3 concertos, G, D, C, 4 vn; T vi, 40:201–3
Concerto, A, 4 vn, ed. W. Friedrich (Mainz, 1951); 40:204

(*for 1 instrument and bc*)
R – 41

Six sonates, vn, bc (Frankfurt, 1715); ed. in *Moecks Kammermusik*, nos.101–3 (Celle, 1948); 41:g1, D1, h1, G1, a1, A1
Kleine Cammer-Music, bestehend aus VI Partien (Frankfurt, 1716); ed. in HM, xlvii (1949); 41:B1, G2, c1, g2, e1, Es1
Sei suonatine, vn, hpd (Frankfurt, 1718); ed. K Schweickert (Mainz, 1938); 41:A2, B2, D2, G3, E1, F1
Solos, vn, bc, op.2 (London, *c*1725); 41:d5, e7, F5, g8, B7, a7

Sonate metodiche, vn/fl, bc (Hamburg, 1728); T i; 41:g3, A3, e2, D3, a2, G4 — 296

18 pieces in Der Getreue Music-Meister (Hamburg, 1728–9/*R*1981): 4 sonatas ed. in HM, vi (1949), 41:C2, F2, fl, B3; 2 sonatas, 4 other pieces ed. in HM, vii (1949), 41:g5, a3, C1, d1, E2, B4; 1 sonata ed. in HM, xiii (1949), 41:D6; 1 sonata ed. in HM, clxxxix (1966), 41:G6; 5 other pieces ed, in HM, viii (1949), 41:D4, D5, E3, G5, h2; Suite, g, ed. in HM, clxxv (1961), 41:g4

Neue Sonatinen, hpd/vn/fl, 2 for rec, bc (Hamburg, 1730–31) [only 1 inst pt. extant]; 41:e3, c2, D7, G7, a4, E4

Continuation des Sonates méthodiques (Hamburg, 1732); T i; 41:h3, c3, E5, B5, d2, C3

3 solo sonatas, fl, vn, ob, in Musique de table (Hamburg, 1733); T xii–xiv, ed. in DDT, lxi–lxii (1927/*R*); 41:h4, A4, g6 — 293, 297–8

XII Solos, vn/fl, bc (Hamburg, 1734), ed. H. Kölbel and E. Meyerolbersleben (Wilhelmshaven, 1972); 41:F3, e4, A5, C4, g7, D8, d3, G8, h5, E6, a5, fis1

10 Sonatas in Essercizii musici (Hamburg, 1739–40): 2 ed. P. Rubardt (Halle, 1953), 41:e5, a6; 2 ed. R. Lauschmann (Hamburg, 1954), 41:B6, e6; 4 ed. H. Ruf (Mainz, 1964–5), 41:C5, d4, F4, A6; 2 ed. H. Ruf (Mainz, 1967), 41:D9, G9; D9 ed. also in NM, clxiii (1953)

21 other solo sonatas, incl. 4, fl, bc, ed. R. Kubik (Stuttgart, n.d.), 41:D10, e9, e11, G12

6 solo sonatas, doubtful; 1 other, frag.

(for 2 instruments and bc)

6 trio, vn, ob, bc (Frankfurt, 1718), ed. K. Schultz-Hauser (Berlin, 1969)

Musique héroïque, ou XII marches (Hamburg, 1728); ed. E. Pätzold (Berlin, n.d.)

Sonata in Der getreue Music-Meister (Hamburg, 1728–9/*R*1981); ed. in HM, x (1949)

III trietti methodici e III scherzi (Hamburg, 1731); ed. M. Schneider (Leipzig, 1948) — 296

Sonates en trio, fl, vns, obs (Paris, 1731–3)

3 sonatas, in Musique de table (Hamburg, 1733); T xii–xiv, ed. in DDT, lxi–lxii (1927/*R*) — 293, 297–8

Six concerts et six suites (Hamburg, 1734); T ix, xi

Scherzi melodichi (Hamburg, 1734); T xxiv

Sonates Corellisantes (Hamburg, 1735); T xxiv

12 trio sonatas in Essercizii musici (Hamburg, 1739–40); 2 ed. in NM, xlvii (1930), cxxxi (1937); 1 ed. R. Lauschmann (Leipzig, 1925); 1 ed. W. Woehl (Leipzig, 1954); 2 ed. H. Töttcher (Hamburg, 1962); 3 ed. H. Ruf (Mainz, 1964–7); 1 ed. W. Woehl (New York, n.d.)

3 trio sonatas, in Sonates en trio, fls, vns etc (Paris, 1738–42) [only fl II extant]

82 other sonatas, incl.: 1 ed. in NM, 1 (1929); 1 ed. in NM, li (1930); 1 ed. in HM, xxv (1940); 1 ed. in NM, cxxxi (1960); 1 ed. in NM, cli (1960); 1 ed. in HM, clxxix (1963); 2 ed. in HM, cxciv–cxcv (1968); 1 ed. in HM, ccxiv (1973); 1 ed, in HM, ccxix (1974); 1 ed. in HM, ccxxiv (1975); 1 ed. J. R. Flexer (Palo Alto, Calif., 1976); 1 ed. G. J. Kinney (Palo Alto, Calif., 1977); 1 ed. F. R. Palmer (Palo Alto, Calif., 1979) [1976, 1977 edns. incl. thematic index, sources, list of edns. of trio sonatas]; for further edns. see also Petzoldt (1967)

(quartets and quintets) — 309

Quadri, vn, fl, va da gamba/vc, bc (Hamburg, 1730); T xviii [= 'Paris' quartets, nos.1–6]

3 quartets in Musique de table (Hamburg, 1733); T xii–xiv, ed. in DDT lxi–lxii (1927/*R*) — 293, 297–8, 309

Six quatuors ou trios, 2 fl/vn, 2 vc/bn (Hamburg, 1733); T xxv

Nouveaux quatuors en six suites (Paris, 1738); T xix [= 'Paris' quartets, nos. 7–12] — 298

Quatrième livre de quatuors, fl, vn, va, bc (Paris, *c*1752)

23 other quartets, incl. 1, 3 vn, bc, ed, in HM, xcvii (1970); 9 quintets, incl. 1 ed. in HM, ccx (1971); for further edns. see Petzoldt (1967)

KEYBOARD — 309

*(principal sources: **B-Bc, D-B/Bds, Dlb, LEm, Mbs, GB-Lbm**)*

R – 30: Fugues; 31: Chorale preludes; 32: Suites; 33: Fantasias, sonatas, concertos; 34: Minuet collections; 35: Individual pieces; 36: Collection of pieces

XX kleine Fugen . . . nach besonderen Modis verfasset, org/hpd (Hamburg, 1731); F ii; ed. in NM, xiii (1928); 30:1–20

Fugues légères et petits jeux, hpd (Hamburg, 1738–9); ed. M. Lange (Kassel, 1929); 30:21–6 — 309–10

2 fughettas, F, D; F ii; 30:27–8

Fugirende und veraendernde Choraele (Hamburg, 1735); F i; ed. A. Thaler (New Haven, Conn., 1965); 31:1–48

3 chorale preludes, ed. in F i, 31:49, 51, 52; 3 others, 31:50, 53, 54

Partia a cembalo solo, G; Ouverture à la polonoise, d: both in Der getreue Music-Meister (Hamburg, 1728–9/*R*1981); ed. in HM, ix (1949); 32:1–2

2 solos, C, F, hpd, in Essercizii musici (Hamburg, 1739–40); ed. H. Ruf (Mainz, 1964); 32:3–4

VI Ouverturen nebst zween Folgesätzen (Nuremberg, before 1750); ed. in *Deutsche Klaviermusik des 17. und 18. Jahrhunderts*, iv–v (Berlin, n.d.); ed. A. Hoffmann (Wolfenbüttel and Zurich, 1964); 32:5–10

Suite, A [= BWV824], ed. in *J. S. Bach: Werke*, xxxvi (1890), 32:14; Ouverture, A, ed. in *Unbekannte Meisterwerke der Klaviermusik* (Kassel, 1930), 32:15; Partie, A [= BW832], ed. in *J. S. Bach: Werke*, xlii (1894), 32:18; 5 other suites, 32:11–13, 16–17

[36] Fantaisies pour le clavessin (Hamburg, 1732–3); ed. M. Seiffert (Kassel, 4/1955); 33:1–36 — 309

Sonata, e, hpd; 33:37

Concerto, b; ed. in *Unbekannte Meisterwerke der Klaviermusik* (Kassel, 1930) [?transcr. of vn conc.]; Anh.33:1

Sept fois sept et un menuet (Hamburg, 1728); ed. I. Amster (Wolfenbüttel and Berlin, 1930); 34:1–50

Zweytes sieben mal sieben und ein Menuet (Hamburg, 1730); 34:51–100

Marche pour M. le Capitaine Weber, F, and Retraite, F; La Poste, B♭: all in Der getreue Music-Meister (Hamburg, 1728–9/*R*1981); ed. in HM, ix (1949); 35:1–2

Menuet; Amoroso; Gigue; Menuet; 35:3–6

Neue auserlesene Arien, Menuetten und Märsche, so mehrentheils von . . . Telemann componiret worden sind [168 pieces, incl. at least 10 authentic], *D-Mbs*; 36:1–168

LUTE

R – 39

(*principal source: PL-Wu*)

Partie polonoise, B♭, ed. as kbd transcr. in Florilegium musicae antiquae, xi (Kraków, 1963); 39:1

Partie, g, 2 lutes, ed. as kbd transcr. in Florilegium musicae antiquae, xxiv (Kraków, 1968); 39:2

1 suite, formerly W. Wolffheim library, lost; Galanteries pour le luth, cited in catalogue of 1733, ? lost

WRITINGS — 294, 311–13

Beschreibung der Augen-Orgel (Hamburg, 1739)

Neues musicalisches System, in L. C. Mizler, Musikalische Bibliothek, iii/4 (Leipzig, 1752/*R*1966), 713; rev. as *Letzte Beschäftigung G. Ph. Telemanns im 86. Lebensjahre, bestehend in einer musikalischen Klang- und Intervallentafel*, in *Unterhaltungen*, iii (Hamburg, 1767) — 312–13

EDITIONS BY TELEMANN

MUSIC

Johann Ernst of Saxe-Weimar: 6 concerts, solo vn, 2 vn, va, hpd/bass viol (Leipzig and Halle, 1718)

Der getreue Music-Meister (Hamburg, 1728–9; facs. repr. ed. G. Fleischhauer, Leipzig, 1981): 1 scene, 1 chorus, 1 duet, 14 arias from operas; 2 songs; 3 sonatas, 1 suite, 2 other pieces for 1/2 insts without bc; 8 sonatas, 1 suite, 9 other pieces for 1 inst, bc; 1 sonata for 2 insts, bc; 2 kbd suites, 3 other kbd pieces; works by 13 other composers — 311

J. Graf: 6 soli, vn, bc (Hamburg and Rudolstadt, 1737)

C. Förster: Sei duetti, 2 vn, bc, op.1 (Paris, 1737)

J. Hövet: Musikalische Probe eines Concerts vors Clavier (Hamburg, 1741)

WRITINGS

C. J. F. Haltmeier: *Anleitung, wie man einen General-Bass . . . in alle Tone transponieren könne* (Hamburg, 1737) — 282

D. Kellner: *Treulicher Unterricht im General-Bass* (Hamburg, 2/1737)

G. A. Sorge: *Anweisung zur Stimmung und Temperatur* (Hamburg, 1744); *Gründliche Untersuchung, ob die . . . Schröterische Clavier-Temperaturen für gleichschwebend passiren können oder nicht* (Hamburg, 1754)

BIBLIOGRAPHY

SPECIALIST STUDIES

1. Magdeburger-Telemann Festtage: Beiträge zu einem neuen Telemannbild. Magedeburg 1962 (Magdeburg, 1963)

R. Petzoldt: *Telemann und seine Zeitgenossen*, Magdeburger Telemann-Studien, i (Magdeburg, 1966)

C. C. J. von Gleich: *Herdenkingstentoonstelling G. Ph. Telemann* (The Hague, 1967) [Gemeentemuseum catalogue]

G. Ph. Telemann, Leben und Werk: Beiträge zur gleichnamigen Ausstellung (Magdeburg, 1967)

K. Zauft: *Telemanns Liedschaffen*, Magdeburger Telemann-Studien, ii (Magdeburg, 1967)

3. Magdeburger Telemann-Festtage: G. Ph. Telemann, ein bedeutender Meister der Aufklärungsepoche. Magdeburg 1967 (Magdeburg, 1969)

I. Allihn: *G. Ph. Telemann und J. J. Quantz*, Magdeburger Telemann-Studien, iii (Magdeburg, 1971)

——: *Telemann-Renaissance: Werk und Wiedergabe*, Magdeburger Telemann-Studien, iv (Magdeburg, 1973)

5. Magdeburger Telemann-Festtage: Telemann und die Musikerziehung. Magdeburg 1973 (Magdeburg, 1975)

I. Allihn: *Telemann und Eisenach*, Magdeburger Telemann-Studien, v (Magdeburg, 1976)

6. Magdeburger Telemann-Festtage: Telemann und seine Dichter. Magdeburg 1977 (Magdeburg, 1978)

Georg Philipp Telemann: Leben—Werk—Wirkung, Konferenz der Zentralen Kommission Musik des Präsidialrates des Kulturbundes der DDR (Berlin, 1980)

Musica, xxxv/1 (1981)

M. Ruhnke: 'In welchem Masse werden zur Zeit in der Musikwissenschaft die Ergebnisse der Quellenforschung benützt und gewürdigt?', *Quellenforschung in der Musikwissenschaft* (Wolfenbüttel, 1982)

BIBLIOGRAPHICAL STUDIES

W. Menke: 'Das Vokalwerk Georg Philipp Telemanns: eine bibliographische Zwischenbilanz', *Mf*, i (1948), 192

M. Ruhnke: 'Zum Stand der Telemann-Forschung', *GfMKB, Kassel 1962*, 161

——: 'Telemann-Forschung 1967', *Musica*, xxi (1967), 6

A. Thaler: 'Der getreue Music-Meister, a "Forgotten" Periodical', *The Consort* (1967), no.24, p.280

M. Ruhnke: 'Telemann als Musikverleger', *Musik und Verlag: Karl Vötterle zum 65. Geburtstag* (Kassel, 1968), 502

——: 'Die Pariser Telemann-Drucke und die Brüder Le Clerc', *Quellenstudien zur Musik: Wolfgang Schmieder zum 70. Geburtstag* (Frankfurt am Main, 1972), 149
J. Schlichte: *Thematischer Katalog der kirchlichen Musikhandschriften des 17. und 18. Jahrhunderts in der Stadt- und Universitätsbibliothek Frankfurt am Main* (*Signaturengruppe Ms. Ff.Mus.*) (Frankfurt am Main, 1979)
M. Ruhnke: Preface to *G. P. Telemann: Musikalische Werke*, xxv (Kassel and Basle, 1981) [incl. section on Telemann's publishing activities]
H. Wettstein: *Georg Philipp Telemann: bibliographischer Versuch zu seinem Leben und Werk 1681–1767* (Hamburg, 1981)
W. Menke: *Thematisches Verzeichnis der Vokalwerke von Georg Philipp Telemann* (Frankfurt am Main, 1982–3)
O. Landmann: *Die Telemann-Quellen der Sächsischen Landesbibliothek* (Dresden, 1983)
M. Ruhnke: *Georg Philipp Telemann: Thematisch-Systematisches Verzeichnis seiner Werke: Instrumentalwerke*, i (Kassel, 1984)

LIFE AND WORKS: GENERAL

WaltherML
J. Mattheson: *Grosse General-Bass-Schule* (Hamburg, 1731) [incl. Telemann's autobiography, 1718]
——: *Grundlage einer Ehren-Pforte* (Hamburg, 1740); ed. M. Schneider (Berlin, 1910/*R*1969) [incl. Telemann's autobiography, 1739]
Herr G Ph. Telemann: Lebenslauf, ed. B. Schmid (Nuremberg, *c*1745)
J. D. Winckler: 'G. Ph. Telemann', *Nachrichten von niedersächsischen berühmten Leuten und Familien*, i (Hamburg, 1768), 342
J. Sittard: *Geschichte des Musik- und Concertwesens in Hamburg* (Altona and Leipzig, 1890)
C. Valentin: *Geschichte der Musik in Frankfurt am Main vom Anfange des XIV. bis zum Anfange des XVIII. Jahrhunderts* (Frankfurt am Main, 1906/*R*1972)
M. Schneider: Preface to DDT, xxviii (1907/*R*)
A. Schering: *Musikgeschichte Leipzigs*, ii (Leipzig, 1926)
E. Valentin: *Georg Philipp Telemann* (Burg, 1931, 3/1952)
W. Kahl: *Selbstbiographien deutscher Musiker des 18. Jahrhunderts* (Cologne and Krefeld, 1948/*R*1970)
E. Valentin: *Telemann in seiner Zeit* (Hamburg, 1960)
A. Hoffmann: *Die Lieder der Singenden Geographie* (Hildesheim, 1962)
H. Grosse: 'Telemanns Aufenthalt in Paris', *HJb 1964–5*, 113
W. Hobohm: 'Telemann und seine Schüler', *GfMKB, Leipzig 1966*, 260

W. Bergmann: 'Telemann in Paris', *MT*, cviii (1967), 1101
O. Büthe: 'Das Frankfurt Telemanns', *Frankfurt, Lebendige Stadt*, xii (1967)
R. Petzoldt: *Georg Philipp Telemann: Leben und Werk* (Leipzig, 1967; Eng. trans., 1974 [with list of edns.])
M. Ruhnke: 'Relationships between the Life and Work of Georg Philipp Telemann', *The Consort* (1967), no.24, p.271
W. Hobohm: 'Zwei Kondolenzschreiben zum Tode Georg Philipp Telemanns', *DJbM*, xiv (1969), 117
K. Grebe: *Georg Philipp Telemann in Selbstzeugnissen und Bilddokumenten* (Reinbeck bei Hamburg, 1970)
L. Füredu and D. Vulpe: *Telemann* (Bucharest, 1971)
H. Grosse and H. R. Jung, eds.: *Georg Philipp Telemann, Briefwechsel* (Leipzig, 1972)
W. Hobohm: 'Drei Telemann-Miszellen', *BMw*, xiv (1972), 237
H. R. Jung: *Georg Philipp Telemann als Eisenacher Kapellmeister und seine weltlichen Festmusiken für den Eisenacher Hof* (diss., U. of Halle, 1975)
C. Oefner: *Das Musikleben in Eisenach 1650–1750* (diss., U. of Halle, 1975)
Georg Philipp Telemann: Autobiographien – 1718, 1729, 1739, ed. G. Fleischhauer, W. Siegmund Schultze and E. Thom (Blankenburg, 1977)
M. Bircher: 'Ein unbekanntes Stammbuchblatt von Georg Philipp Telemann', *Mf*, xxxi (1978), 185
E. Klessmann: *Telemann in Hamburg: 1721–1767* (Hamburg, 1980)
W. Siegmund-Schultze: *Georg Philipp Telemann* (Leipzig, 1980)
E. Rackwitz, ed.: *Georg Philipp Telemann: Singen ist das Fundament zur Music in allen Dingen: eine Dokumentensammlung* (Leipzig, 1981)

WORKS

C. Ottzenn: *Telemann als Opernkomponist: ein Beitrag zur Geschichte der Hamburger Oper* (Berlin, 1902)
M. W. Frey: *Georg Philipp Telemanns Singe-, Spiel- und Generalbass-Übungen* (Zurich, 1922)
H. Graeser: *Georg Philipp Telemanns Instrumental-Kammermusik* (diss., U. of Munich, 1924) [incl catalogue]
R. Meissner: *Georg Philipp Telemanns Frankfurter Kirchenkantaten* (diss., U. of Frankfurt am Main, 1925)
M. Seiffert: 'Georg Philipp Telemanns "Musique de Table" als Quelle für Händel', DDT, *Beihefte*, ii (1927)
H. Hörner: *Georg Philipp Telemanns Passionsmusiken* (Leipzig, 1933)
K. Schäfer-Schmuck: *Georg Philipp Telemann als Klavierkomponist*

(diss., U. of Kiel, 1934; Borna, 1934/*R*1981) [*R* incl. updated bibliography and list of edns.]
H. Büttner: *Das Konzert in den Orchestersuiten Georg Philipp Telemanns* (Wolfenbüttel and Berlin, 1935)
W. Menke: *Das Vokalwerk Georg Philipp Telemanns* (Kassel, 1942)
A. Dürr: 'Zur Echtheit einiger Bach zugeschriebener Kantaten', *BJb*, xxxix (1951–2), 30
F. D. Funk: *The Trio Sonatas of Georg Philipp Telemann* (diss., George Peabody College, Nashville, 1954)
L. Hoffmann-Erbrecht: *Deutsche und italienische Klaviermusik zur Bachzeit* (Leipzig, 1954)
W. Braun: 'B. H. Brockes' "Irdisches Vergnügen in Gott" in den Vertonungen G. Ph. Telemanns und G. Fr. Händels', *HJb 1955*, 42
G. Hausswald: Prefaces to *G. P. Telemann: Musikalische Werke*, vi–viii (Kassel and Basle, 1955)
F. Stein: 'Eine komische Schulmeister-Kantate', *Festschrift Max Schneider* (Leipzig, 1955), 183
H. C. Wolff: *Die Barockoper in Hamburg* (Wolfenbüttel, 1957)
C. H. Rhea: *The Sacred Oratorios of Georg Philipp Telemann* (diss., Florida State U., 1958)
W. S. Newman: *The Sonata in the Baroque Era* (Chapel Hill, 1959, rev. 4/1983)
P. Pisk: *Telemann's Menuet Collection of 1728* (diss., U. of Texas, 1960)
G. Godehart: 'Telemanns Messias', *Mf*, xiv (1961), 139
J. Birke: 'J. Hübners Text zu einer unbekannten Festmusik Telemanns', *Mf*, xvii (1964), 402
M. Ruhnke: 'G. Ph. Telemanns Klavierfugen', *Musica*, xviii (1964), suppl., 103
——: Preface to *G. P. Telemann: Musikalische Werke*, xv (Kassel and Basle, 1964)
W. Bergmann: Prefaces to *G. P. Telemann: Musikalische Werke*, xviii–xix (Kassel and Basle, 1965)
A. Briner: 'Die neuentdeckte Matthäuspassion von Telemann', *SMz*, cv (1965), 295
B. Baselt: 'Georg Philipp Telemann und die protestantische Kirchenmusik', *Musik und Kirche*, xxxvii (1967), 196
——: Preface to *G. P. Telemann: Musikalische Werke*, xx (Kassel and Basle, 1967)
C. P. Gilbertson: *The Methodical Sonatas of Georg Philipp Telemann* (diss., U. of Kentucky, 1967)
W. Maertens: 'Georg Philipp Telemanns Hamburger Kapitänsmusiken', *Festschrift für Walter Wiora* (Kassel, 1967), 335
W. C. Metcalfe: 'The Recorder Cantatas of Telemann's Harmoni-

scher Gottesdienst', *American Recorder*, viii/4 (1967), 113
T. Nishi: 'Georg Philipp Telemann no Junankyoku ni tsuite' [G. P. Telemann's St John Passion], *Ongaku gaku*, xiv/3 (1968)
H.-J. Schulze: 'Das "Kleine Magnificat" BWV Anh.21 und sein Komponist', *Mf*, xxi (1968), 44
C. Annibaldi: 'L'ultimo oratorio di Telemann', *NRMI*, iii (1969), 221
B. Baselt: Preface to *G. P. Telemann: Musikalische Werke*, xxi (Kassel and Basle, 1969)
G. Frum: *The Dramatic-dualistic Style Element in Keyboard Music Published before 1750* (diss., Columbia U., 1969)
A. Hoffmann: *Die Orchestersuiten Georg Philipp Telemanns* (Wolfenbüttel and Zurich, 1969)
S. Kross: *Das Instrumentalkonzert bei Georg Philipp Telemann* (Tutzing, 1969)
M. A. Peckham: *The Opera of Georg Philipp Telemann* (diss., Columbia U., 1969)
G. Fleischhauer: 'Zur instrumentalen Kammermusik Georg Philipp Telemanns', *Colloquium Musica cameralis Brno 1971*, 345
L. Finscher: 'Corelli und die "Corellisierenden" Sonaten Telemanns', *Studi corelliani* (Florence, 1972), 75
G. Fleischhauer: 'Zum Konzertschaffen Georg Philipp Telemanns', *Telemann-Renaissance: Werk- und Wiedergabe* (Magdeburg, 1973), 21
——: 'Telemann, G. Ph.: Orchestersuiten und Instrumentalkonzerte', *Konzertbuch*, iii, ed. H. J. Schaefer (Leipzig, 1974), 469
A. Hoffmann: Preface to *G. P. Telemann: Musikalische Werke*, xxiv (Kassel and Basle, 1974)
K. Koch: 'Eine "Sarrois" von Telemann', *HJb 1974*, 135
H. Friedrichs: *Das Verhältnis von Text und Musik in den Brockespassionen Keisers, Händels, Telemanns und Matthesons* (Munich, 1975)
W. Maertens: *Telemanns Kapitänsmusiken* (diss., U. of Halle, 1975)
J. A. Westrup: 'Telemann and his Contemporaries', *NOHM*, v (Oxford, 1975)
M. Ruhnke: 'Das italienische Rezitativ bei den deutschen Komponisten des Spätbarock', *AnMc*, no.17 (1976), 79
B. Baselt: 'G. Ph. Telemanns Serenade "Don Quichotte auf der Hochzeit des Comacho" ', *Hamburger Jb für Musikwissenschaft*, iii (Hamburg, 1978), 85
M. Ruhnke: 'Zur Hamburger Umtextierung von Telemanns Passionsoratorium "Seliges Erwägen" ', *Festschrift Georg von Dadelsen zum 60. Geburtstag* (Neuhausen-Stuttgart, 1978), 255
P. Drummond: *The German Concerto: five Eighteenth-century Studies* (Oxford, 1980)

N. Anderson: 'Georg Philipp Telemann: a Tercentenary Reassessment', *Early Music*, ix (1981), 499
G. Fleischhauer: 'Zum Instrumentalkonzertschaffen Telemanns', *ÖMz*, xxxvi (1981), 148
C. Lawson: 'Telemann and the chalumeau', *Early Music*, ix (1981), 312
R. Lynch: 'Händels "Ottone": Telemanns Hamburger Bearbeitung', *HJb 1981*, 117
M. Ruhnke: 'Komische Elemente in Telemanns Opern und Intermezzi', *GfMKB, Bayreuth 1981*, 94
——: Preface to *G. P. Telemann: Musikalische Werke*, xxv (Kassel and Basle, 1981)
——: 'Telemanns Hamburger Opern und ihre italienischen und französischen Vorbilder', *Hamburger Jb für Musikwissenschaft*, v (Hamburg, 1981), 9 [incl. discussion of Die wunderbare Beständigkeit der Liebe]
H. C. Wolff: ' "Pimpinone" von Albinoni und Telemann: ein Vergleich', *Hamburger Jb für Musikwissenschaft*, v (Hamburg, 1981), 29
E. Thom and others: *Telemanns Beschreibung einer Augen-Orgel (1739)*, Studien zur Aufführungspraxis und Interpretation von Instrumentalmusik des 18. Jahrhunderts, xviii (Blankenburg, 1983)

STYLE

R. Rolland: *Voyage musical au pays du passé* (Paris, 1919)
W. Serauky: 'Bach-Händel-Telemann in ihrem musikalischen Verhältnis', *HJb 1955*, 72
H. Pohlmann: *Die Frühgeschichte des musikalischen Urheberrechts* (Kassel, 1962)
M. Ruhnke: 'Telemann im Schatten von Bach?', *Hans Albrecht in memoriam* (Kassel, 1962), 143
W. Maertens: 'Georg Philipp Telemann und die Musikerziehung', *Musik in der Schule*, xv (1964), 498
P. Beaussant: 'Situation de Telemann', *La table ronde* (1965), no.207, p.121; (1966), no.208, p.115
G. Carleberg: *Buxtehude, Telemann och Roman, Mus. och biogr. skisser* (Stockholm, 1965)
G. C. Ballola: 'Telemann dotto e galante', *Lo spettatore musicale*, ii (1967)
G. Fleischhauer: 'Die Musik Georg Philipp Telemanns im Urteil seiner Zeit', *HJb 1967–8*, 173; *HJb 1969–70*, 23–73
A. Dürr: 'Eine Handschriftensammlung des 18. Jahrhunderts', *AMw*, xxv (1968), 308
G. von Dadelsen: 'Telemann und die sogenannte Barockmusik',

Musik und Verlag: Karl Vötterle zum 65. Geburtstag (Kassel, 1968), 197

A. Dekker: 'J. S. Bach en G. P. Telemann', *Mens en melodie*, xxiv (1969), 304

L. Finscher: 'Der angepasste Komponist', *Musica*, xxiii (1969), 549

H. C. Wolff: 'Das Tempo bei Telemann', *BMw*, xi (1969), 41

M. Ruhnke: 'Zu L. Finschers neuestem Telemann-Bild', *Musica*, xxiv (1970), 340

S. Kross: 'Telemann und die Aufklärung', *Musicae scientiae collectanea: Festschrift Karl Gustav Fellerer* (Cologne, 1973), 284

C. Wolff: 'Ein Gelehrten-Stammbuch aus dem 18. Jahrhundert', *Mf*, xxvi (1973), 217

G. Fleischhauer: 'Einflüsse polnischer Musik im Schaffen Georg Philipp Telemanns', *Wissenschaftliche Zeitschrift der Martin-Luther-Universität*, xxv/3 (1976), 77

H. C. Wolff: 'Georg Philipp Telemann – 300 Jahre (zum 13. März 1981): Eleganz und Grazie – Symmetrie und Witz', *Mf*, xxxiv (1981), 40

Index